AF816212

MYOCARDIAL ISCHEMIA AND REPERFUSION

Myocardial Ischemia and Reperfusion

Edited by

MICHAEL V. COHEN

University of South Alabama
College of Medicine
Mobile, AL 36688
USA

JAMES M. DOWNEY

University of South Alabama
College of Medicine
Mobile, AL 36688
USA

RICARDO J. GELPI

Departamento de Patologia
Facultad de Medicina
Uriburu 950 – Piso 2
1114 Buenos Aires
Argentina

and

JAN SLEZAK

Institute for Heart Research
Slovak Academy of Sciences
Dubravska Cesta 9.84233
Bratislava
Slovak Republic

Reprinted from *Molecular and Cellular Biochemistry*, Volume 186 (1998)

Springer-Science+Business Media, B.V.

Library of Congress Cataloging-in-Publication Data

Myocardial ischemia and reperfusion/edited by Michael V. Cohen . . . [et al.].
p. cm. -- (Developments in molecular and cellular biochemistry)
ISBN 978-0-7923-8173-0
1. Coronary heart disease – Pathophysiology. 2. Myocardial reperfusion. I. Cohen, Michael V. (Michael Victor), 1944– .II. Series.
[DNLM: 1. Myocardial Ischemia – physiopathology congresses. 2. Myocardial Reperfusion congresses. WG 300 M99763 1998]
RC685.C6M958 1998
616.1'2307 – dc21
DMLM/DLC
for Library of Congress 98-7849
CIP

ISBN 978-0-7923-8173-0 ISBN 978-1-4615-4979-6 (eBook)
DOI 10.1007/978-1-4615-4979-6

Printed on acid-free paper

All rights reserved

© Springer Science+Business Media Dordrecht 1998
Originally published by Kluwer Academic Publishers 1998

No part of the material protected by this copyright notice may be reproduced or utilized in any form or by any means, electronic or mechanical, including photocopying, recording or by any information storage and retrieval system, without written permission from the copyright owner

Molecular and Cellular Biochemistry:

An International Journal for Chemical Biology in Health and Disease

CONTENTS VOLUME 186, Nos. 1 & 2, September 1998

MYOCARDIAL ISCHEMIA AND REPERFUSION

Michael V. Cohen, James M. Downey, Ricardo J. Gelpi and Ján Slezák, guest editors

Molecular and Cellular Biochemistry **186:** 1, 1998.

Preface

It is a pleasure to present to you this focused issue of *Molecular and Cellular Biochemistry*. The articles collected here largely arose from two related meetings held last spring on almost opposite sides of the world. The first was held in Buenos Aires, Argentina on April 11–12, 1997 and was titled Nuevos Avances en el Fenómeno de Isquemia y Reperfusión (New advances in the Phenomenon of Ischemia and Reperfusion). Ricardo Gelpi, the meeting's organizer, attracted over 500 attendees for the 2-day international symposium which had the sponsorship of multiple organizations including the International Society and Federation of Cardiology, The International Society for Heart Research and The American Heart Association, Circulation Council. Much of the financial support was from a generous donation from GEMA of Argentina. Twenty internationally recognized speakers presented data and led the discussions.

The second meeting took place in Stará Lesná located in the beautiful High Tatras Mountains of the Slovak Republic on June 27–30. One hundred heart researchers gathered at this mountain resort to discuss the subject of Myocardial Adaptation to Ischemia. Ján Slezák organized the meeting along with his colleagues at the Institute for Heart Research at the Slovak Academy of Sciences. This meeting, too, was sponsored by several organizations including the International Society and Federation of Cardiology and the International Society for Heart Research.

All participants of both meetings were urged to submit manuscripts, and original research reports were particularly encouraged. In addition there was a world-wide call for submission of papers for the issue. The response was excellent. Special thanks must also go to Naranjan Dhalla whose efforts and encouragement made publication in *Molecular and Cellular Biochemistry* possible. This focused issue on 'Myocardial Ischemia and Reperfusion' is the result of these activities. We hope that you will find some exciting science within these pages.

MICHAEL V. COHEN, Mobile, USA;
JAMES M. DOWNEY, Mobile, USA;
RICARDO J. GELPI, Buenos Airés, Argentina;
JAN SLEZAK, Bratislava, Slovak Republic.

Molecular and Cellular Biochemistry **186:** 3–12, 1998.
© 1998 *Kluwer Academic Publishers.*

Opioid receptor contributes to ischemic preconditioning through protein kinase C activation in rabbits

Takayuki Miki, Michael V. Cohen and James M. Downey
Departments of Physiology and Medicine, University of South Alabama, College of Medicine, Mobile, AL, USA

Abstract

Recent studies have reported that protection from ischemic preconditioning (PC) is blocked by the opioid receptor antagonist naloxone (NAL). We tested whether an opioid agonist could mimic PC in the rabbit heart, whether that protection involved protein kinase C (PKC) activation, and whether opioid receptors act in concert with other PKC-coupled receptors. Rabbit hearts were subjected to 30 min coronary occlusions and were reperfused for either 3 (*in situ*) or 2 (*in vitro*) h. Infarct size was determined by staining with triphenyltetrazolium chloride. In untreated *in situ* hearts 38.5 ± 1.6% of the risk zone infarcted. PC with 5 min ischemia/10 min reperfusion significantly limited infarction to 12.7 ± 2.9% ($p < 0.01$). NAL infusion did not modify infarction (39.6 ± 1.6%) in non-PC hearts, but blocked the effect of one cycle of PC (34.4 ± 3.6% infarction). NAL, however, could not block cardioprotection when PC was amplified with 3 cycles of ischemia/reperfusion (9.9 ± 1.4% infarction, $p < 0.01$ vs. control). Morphine could also mimic ischemic preconditioning, but only at a dose much higher than would be used clinically (3 mg/kg). In isolated hearts pretreatment with morphine (0.3 μM) significantly limited infarction to 9.3 ± 1.2% ($p < 0.01$ vs. 32.0 ± 3.1% in controls). This cardioprotective effect of morphine could be blocked by either the PKC inhibitor chelerythrine (30.4 ± 2.6% infarction) or NAL (34.0 ± 2.6% infarction). Neither chelerythrine nor NAL by itself modified infarction in non-PC hearts. NAL could not block protection from one cycle of PC in isolated hearts indicating that an intact innervation may be required for endogenous opioid production. Thus, opioid receptors, like other PKC-coupled receptors, participate in the triggering of PC in the rabbit heart. (Mol Cell Biochem **186**: 3–12, 1998)

Key words: ischemic preconditioning, opioid receptor, protein kinase C, rabbit, morphine, naloxone

Introduction

Brief ischemia can markedly enhance myocardial resistance to a subsequent ischemic injury. This phenomenon, called ischemic preconditioning, has been widely confirmed in various animal models [1, 2]. Many substances are released by the myocardium during ischemia, and several of them, including adenosine [3, 4], bradykinin [5, 6] and free radicals [7, 8], act to trigger preconditioning. We have proposed that the common mechanism for triggering preconditioning by these various substances is activation of protein kinase C (PKC) through receptor-mediated cell signaling pathways [9, 10]. Opioids, are known to be produced during periods of cardiovascular stress [11, 12], and recently Schultz *et al.* [13] reported that infarct size limitation by preconditioning in the rat could be blocked by naloxone. In the rat opioids are probably the principal endogenous triggers of preconditioning, although adenosine has also recently been demonstrated to initiate protection [14]. Other endogenous agonists produced during ischemia are ineffective [15]. But in the rabbit adenosine and bradykinin clearly play major roles in the initiation of the phenomenon of ischemic preconditioning, and consequently it is not possible to assume that opioids will play an equally significant role. However, it is noteworthy that Chien *et al.* [16] did observe that naloxone also blocked protection in the rabbit. These results suggest that opioids may play a physiological role in the triggering of preconditioning. Indeed the opioid receptor is coupled to G proteins [17, 18]

Address for offprints: J.M. Downey, Department of Physiology, MSB 3024, University of South Alabama, College of Medicine, Mobile, AL 36688, USA

and can activate PKC in neuron cultures [19]. Thus it is possible that endogenously released opioids may simply act in concert with adenosine and bradykinin to activate PKC during ischemic preconditioning.

The present study has tested whether morphine, an opioid receptor agonist, can mimic preconditioning and if the protective effect of morphine, if any, can be blocked by a PKC inhibitor. We also investigated the hypothesis that opioid receptors act in concert with the other PKC-coupled receptors. If opioid receptors act in parallel with other receptors, then blockade of the former should only raise the threshold for preconditioning the heart but not completely block the heart's ability to be preconditioned.

Materials and methods

This study was conducted in accordance with the Guide for the Care and Use of Laboratory Animals published by the US National Institutes of Health (NIH publication No. 85–23, revised 1985).

Surgical preparation: In situ *experiments*

New Zealand White rabbits of either sex, weighing 1.7–2.6 kg, were anesthetized with intravenous sodium pentobarbital (30 mg/kg). The neck was opened with a ventral midline incision and a tracheotomy was performed. The rabbits were ventilated with 100% oxygen via a positive pressure respirator (MD Industries, Mobile, AL, USA). The respiratory rate and tidal volume were adjusted to keep the arterial blood pH in the physiological range. Body temperature was maintained near 38°C with a heating pad. A catheter was inserted into the carotid artery for monitoring of blood pressure. Another catheter was placed in the jugular vein for infusion of drugs. A left thoracotomy was performed in the fourth intercostal space, and the pericardium was opened to expose the heart. A 2-0 silk suture on a curved taper needle was passed around a prominent branch of the left coronary artery, and the ends were pulled through a small vinyl tube to form a snare. The coronary branch was occluded by pulling the snare, which was then fixed by clamping the tube with a small hemostat. Rabbits were anticoagulated with 500 units of heparin after the surgical preparation was completed, and stabilized for 10 min. The rabbit was then treated as described below and myocardial ischemia following traction on the snare was confirmed by appearance of regional cyanosis. Reperfusion was achieved by releasing the snare and was confirmed by visible hyperemia on the ventricular surface. After 3 h of reperfusion, the rabbit was given 2000 units of heparin, and killed with a pentobarbital overdose. The heart was then excised for postmortem analysis.

Surgical preparation: In vitro *experiments*

Rabbits (1.4–2.7 kg) were anesthetized and intubated as described above. After a left thoracotomy and placement of a suture around the coronary artery, the hearts were quickly excised, mounted on a Langendorff apparatus, and perfused at 75 mmHg pressure with non-recirculating Krebs buffer containing (mM) NaCl 118.5, KCl 4.7, $MgSO_4$ 1.2, KH_2PO_4 1.2, $NaHCO_3$ 24.8, $CaCl_2$ 2.5, and glucose 10. The Krebs buffer was gassed with 95% O_2/5% CO_2 , resulting in a pH of 7.4–7.5. The temperature of the perfusate was maintained at 37°C. A fluid-filled latex balloon connected to a transducer with PE240 tubing was inserted into the left ventricle. Balloon volume was adjusted to set the left ventricular systolic pressure near 100 mmHg with a diastolic pressure not exceeding 15 mmHg at the beginning of the experiment. Atrial pacing was performed at 200 beats/min if the spontaneous rate was lower. Total coronary artery flow was measured by timed collection of perfusate dripping from the heart. Following the experimental protocols described below, all hearts were perfused for 2 h before delineation of the risk zone.

Measurement of infarct and risk zone

As recently demonstrated by Birnbaum *et al.* [20] 3 h of reperfusion in *in situ* preparations is sufficient to yield reliable infarct determinations with supravital stains. Furthermore, as shown by Ytrehus *et al.* [21], infarcts are not appreciably different whether evaluated with supravital stains after 2 h of reperfusion in *in vitro* hearts or 3 h of reperfusion in *in situ* hearts or with histological techniques after 72 h of reflow. At the end of the present experiments, *in situ* hearts were quickly removed from the chest and mounted on a Langendorff apparatus and the coronary arteries were perfused with saline to wash out blood. Then in these hearts and those *in vitro* hearts already on the Langendorff apparatus the coronary artery was reoccluded, and 1–10 μm zinc cadmium sulfide fluorescent particles (Duke Scientific Corp., Palo Alto, CA, USA) were infused into the perfusate to demarcate the risk zone as the tissue without fluorescence. The heart was weighed, frozen, and cut into 2 mm-thick slices. The slices were incubated in 1% triphenyltetrazolium chloride (TTC) in pH 7.4 buffer for 20 min at 37°C. The areas of infarct (TTC negative) and risk zone (nonfluorescent under ultraviolet light) were determined by planimetry. Infarct and risk zone volumes were then calculated by multiplying each area by the slice thickness and summing the products. Infarct size was expressed as a percentage of the risk zone.

Experimental protocols: In situ *experiments*

Eight groups of rabbits were studied, and the protocols are summarized in Fig. 1. All rabbits experienced 30 min of regional ischemia followed by 3 h of reperfusion. The control group of rabbits was subjected only to the above ischemia/reperfusion sequence. The second group was the one-cycle ischemic preconditioning group (PC) in which rabbit hearts were subjected to 5 min of regional ischemia and 10 min of reperfusion before the 30 min ischemic period. The third group of rabbits was treated with an intravenous bolus injection of naloxone (NAL) (3 mg/kg), followed by a 70 μg/kg/min infusion starting 55 min before and extending to the end of the 30 min ischemic interval. In the fourth group (NAL + PC), the naloxone dosing of group 3 and preconditioning protocol of group 2 were combined. The fifth group (NAL + 3 × PC) was preconditioned with three cycles of 5 min ischemia/10 min reperfusion. These hearts were also given the same dose and schedule of naloxone as in the NAL group. The last three groups (MOR0.3, MOR0.8 and MOR3.0) were pretreated with intravenous morphine at doses of 0.3, 0.8 and 3 mg/kg, respectively, 15 min before the 30 min ischemia.

Experimental protocols: In vitro *experiments*

In all experiments, infarcts were induced by 30 min of regional ischemia followed by 2 h of reperfusion. Animals were divided into seven groups (Fig. 2). After 20 min of equilibration, the control group experienced only ischemia and reperfusion. In the second group, morphine (MOR) was included in the perfusate (0.3 μM) for 15 min preceding the 30 min regional ischemia. This dose was chosen because it was approximately 10 times the K_i for the μ receptor. In rabbits treated with naloxone (NAL) (1 μM), the latter was included in the perfusate for 30 min, starting 25 min before the regional ischemia. In the NAL + MOR group, naloxone and morphine were present as noted above. In the chelerythrine group (CHE), hearts were perfused with buffer containing chelerythrine (5 μM), a PKC inhibitor, for 15 min starting 5 min before onset of regional ischemia. In the MOR + CHE group, morphine and chelerythrine were given as already described. In the NAL + PC group, rabbits were preconditioned with 5 min of global ischemia and 10 min of reperfusion before the 30 min ischemia, and naloxone was infused as described above.

Exclusion criteria

We excluded rabbits for the following reasons: (1) risk zone area below 0.5 cm^3 [21], (2) ventricular fibrillation lasting more than 2 min, (3) coronary flow below 15 ml/min during reperfusion in *in vitro* hearts. No pharmacologic agents were used for defibrillation.

Chemicals

Naloxone was purchased from Sigma (St. Louis, MO, USA), and chelerythrine chloride was purchased from Research

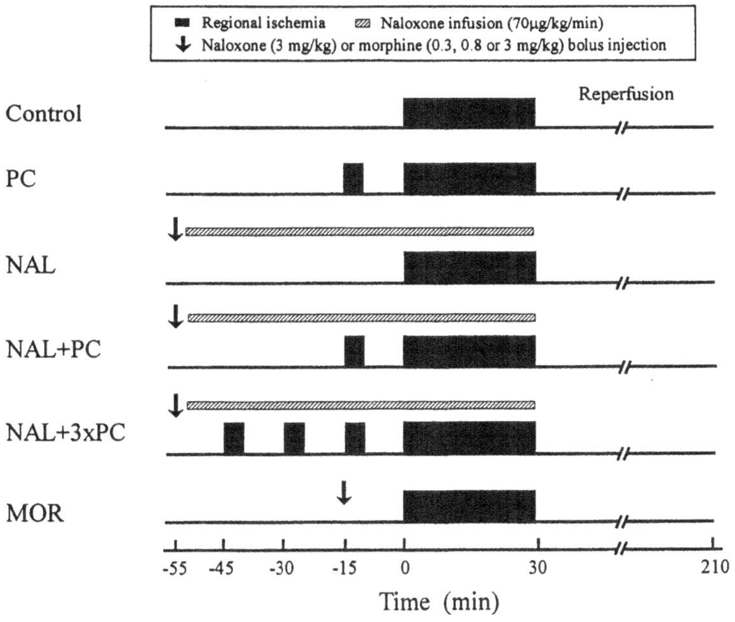

Fig. 1. Experimental protocols for *in situ* hearts. Timing of interventions is indicated by timing line at bottom. Abbreviations: MOR – morphine, NAL – naloxone, PC – ischemic preconditioning

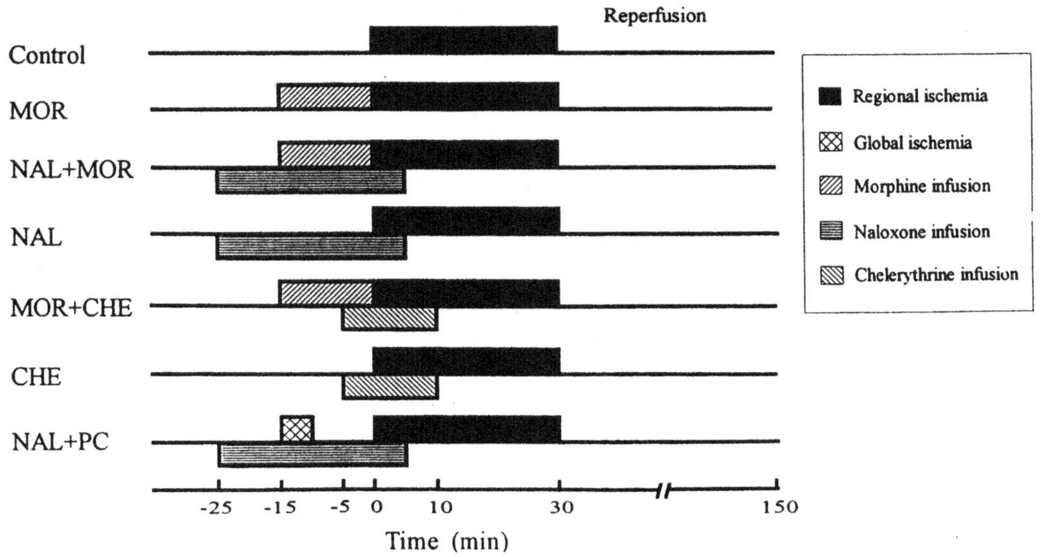

Fig. 2. Experimental protocols for isolated hearts. Timing of interventions is indicated by timing line at bottom. Abbreviations: see Fig. 1, CHE – chelerythrine

Biochemical Inc. (Natick, MA, USA). Morphine sulfate was obtained from Bergen Brunswig (Mobile, AL, USA).

Statistics

All data are presented as mean ± S.E.M. One-way ANOVA combined with Scheffé's *post hoc* test was used to test for differences in infarct size between groups. ANOVA with replication was used to test for differences in hemodynamics in any given group. The difference was considered significant if the p value was less than 0.05.

Results

In situ *experiments*

We excluded two rabbits from analysis. The risk zone of one rabbit in the control group was below 0.5 cm^3, while one rabbit in the PC group experienced prolonged ventricular fibrillation. Additionally one rabbit in the NAL + 3 × PC group fibrillated during the preconditioning ischemia, and one rabbit in the MOR0.3 group fibrillated during the 30 min ischemic period. One other rabbit in the PC group and one in the MOR3.0 group fibrillated during reperfusion. However, these latter 4 hearts spontaneously converted to sinus rhythm within 2 min.

Hemodynamic parameters in the seven groups are summarized in Table 1. Heart rate and mean blood pressure were comparable in all groups under baseline conditions. Administration of naloxone significantly reduced mean blood pressure without changing the heart rate. Administration of morphine at doses of 0.3 and 0.8 mg/kg slightly reduced mean blood pressure, but pressure returned to the baseline level before the coronary occlusion. However, the highest dose of morphine significantly reduced both heart rate and mean blood pressure.

Table 2 presents animal body weight, heart weight and risk zone size data. There were no significant differences in these parameters among the groups. Table 2 also summarizes infarct size data for the groups while Fig. 3 shows the infarct size for each animal. Infarct size expressed as a percentage of risk zone was 38.5 ± 1.6% in the control group, whereas preconditioning with 5 min ischemia/10 min reperfusion (PC group) significantly reduced infarct size to 12.7 ± 2.9% ($p < 0.01$ vs. control group). Pretreatment with naloxone blocked the cardioprotective effect of one cycle of preconditioning. Infarct size in the NAL + PC group was 34.4 ± 3.6% (p = NS vs. control group), while naloxone alone did not modify infarct size. However, when the hearts were preconditioned with three cycles of 5 min ischemia/10 min reperfusion (NAL + 3 × PC), naloxone failed to block protection (9.9 ± 1.4% infarction). Pretreatment with morphine at 0.3 and 0.8 mg/kg (MOR0.3 and MOR0.8 groups) had no effect on infarct size (41.0 ± 3.6% and 38.4 ± 3.4% infarction, respectively). Only the 3 mg/kg dose (MOR3.0) protected the hearts (20.3 ± 3.3% infarction, $p < 0.05$ vs. control group).

In vitro *experiments*

We excluded 3 rabbits: 1 in the NAL + MOR group, 1 in the MOR + CHE group, and 1 in the NAL + PC group. All failed

Table 1. Hemodynamic parameters for *in situ* hearts

	Baseline	Treatment	Ischemia	Reperfusion
Heart Rate (bpm)				
Control	267 ± 14	–	266 ± 12	264 ± 11
PC	271 ± 11	271 ± 13	281 ± 11	273 ± 12
NAL	260 ± 7	262 ± 9	250 ± 11	245 ± 13
NAL + PC	279 ± 8	274 ± 9	271 ± 9	259 ± 10
NAL + 3 × PC	282 ± 13	271 ± 16	271 ± 7	262 ± 10
MOR0.3	274 ± 5	268 ± 4	262 ± 9	262 ± 12
MOR0.8	257 ± 11	256 ± 13	268 ± 11	266 ± 11
MOR3.0	258 ± 5	239 ± 5**	243 ± 6*	234 ± 7**
Mean Blood Pressure (mmHg)				
Control	80 ± 5	–	74 ± 6*	71 ± 5*
PC	89 ± 2	87 ± 2	83 ± 3	78 ± 4
NAL	83 ± 5	77 ± 5*	66 ± 6**	66 ± 7**
NAL + PC	89 ± 4	82 ± 3*	76 ± 2*	67 ± 3**
NAL + 3 × PC	89 ± 3	76 ± 3**	72 ± 4**	64 ± 6**
MOR0.3	83 ± 2	79 ± 3	73 ± 3**	70 ± 3**
MOR0.8	87 ± 7	82 ± 7	76 ± 5	67 ± 6
MOR3.0	83 ± 4	60 ± 4**	63 ± 4**	57 ± 4**

Values are mean ± S.E.M. *p < 0.05 vs. Baseline, **p < 0.01 vs. Baseline. Measurements were made 1 min before coronary occlusion (Treatment), 30 min after coronary occlusion (Ischemia), 30 min after reperfusion (Reperfusion) Abbreviations: MOR – morphine, NAL – naloxone, PC – ischemic preconditioning

Table 2. Infarct size data for *in situ* hearts

	N	Body weight (kg)	Heart weight (g)	Risk zone (cm^3)	Infarct (cm^3)	Infarct size % of risk zone
Control	7	2.0 ± 0.1	6.7 ± 0.3	1.09 ± 0.08	0.43 ± 0.05	38.5 ± 1.6
PC	6	2.1 ± 0.1	8.1 ± 0.4	1.18 ± 0.09	0.15 ± 0.04	12.7 ± 2.9**
NAL	5	2.3 ± 0.1	7.5 ± 0.2	1.05 ± 0.15	0.42 ± 0.07	39.6 ± 1.6
NAL + PC	7	2.2 ± 0.0	8.0 ± 0.4	0.91 ± 0.10	0.34 ± 0.07	34.4 ± 3.6
NAL + 3 × PC	6	2.1 ± 0.1	7.6 ± 0.6	1.13 ± 0.10	0.11 ± 0.02	9.9 ± 1.4**
MOR0.3	5	2.1 ± 0.0	6.5 ± 0.3	0.95 ± 0.10	0.40 ± 0.06	41.0 ± 3.6
MOR0.8	5	2.1 ± 0.1	7.3 ± 0.7	1.10 ± 0.09	0.42 ± 0.04	38.4 ± 3.4
MOR3.0	7	2.2 ± 0.1	7.1 ± 0.2	1.09 ± 0.12	0.24 ± 0.06	20.3 ± 3.3*

Values are mean ± S.E.M. *p < 0.05 vs. Control **p < 0.01 vs. Control. Abbreviations: See Table 1, N – number of rabbits in each group.

to maintain coronary flow above 15 ml/min during reperfusion. Table 3 presents hemodynamic data. Baseline heart rates, left ventricular developed pressures, and coronary flows were comparable in all groups. Administration of morphine tended to decrease developed pressure and coronary flow, but these changes were not statistically significant. Pretreatment with naloxone significantly decreased both developed pressure and coronary flow. On the other hand, administration of chelerythrine increased coronary flow without changing developed pressure. In all groups coronary flow and developed pressure were depressed during ischemia. Developed pressure and coronary flow recovered somewhat over the 2 h reperfusion period in all groups except those treated with chelerythrine. None of the hemodynamic changes correlated with the effects of the drugs on infarct size.

As noted in Table 4 there were no significant differences in body weight, heart weight, or risk zone size among the groups. Infarct size as a percentage of risk zone was 32.0 ± 3.1% in the control group (Fig. 4). A 15 min infusion of 0.3 μM morphine significantly limited infarct size to 9.3 ± 1.2%. This cardioprotective effect of morphine was completely blocked by either naloxone (34.0 ± 2.6% infarction in the NAL + MOR group) or chelerythrine (30.4 ± 2.6% infarction in the MOR + CHE group). Neither naloxone nor chelerythrine modified infarct size in non-preconditioned hearts. Unlike our finding in the *in situ* hearts, naloxone could not block the effect of one cycle of preconditioning in the isolated rabbit heart (10.8 ± 1.0% infarction).

Discussion

In the present study naloxone blocked the protection from one cycle of ischemic preconditioning in *in situ* hearts, confirming

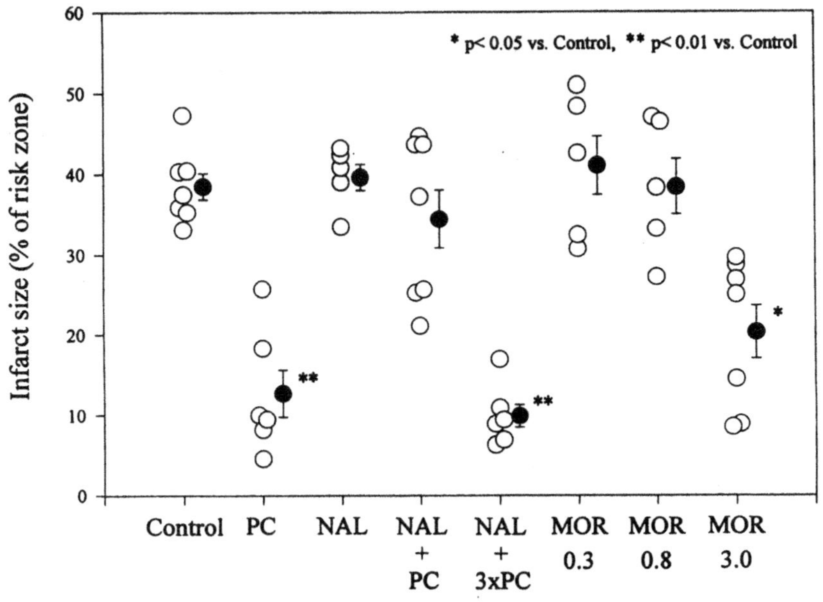

Fig. 3. Graph showing the effect of ischemic preconditioning (PC), naloxone (NAL) and morphine (MOR) on infarct size measured as a percentage of risk zone in *in situ* hearts. Naxolone was able to block the protection of a single cycle of PC (NAL + PC). However, it could not block protection when PC was amplified with three cycles of ischemia/reperfusion (NAL + 3 × PC). Morphine at doses of 0.3 and 0.8 mg/kg (MOR0.3 or MOR0.8) could not mimic ischemic preconditioning. However, a higher dose of 3.0 mg/kg (MOR.3.0) protected the hearts. Open and closed circles represent individual animal infarct sizes and group means, respectively.

Table 3. Hemodynamic parameters for *in vitro* hearts

	Baseline	Treatment	Ischemia	Reperfusion
Heart rate (bpm)				
Control	210 ± 9	–	210 ± 7	207 ± 5
MOR	209 ± 4	209 ± 6	204 ± 4	207 ± 4
NAL + MOR	219 ± 6	210 ± 7	210 ± 7*	206 ± 5*
NAL	207 ± 5	205 ± 3	198 ± 1	200 ± 0
MOR + CHE	221 ± 10	208 ± 6	212 ± 8	203 ± 3
CHE	211 ± 5	210 ± 5	202 ± 2	200 ± 0
NAL + PC	207 ± 4	200 ± 0	200 ± 0	200 ± 0
Left ventricular developed pressure (mmHg)				
Control	106 ± 5	–	60 ± 7**	79 ± 8**
MOR	107 ± 7	90 ± 7	63 ± 10*	81 ± 8*
NAL + MOR	118 ± 4	102 ± 6**	61 ± 7**	77 ± 7**
NAL	115 ± 5	97 ± 9*	59 ± 7**	76 ± 7**
MOR + CHE	116 ± 2	107 ± 3	59 ± 6**	50 ± 3**
CHE	122 ± 5	118 ± 6	64 ± 10**	57 ± 6**
NAL + PC	121 ± 3	90 ± 7**	63 ± 6**	80 ± 4**
Coronary flow (ml/min/g)				
Control	8.7 ± 0.9	–	6.0 ± 1.1**	7.1 ± 1.1 *
MOR	8.3 ± 0.5	7.7 ± 0.6	4.6 ± 0.6**	6.4 ± 0.5*
NAL + MOR	9.4 ± 0.4	7.4 ± 0.3*	4.2 ± 0.4**	5.1 ± 0.7**
NAL	9.8 ± 0.7	7.9 ± 0.6**	4.9 ± 0.5**	6.5 ± 0.7**
MOR + CHE	9.9 ± 0.8	9.1 ± 0.5	4.9 ± 1.1**	4.7 ± 0.8**
CHE	10.6 ± 0.6	11.9 ± 1.0*	6.3 ± 1.1**	5.2 ± 1.0**
NAL + PC	9.8 ± 0.6	8.1 ± 0.8**	4.8 ± 0.6**	5.9 ± 0.6**

Values are mean ± S.E.M. *p < 0.05 vs. Baseline, **p < 0.01 vs. Baseline. Measurements were made 1 min before coronary occlusion (Treatment), 30 min after coronary occlusion (Ischemia), 30 min after reperfusion (Reperfusion) Abbreviations: CHE – chelerythrine, MOR – morphine, NAL – naloxone, PC – ischemic preconditioning

Table 4. Infarct size data for *in vitro* hearts

	N	Body weight (kg)	Heart weight (g)	Risk zone (cm^3)	Infarct (cm^3)	Infarct size % of risk zone
Control	6	2.1 ± 0.2	6.3 ± 0.4	0.98 ± 0.09	0.32 ± 0.06	32.0 ± 3.1
MOR	6	2.3 ± 0.1	6.7 ± 0.3	0.96 ± 0.06	0.09 ± 0.01	9.3 ± 1.2*
NAL + MOR	6	2.2 ± 0.1	7.1 ± 0.3	0.92 ± 0.10	0.33 ± 0.06	34.0 ± 2.6
NAL	5	2.2 ± 0.1	6.6 ± 0.4	1.07 ± 0.11	0.36 ± 0.07	32.9 ± 2.5
MOR + CHE	6	2.1 ± 0.1	6.6 ± 0.3	1.03 ± 0.14	0.33 ± 0.07	30.4 ± 2.6
CHE	5	2.0 ± 0.1	5.7 ± 0.2	1.02 ± 0.08	0.39 ± 0.07	36.7 ± 4.1
NAL + PC	6	2.1 ± 0.1	6.2 ± 0.3	0.90 ± 0.09	0.10 ± 0.02	10.8 ± 1.0*

Values are mean ± S.E.M. *p < 0.01 vs. Control. Abbreviations: See Tables 2 and 3

once again that opioid receptors are also involved in the preconditioning phenomenon. Protection could not be blocked, however, when hearts were preconditioned by three cycles of 5 min ischemia/10 min reperfusion suggesting that opioid receptors act in parallel with other receptors, including those triggered by adenosine and bradykinin, to initiate protection. Morphine could mimic ischemic preconditioning in *in situ* hearts only if a dose almost 20 times higher than the usual clinical dose was used (3 mg/kg). In isolated hearts, morphine (0.3 μM) could mimic the protection of ischemic preconditioning, and that protection could be blocked by naloxone confirming a direct role for opioid receptors. Chelerythrine also blocked the cardioprotective effect of morphine supporting the hypothesis that opioids produced during ischemia help to trigger protection by PKC activation.

In these experiments naloxone decreased mean blood pressure *in situ* and also decreased developed pressure and coronary flow *in vitro*. Several previous studies in rats found no effect of intravenous naloxone on blood pressure [12, 22]. We cannot explain the differences, but these hemodynamic changes appear to be too small to influence infarct size. The highest dose of morphine (3 mg/kg) decreased both heart rate and mean blood pressure in *in situ* preparations, but these changes likely had no effect on infarct size [23].

During myocardial ischemia, many biologically active substances are released by the ischemic tissue, and recent studies have shown that several of them, including adenosine [3, 4], bradykinin [5, 6] and free radicals [7, 8], contribute to

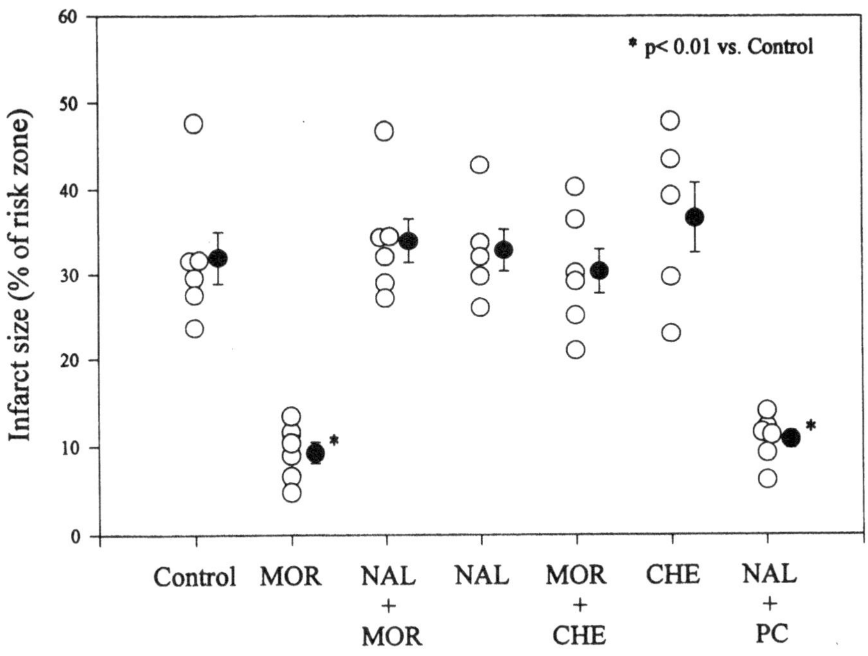

Fig. 4. Graph showing the effect of morphine (MOR), naloxone (NAL), chelerythrine (CHE), and ischemic preconditioning (PC) on infarct size measured as a percentage of risk zone in isolated hearts. The protective effect of MOR was blocked by either NAL or CHE. NAL could not block the protection of PC in the isolated heart. Open and closed circles represent individual animal infarct sizes and group means, respectively.

ischemic preconditioning. The present data support a role for opioids in that phenomenon as well. It has been reported that endogenous opioid peptides are released from the pituitary gland and adrenal medulla during various cardiovascular stress situations, such as shock [12], congestive heart failure [24] and myocardial ischemia [11].

Not all studies report a beneficial effect of activation of opioid receptors. Several reports suggest that naloxone attenuates ischemia-induced arrhythmias in both rats [25] and dogs [26] suggesting participation of endogenous opioid peptides in cardiac arrhythmogenesis. But Chien *et al.* [27] reported that opioid agonists increased tissue preservation and survival time before organ transplantation. The possible involvement of opioids in the protective effect of ischemic preconditioning has been investigated in the rat by Schultz *et al.* [13] and in the rabbit by Chien *et al.* [16]. Our results not only confirm the previous observations that opioids can act as a trigger for ischemic preconditioning but also reveal a part of the mechanism.

In the present study we tested whether a PKC inhibitor could block the effect of morphine. We have previously proposed that PKC activation is a critical part of the mechanism of preconditioning in the rabbit. We noted that PKC inhibitors abolished the infarct size-limiting effect of preconditioning [9]. Furthermore, stimulation of any receptor which can activate PKC was found to induce a preconditioned state. Indeed, adenosine [4], bradykinin [6], catecholamines [28, 29], angiotensin II [30], free radicals [7], and endothelin [31] have all been shown to mimic ischemic preconditioning. Consistent with our hypothesis chelerythrine selectively abolished the infarct size-limiting effect of morphine. These results reveal that morphine, like ischemic preconditioning and pharmacological preconditioning with PKC-coupled agonists, protects the myocardium by a mechanism that involves PKC activation.

Opioid receptors have at least three known subtypes, μ, δ, and κ. These receptors are known to couple to pertussis toxin-sensitive G proteins such as G_i or G_o [17, 18]. If opioid receptors act by activation of PKC, then they must couple to a phospholipase. Ventura *et al.* demonstrated that δ and κ receptors are present in rat heart [32] and δ and κ (not μ) receptor stimulation could increase the level of inositol 1, 4, 5-trisphosphate, suggesting phospholipase C-mediated turnover of phosphatidylinositol in the rat cardiomyocytes [33]. However, Mangoura *et al.* [19] showed that μ-opioid receptor agonists activate phospholipase D which has also been shown to be an important activation pathway for preconditioning in the rabbit heart [34]. In the present study, we used naloxone and morphine, both of which have high affinity for the μ-receptor compared with others [35]. However, at high doses these agents affect other receptor types. Therefore, we cannot specify which receptor type is responsible for the observed cardioprotection.

Although the contribution of PKC to preconditioning in small animals has been confirmed by many [4, 9, 29, 36], a similar role in large animals has been difficult to demonstrate [37, 38]. Kita *et al.* [39] reported that opioid receptors participate in preconditioning against reperfusion arrhythmias in the rat, but a role for PKC activation in that response could not be demonstrated. Schultz *et al.* [40] then showed that the cardioprotection of morphine could be blocked by glibenclamide, a K_{ATP} channel blocker, in *in situ* rat hearts. The suggestion was that the cardioprotection of opioid receptor activation may be effected by the K_{ATP} channel. The evidence that μ- and δ-opioid receptors are linked via a G protein to K^+ channels [41] supports this idea. However, the role of K_{ATP} channels during preconditioning has been controversial in rabbits [42, 43], and, therefore, we did not test any K_{ATP} channel blocking agents in these experiments.

The initial dose of morphine used in the *in situ* studies was 0.3 mg/kg which was the same as that used by Schultz [40] to mimic ischemic preconditioning in the *in situ* rat heart. However, neither this dose nor the higher dose of 0.8 mg/kg limited infarct size in *in situ* rabbit hearts. Only 3 mg/kg which is more than 20 times the usual clinical dose yielded a significant limitation of infarct size. We measured the morphine concentration in arterial blood 5 min after administration in two rabbits treated with 3 mg/kg (National Medical Services Inc, Willow Grove, PA, USA). The concentrations were 0.57 and 0.94 μM. Using morphine's known pharmacokinetics in humans, i.e., a volume of distribution of 3.2 liters/kg and 35% protein binding [44, 45], we calculated that an intravenous injection of 3 mg/kg should have yielded a plasma concentration of 0.91 μM. Although the rabbits in the MOR3.0 group should have had approximately 2–3 times the concentration of morphine in their plasma than was present in the perfusate for the *in vitro* hearts, protection appeared to be less (20.3 ± 3.3% vs. 9.3 ± 1.2%) although the difference was not significant. In light of these observations it should not be assumed that patients receiving a clinical dose of morphine are all preconditioned.

In *in situ* hearts naloxone blocked the protection from one cycle of preconditioning but not that from three cycles. A similar phenomenon has been documented with bradykinin receptor antagonists [6] and free radical scavengers [7]. Four substances (adenosine, bradykinin, free radicals, and opioids) are produced during 5 min of ischemia, and all appear to converge on PKC to trigger preconditioning. We propose that elimination of any one of these triggers results in insufficient stimulation of PKC by the remaining 3 agonists to induce the preconditioned state from a single 5 min period of ischemia. However, when the preconditioning stimulus is amplified with multiple cycles of preconditioning, protection returns, apparently because increased release of the remaining agonists produces enough additional stimulation of PKC. The presence of multiple mediators of preconditioning makes the pathway very redundant which ensures that the heart will be preconditioned by a transient period of ischemia.

Infarct size of preconditioned isolated hearts in our previous studies has been approximately 10% of the risk zone [6, 7, 29]. It is important to note that naloxone failed to block the infarct size-limiting effect of ischemic preconditioning in isolated hearts. Opioid peptides are known to be produced not only in the brain but also by the peripheral autonomic nerves. It has been reported that opioid peptides exist in canine [46] and guinea pig hearts [47, 48]. However, the present study suggests that an intact innervation may be required for endogenous opioid release sufficient to trigger preconditioning. Another possible explanation is that opioids responsible for triggering protection may be produced by non-cardiac tissue, and, therefore, reach the heart through the circulation. A similar effect was seen with bradykinin because HOE 140, a bradykinin receptor antagonist, failed to block preconditioning's protection in isolated hearts [6]. Thus it appears that in the isolated heart only adenosine and free radicals make a measurable contribution to protection.

In summary, this study demonstrates that opioid receptors participate in the triggering effect of ischemic preconditioning through activation of PKC. Also, because multiple cycles of preconditioning could not be blocked by naloxone in the *in situ* heart, opioid receptors appear to act in concert with other PKC-coupled receptors in the heart, especially those triggered by adenosine and bradykinin.

Acknowledgement

This study was supported in part by grants from the National Institutes of Health Heart, Lung, and Blood Institute HL-20648 and HL-50688.

References

1. Miura T, Iimura O: Infarct size limitation by preconditioning: Its phenomenological features and the key role of adenosine. Cardiovasc Res 27: 36–42, 1993
2. Downey JM, Miura T: The role of adenosine in ischemic preconditioning. In: M Hori, Y Maruyama, RS Reneman (eds). Cardiac Adaptation and Failure. Springer-Verlag, Tokyo, 1994, pp 147–166
3. Liu GS, Thornton J, Van Winkle DM, Stanley AWH, Olsson RA, Downey JM: Protection against infarction afforded by preconditioning is mediated by A_1 adenosine receptors in rabbit heart. Circulation 84: 350–356, 1991
4. Sakamoto J, Miura T, Goto M, Iimura O: Limitation of myocardial infarct size by adenosine A_1 receptor activation is abolished by protein kinase C inhibitors in the rabbit. Cardiovasc Res 29: 682–688, 1995
5. Wall TM, Sheehy R, Hartman JC. Role of bradykinin in myocardial preconditioning. J Pharmacol Exp Ther 270: 681–689, 1994
6. Goto M, Liu Y, Yang X-M, Ardell JL, Cohen MV, Downey JM: Role of bradykinin in protection of ischemic preconditioning in rabbit hearts. Circ Res 77: 611–621, 1995
7. Baines CP, Goto M, Downey JM: Oxygen radicals released during ischemic preconditioning contribute to cardioprotection in the rabbit myocardium. J Mol Cell Cardiol 29: 207–216, 1997
8. Tanaka M, Fujiwara H, Yamasaki K, Sasayama S: Superoxide dismutase and N-2-mercaptopropionyl glycine attenuate infarct size limitation effect of ischaemic preconditioning in the rabbit. Cardiovasc Res 28: 980–986, 1994
9. Ytrehus K, Liu Y, Downey JM: Preconditioning protects ischemic rabbit heart by protein kinase C activation. Am J Physiol 266: H1145–H1152, 1994
10. Liu Y, Ytrehus K, Downey JM: Evidence that translocation of protein kinase C is a key event during ischemic preconditioning of rabbit myocardium. J Mol Cell Cardiol 26: 661–668, 1994
11. Chen Y-T, Lin C-J, Lee AY-S: Plasma levels of endogenous opioid peptides in patients with acute myocardial infarction. Jpn Heart J 36: 421–427, 1995
12. Holaday JW: Cardiovascular effects of endogenous opiate systems. Annu Rev Pharmacol Toxicol 23: 541–594, 1983
13. Schultz JEJ, Rose E, Yao Z, Gross GJ: Evidence for involvement of opioid receptors in ischemic preconditioning in rat hearts. Am J Physiol 268: H2157–H2161, 1995
14. Headrick JP: Ischemic preconditioning: Bioenergetic and metabolic changes and the role of endogenous adenosine. J Mol Cell Cardiol 28: 1227–1240, 1996
15. Liu Y, Downey JM: Ischemic preconditioning protects against infarction in rat heart. Am J Physiol 263: H1107–H1112, 1992
16. Chien GL, Van Winkle DM: Naloxone blockade of myocardial ischemic preconditioning is stereoselective. J Mol Cell Cardiol 28: 1895–1900, 1996
17. Aghajanian GK, Wang Y-Y: Pertussis toxin blocks the outward currents evoked by opiate and α_2-agonists in locus coeruleus neurons. Brain Res 371: 390–394, 1986
18. Burns DL, Hewlett EL, Moss J, Vaughan M: Pertussis toxin inhibits enkephalin stimulation of GTPase of NG108-15 cells. J Biol Chem 258: 1435–1438, 1983
19. Mangoura D, Dawson G: Opioid peptides activate phospholipase D and protein kinase C-ε in chicken embryo neuron cultures. Proc Natl Acad Sci USA 90: 2915–2919, 1993
20. Birnbaum Y, Hale SL, Kloner RA: Differences in reperfusion length following 30 minutes of ischemia in the rabbit influence infarct size, as measured by triphenyltetrazolium chloride staining. J Mol Cell Cardiol 29: 657–666, 1997
21. Ytrehus K, Liu Y, Tsuchida A, Miura T, Liu GS, Yang X-M, Herbert D, Cohen MV, Downey JM: Rat and rabbit heart infarction: Effects of anesthesia, perfusate, risk zone, and method of infarct sizing. Am J Physiol 267: H2383–H2390, 1994
22. Lee AY-S, Chen Y-T, Kan M-N, P'eng F-K, Chai C-Y, Kuo J-S: Consequences of opiate agonist and antagonist in myocardial ischaemia suggest a role of endogenous opioid peptides in ischaemic heart disease Cardiovasc Res 26: 392–395, 1992
23. Tsuchida A, Miura T, Miki T, Shimamoto K, Iimura O: Role of adenosine receptor activation in myocardial infarct size limitation by ischaemic preconditioning. Cardiovasc Res 26: 456–461, 1992
24. Liang C-S, Imai N, Stone CK, Woolf PD, Kawashima S, Tuttle RR: The role of endogenous opioids in congestive heart failure: Effects of nalmefene on systemic and regional hemodynamics in dogs. Circulation 75: 443–451, 1987
25. Kan MN, Chen YT, Lee AYS: Naloxone reversal of ischemic arrhythmia is stereospecific and suggests role of endogenous opioid peptides in ischemic heart disease. Proc Soc Exp Biol Med 200: 518–521, 1992
26. Lee AY-S: Stereospecific antiarrhythmic effects of naloxone against myocardial ischaemia and reperfusion in the dog. Br J Pharmacol 107: 1057–1060, 1992

27. Chien S, Oeltgen PR, Diana JN, Salley RK, Su T-P: Extension of tissue survival time in multiorgan block preparation with a delta opioid DADLE ([D-Ala^2, D-Leu^5]-enkephalin). J Thorac Cardiovasc Surg 107: 964–967, 1994
28. Thornton JD, Daly JF, Cohen MV, Yang X-M, Downey JM: Catecholamines can induce adenosine receptor-mediated protection of the myocardium but do not participate in ischemic preconditioning in the rabbit. Circ Res 73: 649–655, 1993
29. Tsuchida A, Liu Y, Liu GS, Cohen MV, Downey JM: α_1-Adrenergic agonists precondition rabbit ischemic myocardium independent of adenosine by direct activation of protein kinase C. Circ Res 75: 576–585, 1994
30. Liu Y, Tsuchida A, Cohen MV, Downey JM: Pretreatment with angiotensin II activates protein kinase C and limits myocardial infarction in isolated rabbit hearts. J Mol Cell Cardiol 27: 883–892, 1995
31. Wang P, Gallagher KP, Downey JM, Cohen MV: Pretreatment with endothelin-1 mimics ischemic preconditioning against infarction in isolated rabbit heart. J Mol Cell Cardiol 28: 579–588, 1996
32. Ventura C, Bastagli L, Bernardi P, Caldarera CM, Guarnieri C: Opioid receptors in rat cardiac sarcolemma: Effect of phenylephrine and isoproterenol. Biochim Biophys Acta 987: 69–74, 1989
33. Ventura C, Spurgeon H, Lakatta EG, Guarnieri C, Capogrossi MC: κ and δ opioid receptor stimulation affects cardiac myocyte function and Ca^{2+} release from an intracellular pool in myocytes and neurons. Circ Res 70: 66–81, 1992
34. Cohen MV, Liu Y, Liu GS, Wang P, Weinbrenner C, Cordis GA, Das DK, Downey JM: Phospholipase D plays a role in ischemic preconditioning in rabbit heart. Circulation 94: 1713–1718, 1996
35. Paterson SJ, Robson LE, Kosterlitz HW: Classification of opioid receptors. Br Med Bull 39: 31–36, 1983
36. Mitchell MB, Meng X, Ao L, Brown JM, Harken AH, Banerjee A: Preconditioning of isolated rat heart is mediated by protein kinase C. Circ Res 76: 73–81, 1995
37. Przyklenk K, Sussman MA, Simkhovich BZ, Kloner RA: Does ischemic preconditioning trigger translocation of protein kinase C in the canine model? Circulation 92: 1546–1557, 1995
38. Vahlhaus C, Schulz R, Post H, Onallah R, Heusch G: No prevention of ischemic preconditioning by the protein kinase C inhibitor staurosporine in swine. Circ Res 79: 407–414, 1996
39. Kita H, Miura T, Tsuchida A, Sakamoto J, Iimura O: Roles of opioid receptor and protein kinase C in suppression of reperfusion arrhythmia by preconditioning. J Mol Cell Cardiol 27: (abstr) A520, 1995
40. Schultz JEJ, Hsu AK, Gross GJ: Morphine mimics the cardioprotective effect of ischemic preconditioning via a glibenclamide-sensitive mechanism in the rat heart. Circ Res 78: 1100–1104, 1996
41. North RA, Williams JT, Surprenant A, Christie MJ: μ and δ receptors belong to a family of receptors that are coupled to potassium channels. Proc Natl Acad Sci USA 84: 5487–5491, 1987
42. Miura T, Goto M, Miki T, Sakamoto J, Shimamoto K, Iimura O: Glibenclamide, a blocker of ATP-sensitive potassium channels, abolishes infarct size limitation by preconditioning in rabbits anesthetized with xylazine/pentobarbital but not with pentobarbital alone. J Cardiovasc Pharmacol 25: 531–538, 1995
43. Walsh RS, Tsuchida A, Daly JJF, Thornton JD, Cohen MV, Downey JM: Ketamine-xylazine anaesthesia permits a K_{ATP} channel antagonist to attenuate preconditioning in rabbit myocardium. Cardiovasc Res 28: 1337–1341, 1994
44. Heel RC, Avery GS: Appendix A. Drug data information. In: GS Avery (ed). Drug Treatment. Principles and Practice of Clinical Pharmacology and Therapeutics. Second edition. Adis Press, Sydney, 1980, 1211–1222
45. Stanski DR, Greenblatt DJ, Lowenstein E: Kinetics of intravenous and intramuscular morphine. Clin Pharmacol Ther 24: 52–59, 1978
46. Barron BA, Gu H, Gaugl JF, Caffrey JL: Screening for opioids in dog heart. J Mol Cell Cardiol 24: 67–77, 1992
47. Lang RE, Hermann K, Dietz R, Gaida W, Ganten D, Kraft K, Unger T: Evidence for the presence of enkephalins in the heart. Life Sci 32: 399–406, 1983
48. Weihe E, McKnight AT, Corbett AD, Kosterlitz HW: Proenkephalin- and prodynorphin-derived opioid peptides in guinea-pig heart. Neuropeptides 5: 453–456, 1985

Molecular and Cellular Biochemistry **186:** 13–18, 1998.
© 1998 *Kluwer Academic Publishers.*

Equal reduction in infarct size by ethylisopropyl-amiloride pretreatment and ischemic preconditioning in the *in situ* rabbit heart

Jens Munch-Ellingsen, Jan Eirik Løkebø, Einar Bugge and Kirsti Ytrehus

Department of Medical Physiology, University of Tromsø, Norway

Abstract

Inhibition of Na^+/H^+ exchange with amiloride analogues has been shown to provide functional protection during ischemia and reperfusion and to reduce infarct size in isolated rat hearts. In rat hearts, treatment with ethylisopropyl-amiloride (EIPA, a selective Na^+/H^+ exchange inhibitor) was additive to the protection afforded by ischemic preconditioning. In addition, such compounds have been demonstrated to reduce infarct size in *in situ* rabbit hearts. The aim of the present study was to determine to what extent preischemic treatment with EIPA could reduce infarct size in an *in situ* rabbit model of regional ischemia and reperfusion. We also wished to determine if this effect was additive to the infarct reducing effect of ischemic preconditioning. Anaesthetized, open chest rabbits, were subjected to 45 min of regional ischemia and 150 min of reperfusion. The risk zone was determined by fluorescent particles and infarct size was determined by TTC staining. Four groups were investigated: control, ischemic preconditioned (IP) (5 min of ischemia followed by 10 min reperfusion), EIPA (0.65 mg/kg iv given preischemically) and EIPA + IP. The main results expressed as percent infarction of the risk zone ± S.E.M. for the different groups were: control 59.2 ± 3.3% (n = 6), IP 16.3 ± 2.1% (n = 6) ($p < 0.001$ vs. control), EIPA 16.9 ± 4.1% (n = 5) ($p < 0.001$ vs. control), EIPA + IP 22.5 ± 9.5% (n = 6) ($p < 0.001$ vs. control). In conclusion: EIPA, when administered prior to ischemia, caused a reduction in infarct size in the *in situ* rabbit heart which was similar to that seen with ischemic preconditioning, however, the effect was not additive to ischemic preconditioning. (Mol Cell Biochem **186**: 13–18, 1998)

Key words: TTC, EIPA, anaesthetized rabbit, myocardial infarction, ischemic preconditioning, sodium hydrogen exchange blockers

Introduction

Several interventions reduce (or delay) cell death in the ischemic heart provided they are used as pretreatment. Ischemic preconditioning and pretreatment with a Na^+/H^+-exchange blocker are two such interventions which, among others, have been intensively investigated in recent years [1, 2].

The mechanism behind protection by ischemic preconditioning is not clear at present although involvement of protein kinase C and ATP dependent potassium channels seems certain in several species [3, 4]. The mechanism responsible for protection with Na^+/H^+-exchange blockers is proposed to be provided by a secondary decrease in intracellular Ca^{2+} loading during ischemia and reperfusion. Activation of the exchanger, as occurs during ischemia [5, 6], will tend to increase intracellular Na^+-concentration ($[Na^+]_i$) [7], which, secondarily, could lead to an increase in $[Ca^{2+}]_i$ through the Na^+/Ca^{2+} exchange mechanism [8]. Accordingly, Tani and Neely showed that $[Na^+]_i$ increases in the ischemic rat heart and that the elevated $[Na^+]_i$ seen towards the end of the ischemic period is linearly correlated with both $^{45}Ca^{2+}$ uptake and depression of ventricular function during reperfusion [8]. Furthermore, over the past years an

Address for offprints: J. Munch-Ellingsen, Department of Medical Physiology, Institute of Medical Biology, University of Tromsø, N-9037 Tromsø, Norway

increasing amount of data has emerged showing that inhibition of Na^+/H^+ exchange with amiloride or amiloride analogues may protect the heart during ischemia and reperfusion [2, 9, 10].

Ethylisopropyl-amiloride (EIPA) is reported to be a specific and potent inhibitor of the Na^+/H^+ exchanger [11]. In a recent study we demonstrated that EIPA significantly decreased infarct size after 30 min ischemia in an *in situ* rabbit model when given before ischemia, but that it had no effect when given at the time of reperfusion [2].

The infarct reducing effect of EIPA treatment is comparable in magnitude to ischemic preconditioning in the isolated perfused rat heart, and we have previously shown that these two infarct reducing mechanisms are additive in the buffer-perfused rat heart [12]. Recently a preliminary report indicated that the Na^+/H^+ exchange blocker, dimethyl-amiloride, prevented ischemic preconditioning in canine heart [13]. It was proposed that protein kinase C (PKC) stimulates the Na^+/H^+ exchanger [14] and furthermore that PKC is closely linked with ischemic preconditioning [15]. We wished to examine the effect of combined treatment with EIPA and ischemic preconditioning and to see whether these are two distinct mechanism for protecting the myocardial cells against ischemic injury. If this is the case, we would expect that a combined treatment would result in an additive effect. In the present study the duration of ischemia was prolonged from 30–45 min compared to our previous study [2] in order to give a more severe ischemic model which would better enhance the possible separation of these two protective mechanisms.

Materials and methods

Surgical preparation of animals

New Zealand white rabbits of either sex, weighting between 2.0–3.0 kg, were anaesthetized with intravenous sodium-pentobarbital (35–45 mg/kg) administered via a marginal ear vein. The trachea was intubated through an midline cervical incision and mechanical ventilation was provided by a positive-pressure respirator (MD industries, Mobile, AL, USA) using 100% O_2, while tidal volume was approximately 15 ml and initial respiration rate was 30 breaths per min. The respiratory rate was adjusted through each experiment in order to maintain blood pH within the physiological range. Positive end-expiratory pressure (2 cm water) was applied to the expiratory outflow of the ventilator to prevent atelectasis.

The left carotid artery was ligated and cannulated for blood pressure monitoring. A catheter was placed in the right jugular vein for administration of additional anaesthesia and EIPA or vehicle. ECG (lead II) was recorded for heart rate monitoring, and to aid in the verification of ischemia. A sternum split was made and a loose pericardial cradle was made to support the heart. A 3-0 silk suture on a curved tapered needle was passed beneath a major branch of the left coronary artery and the ends were drawn through the holes of a small button to form a snare by which the coronary artery could be occluded. A brief occlusion (< 5 sec) was made to ensure and locate tissue ischemia by cyanosis. Reperfusion was achieved by releasing the snare, and was confirmed by visible hyperemia over the surface. Bipolar electrodes were attached to the right atrium for pacing (Medtronic 5325) in order to standardize the experimental conditions.

Drugs

EIPA was dissolved in dimethylsulphoxide (DMSO, 10 mg/ml as a stock solution), and was further diluted in saline to the required concentrations. Ethylisopropyl-amiloride (EIPA) was purchased from Research Biochemicals International, Natick, MA, USA, while DMSO was purchased from Sigma Chemical Company, St. Louis, MO, USA.

Measurement of risk zone and infarct size

At the end of the experiments the hearts were quickly removed and mounted as a retrogradely perfused Langendorff preparation. The hearts were flushed with saline maintained at room temperature for 60 sec, and thereafter the coronary snare was reoccluded. A 0.5% suspension of zinc-cadmium fluorescent particles (1-10 µm in diameter, from Duke Scientific Corp., Palo Alto, CA) was infused into the perfusate to mark the risk zone as non-fluorescent tissue. The hearts were then weighed and frozen overnight before being cut into 2 mm-thick slices parallel with the atrioventricular groove. The slices were then thawed and stained by incubation at 37°C for 20 min in 1% triphenyl-tetrazolium-chloride (TTC) in a phosphate buffer (pH 7.4). Thereafter the slices were immersed in 10% formalin to enhance the contrast of the stain before being placed between two glass plates to a uniform 2 mm thickness. The area of the left ventricle, the infarcted area (tetrazolium negative) and the risk zone (non-fluorescent under UV-light), were determined by computer based planimetry. Planimetry was carried out by an investigator without prior knowledge of experimental group. The volume of the left ventricle, the infarct volume and the risk zone volume were then calculated.

Experimental protocol

Four groups of animals were studied. All animals were subjected to 45 min of regional ischemia and 150 min of reperfusion. The animals in the first group (control, n = 6), served as vehicle (DMSO) treated controls. The animals in the second group (EIPA, n = 5), were given 0.65 mg/kg EIPA intravenously (bolus of 0.5 mg/kg diluted in 2 ml saline followed by a continuous infusion of 0.01 mg/kg/min for 15 min) with regional ischemia being initiated 25 min after the EIPA bolus. The animals in the third group (IP, n = 6) were ischemic preconditioned before the 45 min ischemic period. The ischemic preconditioning consisted of one cycle of 5 min regional ischemia followed by 10 min reperfusion. The animals in the fourth group (EIPA + IP, n = 6) were given the same dose of EIPA as the animals in the EIPA group, but in addition IP was applied. EIPA administration commenced 10 min before the preconditioning ischemia.

Statistical analysis

Values are presented as mean ± S.E.M. One way analysis of variance with the Tukey *post hoc* test was used to test for between group differences. A $p < 0.05$ was accepted as statistically significant.

Results

Infarct size

There were no statistically significant differences between the size of the risk volume among the groups. Infarct volume expressed as percent of the risk volume infarcted is presented in Fig. 1, while infarct volume versus volume of the risk zone in absolute values are presented in Fig. 2. In the vehicle treated control group infarct size was 59.2 ± 3.3% of the risk zone. Treatment with EIPA prior to ischemia significantly reduced infarct size to 16.9 ± 4.1% of the risk zone ($p < 0.001$ vs. control), while the combined treatment with EIPA and ischemic preconditioning (IP) reduced infarct size to 22.5 ± 9.5% ($p < 0.001$ vs. control), and IP alone reduced infarct size to 16.3 ± 2.1% ($p < 0.001$ vs. control).

Hemodynamic parameters

There were no significant differences between measured values at baseline or at any other time points investigated for mean arterial pressure and systolic and diastolic blood pressure. Mean arterial pressure values at selected time points in the protocol are presented in Table 1.

Discussion

The main finding in this study was that the observed reduction in infarct size following EIPA treatment prior to 45 min of ischemia (71.5%) was similar in magnitude to that afforded by ischemic preconditioning (72.4%). No additive protective effect could be detected by combining these two

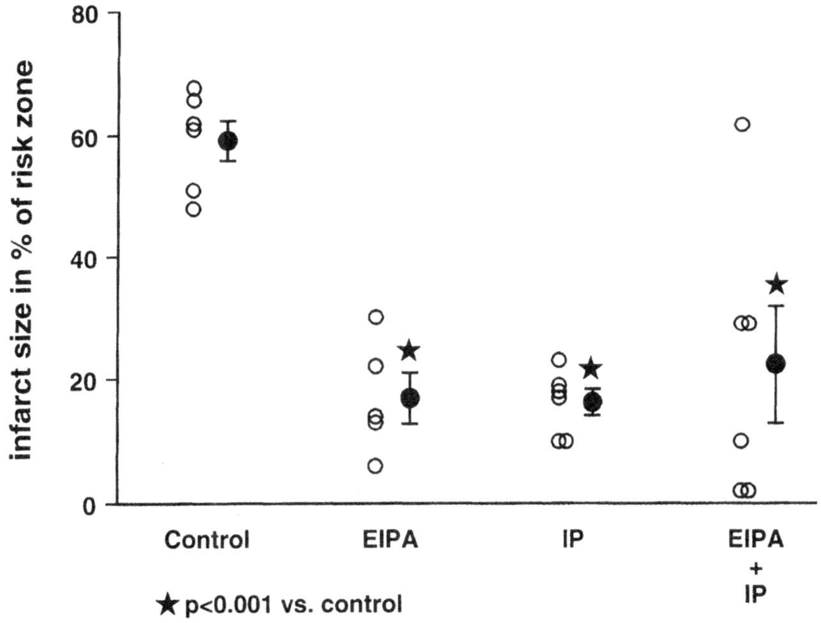

Fig. 1. Infarct size expressed as percent of the risk zone. Open symbols represent single hearts, columns with error bars represent mean of groups ± S.E.M. *p < 0.001 vs. control group.

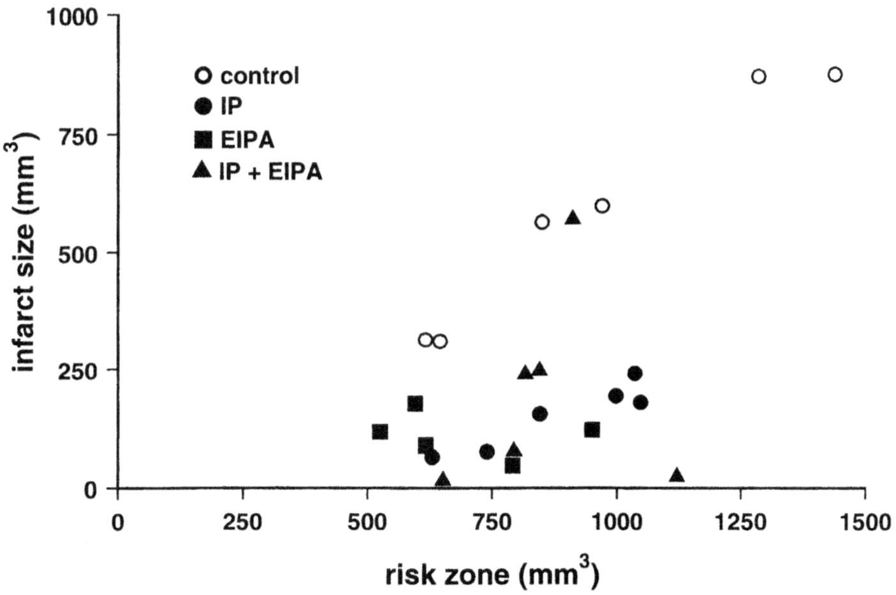

Fig. 2. Absolute infarct volumes plotted against absolute risk zone volumes. Protection is indicated by a downward shift of the relationship.

Table 1. Mean arterial pressure

group	stabilization	preischemic	isch 5	isch 40	rep 30	rep 120
control	75.3 ± 5.8	80.7 ± 6.6	69.2 ± 6.1	65.6 ± 6.4	76.9 ± 5.6	73.3 ± 4.9
EIPA	86.7 ± 4.6	94.2 ± 3.5	65.3 ± 12.5	71.7 ± 12.0	93.3 ± 7.8	76.1 ± 9.6
IP	83.3 ± 3.3	80.6 ± 5.6	50.6 ± 13.5	63.9 ± 9.5	71.7 ± 3.5	77.2 ± 5.3
IP + EIPA	73.9 ± 8.7	75.0 ± 9.8	55.0 ± 11.4	66.7 ± 11.3	84.7 ± 5.1	63.1 ± 8.6

All values are mean ± S.E.M. Stabilization is after surgery and stabilization. Preischemic is after drug/vehicle/ischemic preconditioning. Isch – ischemia; rep – reperfusion

treatments in contrast to what has been observed in the buffer-perfused rat heart [12].

In a recent report [2] we demonstrated that administration of EIPA prior to 30 min regional ischemia reduced infarct size in the *in situ* rabbit heart to the same degree (76%). When comparing IP and EIPA treatments prior to 30 min regional ischemia in the rabbit heart an equal degree of protection could be demonstrated, which was similar to that found in this study using 45 min of ischemia [2, 16]. In recent studies utilizing isolated rat hearts [12, 17] it was demonstrated that the protection afforded by Na^+/H^+ exchange inhibition was additive to that produced by ischemic preconditioning alone, indicating that the two modes of protection may act through different mechanisms. This finding was not confirmed in the present study. We assume that even though the degree of protection by either intervention alone was high in the present study there is ample room for further reduction, as has been demonstrated, for example, in the study of Bugge and Ytrehus [2].

The dose of EIPA used in the present study was based on previous reports [2, 18], which have shown infarct reduction both in rabbit and dog. EIPA is reported to have other effects which could also relate to myocardial protection. These effects of EIPA include inhibition of the Na^+/Ca^{2+} exchange [11] and the atrial sodium current I_{Na} [19], L-type calcium channel blockade [20], and effects on several other ion transport systems, protein kinases and enzymes [11]. However, inhibition of Na^+/Ca^{2+} exchange and I_{Na} as well as calcium channel blockade, protein kinase- and enzyme inhibition is reported to occur at concentrations far above the concentration required for inhibition of Na^+/H^+ exchange [11]. We, therefore, think that inhibition of Na^+/H^+ exchange is the most likely major mechanism responsible for the cardioprotection demonstrated by EIPA in this study although other mechanism can not be excluded.

One important question that needs to be explained is as to why there is a difference between rat and rabbit when it comes to addition of the two treatments. Both differences in experimental protocol as well as differences in species could be involved. The rat study was made in a buffer-perfused rat heart compared to the present study which employed an *in situ* model, while the hearts in the present study were perfused with blood. EIPA is reported to influence red blood cell volume [21] but it is difficult to see how this could explain the difference between the rat and rabbit study. Differences in species should therefore be considered.

Opening of the K_{ATP}-channel during ischemia has been suggested to contribute to maintenance of the diastolic potential at negative values, which subsequently favors the extrusion of Ca^{2+} through the Na^+/Ca^{2+} exchange but results in intracellular Na^+ increase. Na^+/H^+ exchange and the inhibition of the Na^+/K^+-ATPase also promotes Na^+ loading during ischemia. Protection by Na^+/H^+ exchange inhibitors is thought to result from a chain of events starting with lessened Na^+ loading with subsequently lessened Na^+/Ca^{2+} exchange, which in term causes a decreased Ca^{2+} loading during both ischemia and especially during reperfusion. The Na^+/Ca^{2+} exchange is electrogenic and is not only influenced by the chemical gradients of Na^+ and Ca^{2+} but also by the membrane potential. The exchanger, which extrudes Ca^{2+} at polarized membrane potentials, can also operate in the reverse mode at depolarized membrane potentials and result in Ca^{2+} loading [17]. In addition to intracellular Ca^{2+} control, a role of the Na^+/H^+ exchange inhibitors in protection might also be linked to stimulation of other mechanisms for proton extrusion from the ischemic cell such as the proton pump [22].

The activity of the Na^+/Ca^{2+} exchanger is quite different in rat and rabbit. The differences seems to have their origin in differences in intracellular sodium concentrations (Na_i) rather than differences in the exchanger itself. The rat heart has substantially higher [Na_i] than the rabbit heart which has an impact on the reversal potential of the Na^+/Ca^{2+} exchanger [23]. In rat the ventricular action potential resembles atrial action potentials with a short high plateau and a wider low plateau. The Na^+/Ca^{2+} exchanger is believed to cause the low plateau in these cells. In ventricular cells in species other than the rat the plateau is supposed to be mainly caused by opening of Ca^{2+} channels and contribution by the Na^+/Ca^{2+} exchange is supposedly small. This might explain the difference between rat and rabbit heart with respect to ischemic protection. Na^+ loading through Na^+/H^+-exchange might be a more devastating factor in the rat heart compared to the rabbit heart.

In conclusion, we have demonstrated that ethylisopropyl-amiloride, when administered prior to ischemia, causes similar level of protection as ischemic preconditioning but Na^+/H^+-exchange inhibition is not additive to ischemic preconditioning in the rabbit heart *in situ*.

Acknowledgements

The authors thank Randi Skogstad and Fredrik Bergheim for their skilled technical assistance. This study was supported by the Norwegian Research Council and by the Norwegian National Health Association, Council of Cardiovascular Disease.

References

1. Murry CE, Jennings RB, Reimer KA: New insights into potential mechanisms of ischemic preconditioning [editorial]. Circ 84: 442–445, 1991
2. Bugge E, Munch-Ellingsen J, Ytrehus K: Reduced infarct size in the rabbit heart *in vivo* by ethylisopropyl-amiloride. A role for Na^+/H^+ exchange. Basic Res Cardiol 91: 203–209, 1996
3. Ytrehus K, Liu Y, Downey JM: Preconditioning protects ischemic rabbit heart by protein kinase C activation. Am J Physiol 266: H1145–H1152, 1994
4. Munch-Ellingsen J, Ytrehus K, Bugge E: Blockade of the K_{ATP}-channel by glibenclamide aggravates ischemic injury, and counteracts ischemic preconditioning. Basic Res Cardiol 382–388, 1996
5. Karmazyn M, Moffat MP: Role of Na^+/H^+ exchange in cardiac physiology and pathophysiology: Mediation of myocardial reperfusion injury by the pH paradox. Cardiovasc Res 27: 915–924, 1993
6. Duff HJ: Clinical and *in vivo* antiarrhythmic potential of sodium-hydrogen exchange inhibitors. Cardiovasc Res 29: 189–193, 1995
7. Frelin C, Vigne P, Lazdunski M: The role of the Na^+/H^+ exchange system in cardiac cells in relation to the control of the internal Na^+ concentration. A molecular basis for the antagonistic effect of ouabain and amiloride on the heart. J Biol Chem 259: 8880–8885, 1984
8. Tani M, Neely JR : Role of intracellular Na^+ in Ca^{2+} overload and depressed recovery of ventricular function of reperfused ischemic rat hearts. Possible involvement of H^+-Na^+ and Na^+-Ca^{2+} exchange. Circ Res 65: 1045–1056, 1989
9. Moffat MP, Karmazyn M: Protective effects of the potent Na/H exchange inhibitor methylisobutyl amiloride against post-ischemic contractile dysfunction in rat and guinea-pig hearts. J Mol Cell Cardiol 25: 959–971, 1993
10. Scholz W, Albus U, Linz W, Martorana P, Lang HJ, Scholkens BA: Effects of Na^+/H^+ exchange inhibitors in cardiac ischemia. J Mol Cell Cardiol 24: 731–739, 1992
11. Kleyman TR, Cragoe EJ: Amiloride and its analogs as tools in the study of ion transport. J Membrane Biol 105: 1–22, 1988
12. Bugge E, Ytrehus K: Inhibition of sodium-hydrogen exchange reduces infarct size in the isolated rat heart – a protective additive to ischaemic preconditioning. Cardiovasc Res 29: 269–274, 1995
13. Przyklenk A, Hata K, Kloner RA: The Na^+-H^+ exchange inhibitor dimethylamiloride blocks preconditioning in the canine model. Circ 94: (abstr) I-424, 1996
14. Wadsworth S, Wu AM, Spitzer A, Chander A: Protein kinase C in intracellular pH regulation in alveolar type II cells. Am J Physiol Lung Cell Mol Physiol 271: L106–L113, 1996
15. Liu Y, Ytrehus K, Downey JM: Evidence that translocation of protein kinase C is a key event during ischemic preconditioning of rabbit myocardium. J Mol Cell Cardiol 26: 661–668, 1994
16. Munch-Ellingsen J, Løkebø JE, Bugge E, Ytrehus K: Dofetilide. Scand J Clin Lab Invest, 1997

17. Janvier NC, Boyett MR: The role of Na-Ca exchange current in the cardiac action potential. Cardiovasc Res 69–84, 1996
18. Duff HJ, Lester WM, Rahmberg M: Amiloride. Antiarrhythmic and electrophysiological activity in the dog. Circ 78: 1469–1477, 1988
19. Gold MR, Strichartz GR: Use-dependent block of atrial sodium current by ethylisopropylamiloride. J Cardiovasc Pharmacol 17: 792–799, 1991
20. Pierce GN, Cole WC, Liu K, Massaeli H, Maddaford TG, Chen YJ, McPherson CD, Jain S, Sontag D: Modulation of cardiac performance by amiloride and several selected derivatives of amiloride. J Pharmacol Exp Ther 265: 1280–1291, 1993
21. Clark JD, Limbird LE: Na^+-H^+ exchanger subtypes: A predictive review. Am J Physiol 261: C945–C953, 1991
22. Gottlieb RA, Gruol DL, Zhu JY, Engler RL: Preconditioning in rabbit cardiomyocytes. J Clin Inv 97: 2391–2398, 1996
23. Shattock MJ, Bers DM: Rat vs. rabbit ventricle: Ca flux and intracellular Na assessed by ion-selective microelectrodes. Am J Physiol 256: C814–C822, 1989

Molecular and Cellular Biochemistry **186**: 19–25, 1998.
© 1998 *Kluwer Academic Publishers.*

Loss of myocardial protection from ischemic preconditioning following chronic exposure to R(–)-N^6-(2-phenylisopropyl)adenosine is related to defect at the adenosine A_1 receptor

M. Wail Hashimi, Jon D. Thornton, James M. Downey and Michael V. Cohen

Departments of Medicine and Physiology, University of South Alabama College of Medicine, Mobile, AL, USA

Abstract

Exogenously administered adenosine agonist will protect myocardium against infarction during ischemia. However, long-term exposure to adenosine agonists is associated with loss of this protection. To determine why this protection is lost, isolated, perfused rabbit hearts were studied after administration of R(–)-N^6-(2-phenylisopropyl)adenosine (PIA), 0.25 mg/h IP, for 3–4 days to intact animals. All hearts experienced 30 min of regional ischemia and 120 min of reperfusion. Control groups 1 and 2 were untreated. In group 1 this ischemia/reperfusion was the only intervention, whereas group 2 hearts were preconditioned with a cycle of 5 min global ischemia/10 min reperfusion preceding the 30 min regional ischemia. Groups 3–5 had been chronically exposed to PIA. Group 3 hearts had 1 preconditioning ischemia/reperfusion cycle before the prolonged ischemia. Group 4 received a 5 min infusion of 0.1 μmol/L phenylephrine in lieu of global ischemia, whereas group 5 was instead treated with 1 μmol/L carbachol. Infarct size averaged 32% of the risk zone in group 1, whereas ischemic preconditioning limited infarction to 8.2% in group 2. Prolonged exposure of group 3 hearts to PIA resulted in the inability of preconditioning with 5 min global ischemia to protect (28.7 ± 4.4% infarction). However, protection was restored by either phenylephrine, an agonist of α_1-adrenergic receptors which couple to G_q and stimulate PKC, or carbachol, an agonist of M_2-muscarinic receptors which couple instead to G_i as do adenosine A_1 receptors (5.2 ± 1.7% and 9.2 ± 2.1% infarction, resp.). Therefore, cross tolerance to ischemic preconditioning develops after chronic PIA infusion. Since both the G_i and the PKC components of the preconditioning pathway were shown to be intact, tolerance must have been related to downregulation or desensitization of the A_1 adenosine receptor. (Mol Cell Biochem **186**: 19–25, 1998)

Key words: adenosine A_1 receptor, carbachol, desensitization, downregulation, ischemic preconditioning, phenylephrine, r-phenylisopropyl adenosine

Introduction

In several species ischemic preconditioning, the ability to render the heart resistant to infarction by brief periods of ischemia, is initiated by A_1 adenosine receptor stimulation following release of endogenous adenosine and can be simulated by infusion of an adenosine agonist [1–3]. Because of the requirement for pretreatment it would be useful if high-risk patients could be kept in a preconditioned state by continuous treatment with an appropriate adenosine agonist. It has been observed, however, that chronic stimulation of adenosine receptors by continuous exposure to an adenosine agonist for 3 days [4] or even by endogenous adenosine released during repetitive coronary occlusions [5] results in the loss of the ability of ischemic preconditioning to protect. The mechanism by which this loss of protection occurs is not clear. Work by several investigators has shown that chronic exposure of adipocytes to R(–)-N^6-(2-phenyl-

Address for offprints: M.V. Cohen, Department of Physiology, MSB 3050, University of South Alabama, College of Medicine, Mobile, Alabama 36688, USA

isopropyl) adenosine (PIA) results in desensitization (uncoupling of the receptor – effector association) of the A_1 and A_2 adenosine receptors as well as downregulation of A_1 adenosine receptors [6, 7] and of the G_i protein to which they couple [8]. Phosphorylation with subsequent internalization and degradation appears to play a more important role in desensitization of the A_1 receptor than A_2 receptor [7]. Similar findings of downregulation and desensitization have been observed with other receptor systems such as the β adrenergic receptor [9, 10]. Although both the A_1 adenosine receptor and G protein are downregulated, data suggest that the downregulation appears to be almost twice as pronounced for the receptor [6]. Furthermore, the G protein may be in such abundance in the cell membrane that a substantial loss of G protein may result in little loss in physiologic activity. Nonetheless in an earlier study [4] rabbits treated for 72 h with the selective A_1 agonist 2-chloro-N6-cyclopentyladenosine (CCPA) demonstrated modest, but significant, attenuation of the bradycardic response to bolus injection of the muscarinic agonist acetylcholine, suggesting possible downregulation of the G protein shared by both adenosine and muscarinic receptors.

There is evidence that occupancy of the cardiac A_1 receptor results in coupling to a G protein [11] and stimulation of protein kinase C (PKC) [12], and PKC stimulation appears to be a critical step in the protection of ischemic myocardium, at least in the rat and rabbit [13–15]. The loss of the ability of ischemic preconditioning to protect hearts chronically exposed to adenosine [5] or one of its agonists [4] may be related to downregulation of the adenosine receptor or G protein, or alternately may be caused by changes in PKC activity since chronic stimulation of PKC is well known to result in enzyme depletion [16, 17]. The present study seeks to identify the point at which tolerance occurs by stimulating the signal transduction cascade at key points between the adenosine receptor and PKC to attempt to activate the preconditioning response in adenosine-tolerant hearts. The proposed strategy tests whether PKC is intact by stimulating α_1-adrenergic receptors which couple to their PKC-activating phospholipase through G_q. Similarly G_i function is to be tested by attempting to precondition with an M_2 agonist since both muscarinic M_2- and adenosine A_1-receptors are thought to couple to the same G protein, G_i. Failure of both agonists to protect the heart would reveal a deficiency of PKC, whereas only a carbachol failure would imply a problem with G_i. If both agonists can mimic ischemic preconditioning's protection, then the defect must be proximal to the G protein, i.e. at the adenosine receptor itself.

Materials and methods

Surgical preparation of animals

New Zealand White rabbits (1.5–2.6 kg) of either sex were anesthetized with intravenous sodium pentobarbital, 30 mg/kg. The rabbits were intubated and ventilated with a positive pressure ventilator (MD Industries, Mobile, AL, USA) and 100% oxygen at a rate of 30–35 breaths per minute and tidal volume of approximately 15 ml. The peritoneal cavity was entered through a midline abdominal incision. An Alzet (Palo Alto, CA, USA) osmotic pump rated to deliver 10 µl/h was placed into the peritoneal cavity and the incision was closed. Immediately following implantation rabbits received an intravenous bolus of 2 mg of PIA and heart rates were recorded. Heart rates were subsequently measured at 24, 48 and 72 h after implantation. All rabbits were treated with 600,000 U of benzathine penicillin IM following surgery.

Preparation of Alzet pump

Fifty mg of PIA were dissolved in a mixture of 1 ml of saline and 1 ml of dimethylsulfoxide. The solution was injected into the osmotic pump which was then incubated for at least 12 h in sterile saline at 38°C before implantation. At a putative delivery rate of 10 µl/h, PIA was infused intraperitoneally at 0.25 mg/h. At the end of the experiment all pumps were removed and sectioned. In every instance the walls surrounding the internal cavity had swollen to obliterate the lumen indicating proper function of the pump.

Isolation of the heart

By 72–96 h following implantation, heart rates had returned to pre-PIA levels. Rabbits were anesthetized with sodium pentobarbital, 30 mg/kg IV. A tracheotomy was performed and the rabbits were ventilated with 100% O_2 at the same rate and volume as outlined above. Response of heart rate to an intravenous bolus of 2 mg PIA was again recorded. A left thoracotomy was performed in the fourth left intercostal space to expose the heart. A 2-0 silk thread on a needle was passed around a branch of the left coronary artery. The ends of the tie were threaded through a vinyl tube to form a snare.

The hearts were rapidly excised and hung on a Langendorff apparatus by the aortic root in less than 1 min and perfused with Krebs-Henseleit bicarbonate buffer. This buffer consisted of 118.5 mmol/L NaCl, 4.7 mmol/L KCl, 1.2 mmol/L $MgSO_4$, 1.2 mmol/L KH_2PO_4, 2.5 mmol/L $CaCl_2$, 24.8 mmol/L $NaHCO_3$ and 10 mmol/L glucose. The perfusate was warmed to 38°C and bubbled with 95% O_2 - 5% CO_2. The pH of the perfusate was maintained at 7.4 by changing the flow of bubbled gas. The reservoir height was set so that the perfusion pressure was 75 mmHg.

After the heart was hung on the Langendorff apparatus, a latex balloon connected to a catheter was inserted into the left ventricular cavity. The balloon was then filled with saline to produce an end-diastolic pressure of 5 mmHg. Demand pacing of the heart at 220 bpm was performed. Experiments were carried out only after the heart had been allowed to stabilize for 10 min.

Measurements of infarct size and risk area

After completion of the experiment the coronary artery snared during the experiment was reoccluded. Fluorescent 1–10 µm ZnCd particles (Duke Scientific Corp, Palo Alto, CA, USA) were added to the perfusate to distinguish the non-fluorescent perfusion field of the occluded artery from the fluorescing, perfused myocardium. The heart was then frozen and cut into 2 mm transverse slices. The slices were stained in a 1% solution of triphenyltetrazolium chloride in sodium phosphate buffer, pH 7.4, for 20 min at 37°C. The risk zone was identified by illuminating slices with UV light and the infarcted tissue was visualized under white light. Planimetry was used to measure the areas of the infarct and risk zones. Volumes were calculated by multiplying areas by the thickness of the tissue section.

Experimental protocol

In all hearts regional ischemia for 30 min was produced by coronary occlusion and was followed by 2 h of reperfusion

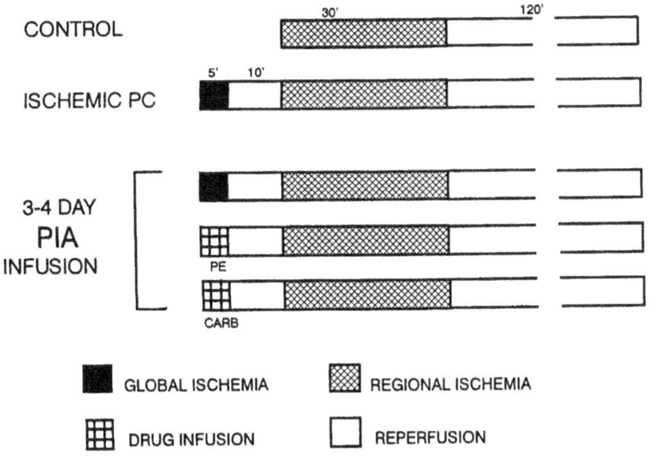

Fig. 1. Experimental protocol. Abbreviations: CARB – carbachol; pc – preconditioning; PE – phenylephrine; PIA – R(–)-N^6-(2-phenylisopropyl)-adenosine.

before risk zone and infarct sizes were measured as detailed above. The hearts were divided into 5 experimental groups (Fig. 1). The first 2 groups had not been exposed to PIA. Control hearts (group 1) were subjected to only the 30 min coronary occlusion/120 min reperfusion. The second group of hearts underwent ischemic preconditioning with 5 min global ischemia followed by 10 min reperfusion. The remaining 3 groups of hearts were removed from animals which had been treated with PIA for 3–4 days. The protocol for group 3 hearts was designed to determine if chronically infused PIA resulted in loss of the protection of ischemic preconditioning. These hearts were preconditioned with 5 min global ischemia/10 min reperfusion prior to 30 min regional ischemia.

The last two protocols were intended to test if down-regulation occurred at the level of the adenosine receptor, G protein or PKC. Group 4 hearts were treated with a 5 min infusion of 0.1 µmol/L phenylephrine (Sigma Chemical Co., St. Louis, MO, USA) in lieu of global ischemia. Then a 10 min drug-free period permitted drug washout before the 30 min regional ischemia. In the fifth group of hearts carbachol (Sigma Chemical Co.) was added to the perfusate for 5 min at a concentration of 1 µmol/L followed by 10 min of drug-free perfusion and the 30 min regional ischemia.

Statistical analysis

All values are given as mean ± S.E.M. Differences in hemodynamics over time were tested by analysis of variance with replication. Group differences in infarct size were evaluated by one way analysis of variance with Tukey's *post hoc* test. A p value of < 0.05 was considered to be significant.

Results

Hemodynamic variables

Twenty-six rabbits were treated with PIA and were then divided into one of 3 experimental groups (groups 3–5). No rabbits died during the chronic infusion phase of the experiment. However, 2 died during isolation of the heart and two other hearts had to be discarded from analysis because of technical problems. For this experiment, 2 control animals (group 1) and 2 rabbits with ischemic preconditioning (group 2) were prepared. Because infarct sizes in these rabbits were comparable to those already observed in similarly prepared animals, these rabbits were added to historical control and ischemic preconditioning groups [2]. For infarct size analysis, groups 1 and 2 represent a composite of current and historical data and serve as reference for groups 3–5.

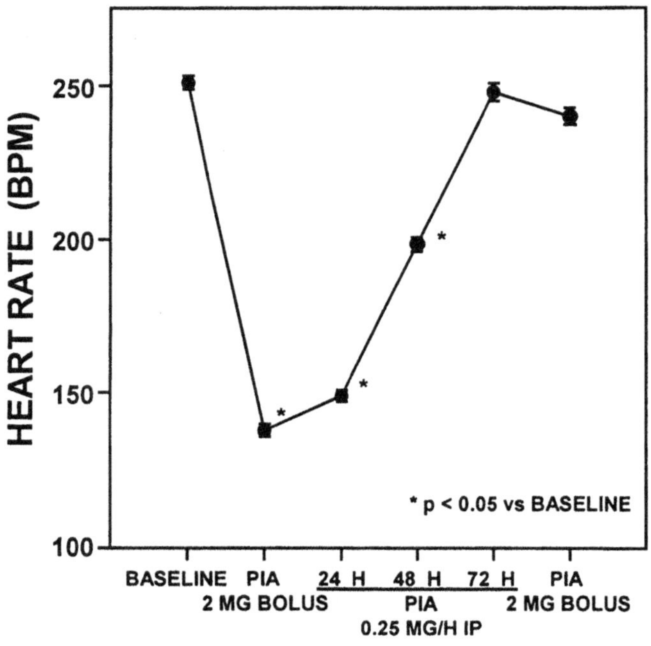

Fig. 2. With initial bolus and early continuous infusion of R(–)-N^6-(2-phenylisopropyl)adenosine (PIA) heart rate declined. After 72 h of continuous infusion of PIA, tolerance to the bradycardic effects of PIA developed. Abbreviation: IP – intraperitoneal

Heart rate (Fig. 2) was measured at baseline, immediately following IV PIA bolus, at 24, 48, 72, and, when necessary, 96 h after the start of chronic PIA infusion, and finally following a 2 mg IV bolus of PIA prior to the left thoracotomy. Baseline heart rate of 251 ± 2 bpm was seen to decline significantly to 137 ± 2 bpm ($p < 0.001$) with injection of 2 mg of PIA. During the initial 24 h there continued to be marked bradycardia with average heart rate of 149 bpm. This rate was still significantly lower than baseline ($p < 0.001$). By 72 h the mean heart rate had returned to the pre-treatment level, and a second intravenous bolus of 2 mg PIA had minimal effect.

Hemodynamic parameters for experimental groups 3–5 are summarized in Table 1. Heart rate, developed left ventricular pressure and coronary flow were comparable in the 3 experimental groups at all phases of the experiment except for obvious differences during global ischemia and infusions of phenylephrine or carbachol. There were no significant differences in variables at baseline, during global ischemia, or following reperfusion in any of the 5 groups.

Infarct size data

Body weights, heart weights and risk zone volumes were similar in the 5 experimental groups and average values for groups 3–5 are presented in Table 2. Individual heart infarct sizes normalized as a percentage of risk zone are presented in Fig. 3. Infarct size of the group receiving chronic PIA infusion prior to ischemic preconditioning (group 3) was similar to that in control hearts (group 1) (28.7 ± 4.4% and 32.0 ± 2.9%, resp.) and significantly higher than in untreated, ischemically preconditioned hearts (group 2) (8.2 ± 1.5%). However, protection was once again observed (5.2 ± 1.7% infarction, pNS vs. group 2) in hearts chronically exposed to PIA if hearts were preconditioned with phenylephrine, an α_1-adrenergic agonist, in lieu of ischemia (group 4). Hearts were equally protected in group 5 in which carbachol, an M_2-muscarinic agonist, was administered for 5 min beginning 15 min prior to the 30 min period of regional ischemia (9.2 ± 2.1% infarction, pNS vs. group 2).

Discussion

Adenosine and its A_1-potent congeners have been shown to protect myocardium in the rabbit by stimulation of the adenosine A_1-receptor (A_1AR) [1–3]. The more recently described A_3 receptor probably works in parallel with the A_1

Table 1. Hemodynamic parameters for hearts from rabbits treated for 3–4 days with continuous infusion of R(–)-N^6-(2-phenylisopropyl)adenosine

	Heart rate (beats/min)			Left ventricular developed pressure (mmHg)			Coronary flow (ml/min)		
	PC	PE	CARB	PC	PE	CARB	PC	PE	CARB
Baseline	234 ± 6	244 ± 4	242 ± 4	96 ± 1	94 ± 1	96 ± 1	59.0 ± 1.8	57.9 ± 1.9	62.3 ± 1.8
Global ischemia/ drug infusion	220*	220*	220*	6 ± 4†	65 ± 2†	66 ± 2†	0†	52.4 ± 2.5	65.0 ± 2.8
10′ Reperfusion	220*	220*	220*	91 ± 1	86 ± 2	93 ± 1	42.0 ± 2.7†	48.1 ± 2.7‡	51.9 ± 3.4‡
30′ Cor occl	220*	220*	220*	66 ± 2†	68 ± 5†	68 ± 2†	28.9 ± 1.5†	33.1 ± 1.6†	30.6 ± 1.5†
5′ Reperfusion	220*	220*	220*	82 ± 3‡	84 ± 2	89 ± 1	35.4 ± 1.5†	37.2 ± 1.4†	32.9 ± 1.7†
30′ Reperfusion	220*	220*	220*	76 ± 3†	84 ± 2	89 ± 2	32.3 ± 2.0†	33.4 ± 0.9†	33.4 ± 1.2†
120′ Reperfusion	220*	220*	220*	66 ± 3†	69 ± 4†	66 ± 3†	30.0 ± 1.4†	31.8 ± 1.0†	32.3 ± 1.5†

*paced rate. Abbreviations: CARB – carbachol; PC – ischemic preconditioning; PE – phenylephrine infusion.
Statistical significance of difference between observation and baseline value: ‡$p < 0.025$; †$p < 0.001$.

Table 2. Heart weights and infarct size data for rabbits treated for 3–4 days with continuous infusion of R(–)-N^6-(2-phenylisopropyl)adenosine

	PC (n = 7)	Phenylephrine (n = 7)	Carbachol (n = 8)
Rabbit weight (kg)	1.9 ±0.0	1.9 ±0.1	1.8 ±0.0
Heart weight (g)	7.4 ±0.5	7.9 ±0.6	7.5 ±0.4
Risk zone (mm^3)	781 ±71	733 ± 149	730 ± 60
Infarct zone (mm^3)	220 ±33	33 ±13	66 ±10
Infarction (% of risk zone)	28.7 ± 4.4	5.2 ±1.7*	9.2 ±2.1*

*$p < 0.001$ vs. PC. Abbreviations: PC – ischemic preconditioning.

receptor [18], although the role of the former in the rabbit is not as firmly established. The adenosine receptor in turn is thought to be coupled to PKC by a G protein [11, 12]. Many G protein subtypes have been identified, but two seem to be involved in the preconditioning cascade. Stimulation of either A_1AR [1–3] or muscarinic M_2 [19] receptors can protect myocardium, and both are believed to interact exclusively with pertussis toxin-sensitive G_i [5]. α_1-Adrenergic agonists such as norepinephrine [20] and phenylephrine [14] also can salvage ischemic myocardium but they couple to PKC through pertussis toxin-insensitive G_q [21, 22]. Unfortunately, several studies have shown that chronic stimulation of the A_1AR will result in loss of the ability to effect ischemic preconditioning [4, 5]. Cross tolerance to ischemic preconditioning has been noted following prolonged (72 h) exogenous administration of CCPA, a selective A_1-agonist [4]. There is also evidence that the heart can become tolerant to its own endogenous adenosine. When the risk zone is repeatedly exposed to endogenous adenosine released during multiple repetitive coronary occlusions over several days, the protection from ischemic preconditioning is again lost, presumably because of the development of adenosine tolerance [5]. This loss of protection is an obvious impediment to projected prophylactic treatment of high-risk patients with an adenosine agonist. Furthermore, patients with a history of recurrent anginal attacks may be resistant to any adenosine-based cardioprotective treatment.

Although the protection of ischemic preconditioning following the triggering stimulus fades after 1–2 h [23], Baxter and colleagues [24] have demonstrated a second window of protection 24 h after either ischemic preconditioning or a bolus injection of CCPA. These researchers have proposed that this second phase of protection is related to expression of protective proteins. Thus it is perhaps surprising that we saw no evidence of protection in our rabbits treated with CCPA for 3 days. There should have been ample time to produce protective stress proteins. This second phase of protection may have been affected by the CCPA-mediated downregulation of some signal transduction pathways required during the prolonged ischemia. This view is supported by a recent report in which less intense exposure of the heart to CCPA, i.e. 0.1 mg/kg every 48 h, did not result in loss of either the hemodynamic effects of CCPA or delayed protection for up to 10 days [25].

Investigators have studied the adenosine pathway in an attempt to determine where desensitization and downregulation following chronic agonist exposure may be occurring. Chronic stimulation by an agonist may affect regulation in one of two ways: a decrease in receptor number

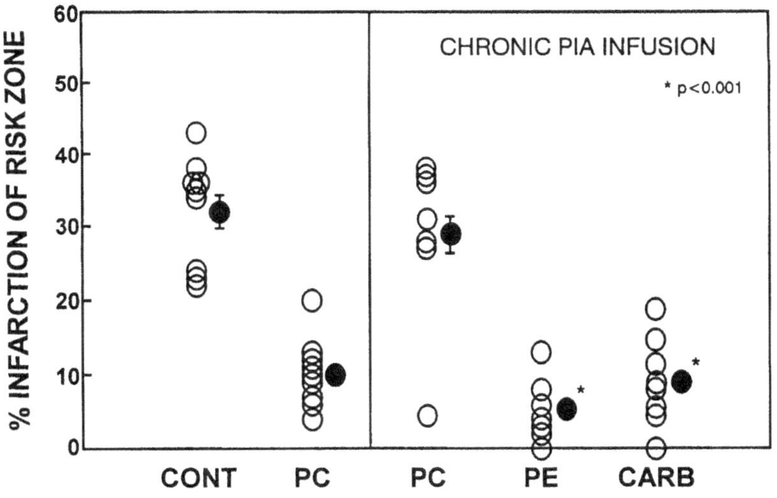

Fig. 3. Infarct size as a percentage of myocardium at risk. Open circles represent infarcts in individual animals, and closed circles are means ± S.E.M. The 2 groups on the left (cont – control and pc – ischemic preconditioning) were untreated, whereas the three groups on the right had received a 3–4 day infusion of R(–)-N^6-(2-phenylisopropyl)adenosine (PIA). After prolonged exposure to PIA the ability of ischemic preconditioning to protect the heart was lost. However, a 5 min infusion of either phenylephrine (PE) or carbachol (CARB) in lieu of ischemia restored protection from infarction.

or downregulation and an uncoupling of the receptor-effector association or desensitization. In many systems both may be occurring. Adenosine receptors may be subdivided into the A_1AR which inhibits and the A_2AR which stimulates adenylyl cyclase. More recently the A_3AR has also been characterized [26]. Cardiac muscle possesses receptors of the A_1 class which couple to PKC [12] by the GTP-dependent regulatory protein G_i [11]. The A_2AR binds G_s [11]. In rat adipocytes *in vivo* and *in vitro* chronic stimulation of the A_1AR results in downregulation of the receptor [6, 7, 27] as well as the associated G protein [6, 8, 27, 28]. Downregulation appears to be much more pronounced in the *in vitro* model with 80% reduction in the A_1AR [6] than in rats receiving PIA *in vivo* with only a 30% reduction in A_1AR [29]. Furthermore, a reduction of G_i of 40–50% [6, 8] and an uncoupling of the A_1AR from G_i [7, 28] in rat adipocytes has also been noted with chronic stimulation of the adenosine receptor. Lee *et al.* [27] also found as much as a 45% decrease in G protein in rat myocytes exposed to PIA for 7 days. Although these data shed light on the biochemical changes occurring with chronic stimulation of the adenosine receptor, they do not indicate which change may have physiologic importance or whether loss of PKC itself may be the limiting step. Therefore, we designed our study to delineate where the physiologically important site of downregulation occurs.

Chronic stimulation of the adenosine receptor was accomplished with an osmotic pump delivering PIA into the peritoneal cavity. Heart rates were monitored as a marker of drug effect. With bolus injection of adenosine agonist in otherwise untreated hearts, bradycardia occurred and this heart rate effect was still evident after 24 h of PIA infusion. Subsequently loss of A_1AR-mediated bradycardia was documented, suggesting tolerance to the infused PIA had developed. This observation was identical to that made with CCPA which was infused intravenously for 4 days with an electric external pump [4].

Several of the rabbits exposed to PIA for 3–4 days were subjected to one cycle of ischemic preconditioning before the standard 30 min period of regional ischemia (group 3, Fig. 3). As predicted by the results in rabbits with multiple coronary occlusions [5] and those treated with prolonged infusions of CCPA [4], ischemic preconditioning no longer had any protective effect. Hence, in addition to developing tolerance to the bradycardic effect of PIA, these hearts also lost the protective effect of ischemic preconditioning known to be initiated by A_1AR stimulation.

When 0.1 μmol/L phenylephrine, an α_1-adrenergic agonist known to stimulate PKC [30, 31] through G_q [21, 22], was administered in lieu of ischemic preconditioning to hearts previously exposed to 3–4 days of PIA, marked protection was again observed. This finding suggested that PKC had not been significantly downregulated and was still physiologically intact. However, because phenylephrine and adenosine do not couple to PKC through the same G protein, it could not be assumed that the G protein which couples to adenosine (G_i) was still intact. Therefore, carbachol, an agonist of M_2-muscarinic receptors which, like A_1AR, also couple to PKC through G_i [11, 12], was given in lieu of either ischemia or phenylephrine. Once again, protection was evident. Therefore, these findings suggest that both G_i and PKC were still functional in the adenosine-tolerant hearts. Apparently enough G_i was still present in the hearts treated chronically with PIA to result in carbachol-mediated protection, even if levels were decreased [8, 27] to the extent that the bradycardic response to acetylcholine in hearts exposed for long periods to an adenosine agonist was attenuated [4]. By deduction, the loss of protection from chronic stimulation must be related to a step before involvement of G_i. Hence it is reasonable to conclude that tolerance is likely to be caused by downregulation of the adenosine receptor and/or uncoupling of the adenosine receptor from the G protein.

Two general mechanisms have been postulated to regulate the number and functional integrity of adenosine receptors: downregulation and desensitization. The adenosine A_1 receptor appears to utilize both of these mechanisms while the A_2 adenosine receptor appears to rely on only desensitization for modulation of activity [7]. Downregulation may be related to internalization and degradation of the receptor, alteration in the cell membrane, and/or diminished resynthesis. Ramkumar *et al.* [7] have found that the A_1AR present in intracellular 'light vesicles' could only account for 5–10% of the receptors lost indicating that either degradation does not play an important role, or that the site of degradation is not known. A decrease in mRNA responsible for the production of A_1AR is an interesting hypothesis which has actually been observed in the β-adrenergic system [10, 32, 33]. Receptor phosphorylation may be important by somehow tagging the receptor for internalization and degradation [7].

Two models of desensitization have been identified: (a) uncoupling of the receptor from its effector molecule, i.e. the adenosine – G protein association and (b) rapid, reversible internalization of the receptor (sequestration). For the A_2 adenosine receptor, sequestration does not appear to have much of a role in regulation, but rather desensitization is dependent on an uncoupling from the G protein with subsequent loss of receptor function [7]. For the A_1 adenosine receptor the relationship between uncoupling and sequestration has not been clearly defined. Interestingly sequestration has also not been found to be important in regulation of the β-adrenergic system [32]. Unfortunately, the present experimental design cannot distinguish between these mechanisms.

Our data clearly indicate that chronic stimulation of the adenosine receptor will result in downregulation and/or desensitization. However, this finding is merely the first step in understanding the process. More studies will be necessary if an attempt is to be made to chronically stimulate the protection cascade in ischemic myocardium of affected patients with resulting diminution of the risk of infarction.

Acknowledgement

This work was supported in part by grants from the National Institutes of Health, Heart, Lung and Blood Institute HL-20648 and HL-50688.

References

1. Liu GS, Thorton J, Van Winkle DM, Stanley AWH, Olsson RA, Downey JM: Protection against infarction afforded by preconditioning is mediated by A_1 adenosine receptors in rabbit heart. Circulation 84: 350–356, 1991
2. Tsuchida A, Liu GS, Wilborn WH, Downey JM: Pretreatment with the adenosine A_1 selective agonist, 2-chloro-N6-cyclopentyladenosine (CCPA), causes a sustained limitation of infarct size in rabbits. Cardiovasc Res 27: 652–656, 1993
3. Thornton JD, Liu GS, Olsson RA, Downey JM: Intravenous pre-treatment with A_1-selective adenosine analogues protects the heart against infarction. Circulation 85: 659–665, 1992
4. Tsuchida A, Thompson R, Olsson RA, Downey JM: The anti-infarct effect of an adenosine A_1-selective agonist is diminished after prolonged infusion as is the cardioprotective effect of ischaemic preconditioning in rabbit heart. J Mol Cell Cardiol 26: 303–311, 1994
5. Cohen MV, Yang X-M, Downey JM: Conscious rabbits become tolerant to multiple episodes of ischemic preconditioning. Circ Res 74: 998–1004, 1994
6. Green A: Adenosine receptor down-regulation and insulin resistance following prolonged incubation of adipocytes with an A_1 adenosine receptor agonist. J Biol Chem 262: 15702–15707, 1987
7. Ramkumar V, Olah ME, Jacobson KA, Stiles GL: Distinct pathways of desensitization of A_1- and A_2-adenosine receptors in DDT_1 MF-2 cells. Mol Pharmacol 40: 639–647, 1991
8. Green A, Johnson JL, Milligan G: Down-regulation of G_i sub-types by prolonged incubation of adipocytes with an A_1 adenosine receptor agonist. J Biol Chem 265: 5206–5210, 1990
9. Lohse MJ, Andexinger S, Pitcher J, Trukawinski S, Codina J, Faure J-P, Caron MG, Lefkowitz RJ: Receptor-specific desensitization with purified proteins: Kinase dependence and receptor specificity of β-arrestin and arrestin in the β_2-adrenergic receptor and rhodopsin systems. J Biol Chem 267: 8558–8564, 1992
10. Hausdorff WP, Caron MG, Lefkowitz RJ: Turning off the signal: Desensitization of β-adrenergic receptor function. FASEB J 4: 2881–2889, 1990
11. Fleming JW, Wisler PL, Watanabe AM: Signal transduction by G proteins in cardiac tissues. Circulation 85: 420–433, 1992
12. Kohl C, Linck B, Schmitz W, Scholz H, Scholz J, Tóth M: Effects of carbachol and (–)-N^6-phenylisopropyladenosine on myocardial inositol phosphate content and force of contraction. Br J Pharmacol 101: 829–834, 1990
13. Ytrehus K, Liu Y, Downey JM: Preconditioning protects ischemic rabbit heart by protein kinase C activation. Am J Physiol 266: H1145–H1152, 1994
14. Tsuchida A, Liu Y, Liu GS, Cohen MV, Downey JM: α_1-Adrenergic agonists precondition rabbit ischemic myocardium independent of adenosine by direct activation of protein kinase C. Circ Res 75: 576–585, 1994
15. Speechly-Dick ME, Mocanu MM, Yellon DM: Protein kinase C: Its role in ischemic preconditioning in the rat. Circ Res 75: 586–590, 1994
16. Huang K-P, Huang FL, Nakabayashi H, Mahoney CW, Chen K-H: Mechanism of protein kinase C-mediated signal transduction. In: S Papa, A Azzi, JM Tager (eds). Adenosine Nucleotides in Cellular Energy Transfer and Signal Transduction. Birkhaüser Verlag, Basel, 1992, pp 219–230
17. Savart M, Letard P, Bultel S, Ducastaing A: Induction of protein kinase C down- regulation by the phorbol ester TPA in a calpain/protein kinase C complex. Int J Cancer 52: 399–403, 1992
18. Liu G-S, Downey JM, Cohen MV: Adenosine, ischemia, and pre-conditioning. In: KA Jacobson, MF Jarvis (eds). Purinergic Approaches in Experimental Therapeutics. Wiley-Liss, New York, 1997, pp 153–172
19. Thornton JD, Liu GS, Downey JM: Pretreatment with pertussis toxin blocks the protective effects of preconditioning: Evidence for a G-protein mechanism. J Mol Cell Cardiol 25: 311–320, 1993
20. Thornton JD, Daly JF, Cohen MV, Yang X-M, Downey JM: Catecholamines can induce adenosine receptor-mediated protection of the myocardium but do not participate in ischemic preconditioning in the rabbit. Circ Res 73: 649–655, 1993
21. Exton JH: Phosphoinositide phospholipases and G proteins in hormone action. Annu Rev Physiol 56: 349–369, 1994
22. LaMorte VJ, Thornburn J, Absher D, Spiegel A, Brown JH, Chien KR, Feramisco JR, Knowlton KU: G_q- and ras-dependent pathways mediate hypertrophy of neonatal rat ventricular myocytes following α_1-adrenergic stimulation. J Biol Chem 269: 13490–13496, 1994
23. Van Winkle DM, Thornton JD, Downey DM, Downey JM: The natural history of preconditioning: cardioprotection depends on duration of transient ischemia and time to subsequent ischemia. Coron Artery Dis 2: 613–619, 1991
24. Baxter GF, Marber MS, Patel VC, Yellon DM: Adenosine receptor involvement in a delayed phase of myocardial protection 24 hours after ischemic preconditioning. Circulation 90: 2993–3000, 1994
25. Dana A, Baxter GF, Walker JM, Yellon DM: Prolonging the delayed phase of myocardial protection: Anti-infarct effects of adenosine A_1 receptor activation are maintained despite repetitive dosing. Heart 77(Suppl 1): P23, 1997 (abstr)
26. Salvatore CA, Jacobson MA, Taylor HE, Linden J, Johnson RG: Molecular cloning and characterization of the human A_3 adenosine receptor. Proc Natl Acad Sci USA 90: 10365–10369, 1993
27. Lee HT, Thompson CI, Hernandez A, Lewy JL, Belloni FL: Cardiac desensitization to adenosine analogues after prolonged *R*-PIA infusion *in vivo*. Am J Physiol 265: H1916–H1927, 1993
28. Milligan G: Agonist regulation of cellular G protein levels and distribution: Mechanisms and functional implications. Trends Pharmacol Sci 14: 413–418, 1993
29. Parsons WJ, Stiles GL: Heterologous desensitization of the inhibitory A_1 adenosine receptor – adenylate cyclase system in rat adipocytes: Regulation of both N_s and N_i. J Biol Chem 262: 841–847, 1987
30. Minneman KP: α_1-Adrenergic receptor subtypes, inositol phosphates, and sources of cell Ca^{2+}. Pharmacol Rev 40: 87–119, 1988
31. Fedida D, Braun AP, Giles WR: α_1-Adrenoceptors in myocardium: Functional aspects and transmembrane signaling mechanisms. Physiol Rev 73: 469–487, 1993
32. Collins S, Caron MG, Lefkowtiz RJ: Regulation of adrenergic receptor responsiveness through modulation of receptor gene expression. Annu Rev Physiol 53: 497–508, 1991
33. Moalic J-M, Bourgeois F, Mansier P, Machida CA, Carré F, Chevalier B, Pitarque P, Swynghedauw B: β_1 Adrenergic receptor and Gαs mRNAs in rat heart as a function of mechanical load and thyroxine intoxication. Cardiovasc Res 27: 231–237, 1993

Molecular and Cellular Biochemistry **186**: 27–34, 1998.
© 1998 *Kluwer Academic Publishers.*

Rapid pacing-induced preconditioning is recaptured by farnesol treatment in hearts of cholesterol-fed rats: Role of polyprenyl derivatives and nitric oxide

Peter Ferdinandy,[1] Csaba Csonka,[1] Tamás Csont,[1] Zoltán Szilvássy[2] and László Dux[1]

[1]*Departments of Biochemistry and* [2]*Medicine I., Albert Szent-Györgyi Medical University, Szeged, Hungary*

Abstract

We have previously shown that hypercholesterolemia leads to the loss of pacing-induced preconditioning (PC), possibly due to the impairment of cardiac nitric oxide (NO) synthesis. It has been shown that excess exogenous cholesterol inhibits formation of several polyprenyl derivatives involved in signal transduction. In the present study, we examined whether PC and cardiac NO synthesis are restored by treatment with the key polyprenyl product, farnesol, in cholesterol-fed rats. Rats fed 2% cholesterol-enriched/control diet for 24 weeks were given i.p. 5 μM/kg farnesol/vehicle, respectively. An hour later, hearts were isolated and prepared for 'working' perfusion, then subjected to PC/non-PC protocols of 3 intermittent periods of pacing of 5 min duration at 10 Hz, followed by a 10 min coronary occlusion to test the effect of PC. PC increased ischemic aortic flow (AF) from its control value of 15.6 ± 1.5 to 27.3 ± 1.7 mL/min ($p < 0.05$). PC was not observed in hearts obtained from hypercholesterolemic rats (AF: 15.7 ± 1.2 mL/min), however, it reappeared in the farnesol-treated hypercholesterolemic group (AF: 31.8 ± 3.4 mL/min, $p < 0.05$). In tissue samples from the left ventricle, cholesterol-diet markedly decreased the intensity of the electron spin resonance spectra of NO obtained after *in vivo* spin trapping with Fe^{2+}-diethyl-dithio-carbamate complex. Farnesol-treatment did not influence cardiac NO content in the cholesterol-fed or in the control group. These results show that the lost PC can be recaptured by farnesol-treatment in hypercholesterolemia, however, farnesol-treatment does not restore cardiac NO synthesis. (Mol Cell Biochem **186**: 27–34, 1998)

Key words: pacing, preconditioning, farnesol, high-cholesterol diet, nitric oxide, electron spin resonance, rat heart

Introduction

Since the original observation by Murry *et al.* [1] more than a decade ago, the phenomenon of ischemic preconditioning has been well characterized, however, there is a considerable controversy in the literature regarding the mechanism of ischemic preconditioning (see for reviews [2, 3]). A variety of substances, i.e. adenosine, bradykinin, nitric oxide (NO), prostacyclin, norepinephrine etc. acting via G-protein and protein-kinase dependent mechanisms have been proposed to be potential mediators of preconditioning (see for reviews [2, 4–6]).

We have recently demonstrated that the early cardioprotective effect of preconditioning induced by rapid pacing is lost in rabbits and rats fed a high-cholesterol diet; and that reduced cardiac NO synthesis may be involved in the impairment of preconditioning in hypercholesterolemia [7–10]. High-cholesterol diet is regarded as an important factor in the development of ischemic heart disease. The heart of hyperlipidemic/atherosclerotic patients adapts poorly to

Address for offprints: P. Ferdinandy, Department of Biochemistry, Albert Szent-Gyögyi Medical University, H-6701 Szeged, P O Box 415, Hungary

physical exercise or other kinds of stress, showing that the endogenous adaptive mechanisms against myocardial stress are impaired (see for review [11]). Therefore, restoration of the effect of preconditioning in hyperlipidemia would be of great importance.

It has been shown that excess exogenous cholesterol inhibits formation of several polyprenyl derivatives through inhibition of hydroxy-methyl-glutaryl-coenzyme-A (HMG-CoA) reductase enzyme (Fig. 1.) [12, 13]. Recent evidences suggest that prenylation of several proteins including nuclear protein lamin B and a number of GTP-binding regulatory proteins (G proteins) responsible for controlling a wide spectrum of signal transduction pathways, is a prerequisite for their physiological function [13, 14]. Ubiquinone (coenzyme Q_{10}) is a key polyprenyl derivative molecule in the mitochondrial electron-transport system and as an endogenous antioxidant, and it was shown to protect the ischemic/reperfused myocardium [15]. Consequently, high-cholesterol diet-induced inhibition of polyprenyl biosynthesis may impair signal-transduction processes and the endogenous antioxidant defense systems, thereby leading to the loss of preconditioning.

Therefore, the present study was devoted to examine whether the cardioprotective effect of preconditioning is recaptured by treatment with the key polyprenyl derivative, farnesol, in high-cholesterol-fed rats; and if farnesol treatment alleviates reduction of basal cardiac NO content due to high-cholesterol feeding. The results presented here suggest that farnesol treatment does recapture preconditioning, however, farnesol does not restore basal cardiac NO synthesis in hypercholesterolemia.

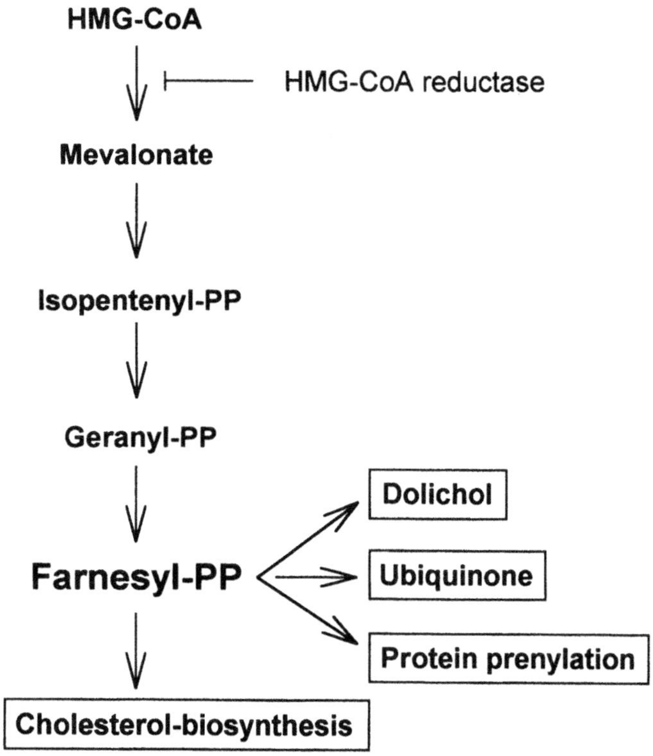

Fig. 1. Schematic diagram of isoprenoid biosynthesis. HMG-CoA: hydroxy-methyl-glutaryl-coenzyme-A.

Materials and methods

Animals and treatments

Male Wistar rats (8 weeks old), housed in a room maintained at 12 h light-dark cycles and a constant temperature of 22 ± 2°C, were fed laboratory chow enriched with 2% cholesterol (Sigma) or standard chow for 24 weeks. At the end of the diet period, body weight of the animals were 420–500 g, and there were no significant difference between groups. Cholesterol-fed and control rats were injected intraperitoneally 5 μM/kg farnesol (Sigma, St. Louis, MO, USA) or physiological saline at the end of the 24 weeks, respectively. An hour later, 14 rats from each group were used for isolated heart preparations, and 4 animals from each group were used for NO measurement in separate experiments. When the thoraxes were opened for excision of the hearts, blood samples were taken from the thoracic aorta, and serum cholesterol and triglycerides were assayed by means of an automatic analyzer (Hitachi-911) using Boehringer-Mannheim (Mannheim, Germany) kits. Hearts of Male Wistar rats were chosen for the study, since this species shows a moderate increase in serum cholesterol when on a cholesterol diet. While no substantial atherosclerosis develops, the increased tissue cholesterol leads to strong metabolic effects [16, 17]. The 24-week cholesterol diet moderately increased serum cholesterol and triglyceride from 1.64 ± 0.06 and 0.49 ± 0.04 to 1.92 ± 0.09 ($p < 0.05$) and 1.34 ± 0.17 ($p < 0.05$) mM/L, respectively.

Isolated heart preparation

At the end of the 24 week diet period, 1 h after farnesol/vehicle treatment, hearts were excised after light anesthesia with diethylether, and mounted on a working heart apparatus. Hearts were perfused through the left atrium with 37°C Krebs-Henseleit bicarbonate buffer containing (in mM) NaCl 118, KCl 4.3, $CaCl_2$ 2.4, $NaHCO_3$ 25, KH_2PO_4 1.2, $MgSO_4$ 1.2 and glucose 11.1, gassed with 95% O_2 and 5% CO_2 as described [18, 19]. The perfusion system allowed switching the perfusion line between retrograde Langendorff and left atrial 'working' perfusion. A suture was placed around the left main coronary artery close to its origin, allowing induction of regional ischemia

[20]. The coronary occlusion results in an ischemic zone comprising 37.4 ± 1.3 % of the left ventricle, which can not be altered even by strong coronary dilator/constrictor agents [19] since the rat is a coronary collateral-deficient species. Pacing at 10 Hz (600 bpm) with double threshold square impulses (impulse duration: 5 ms) was performed by an electric stimulator (Experimetria, Budapest, Hungary) through silver electrodes attached directly to the surface of the right ventricle. Heart rate (HR) derived from the left ventricular pressure curve, coronary flow (CF) measured by collecting effluent from the right atrium in a measuring cylinder for a timed period, aortic flow (AF) measured by a calibrated rotameter (KDG Flowmeters, Sussex, England), and left ventricular end-diastolic pressure (LVEDP) were measured. Preload (1.7 kPa) and afterload (9.8 kPa) were kept constant throughout the experiments. Intraventricular pressure was measured by means of a pressure transducer (B. Braun, Melsungen, Germany) connected to a small polyethylene catheter inserted into the left ventricle through the left atrial cannula [20]. Ventricular pressure was on-line digitized with a 200 Hz sampling frequency, recorded, and stored on an IBM computer.

Design of isolated heart experiments

Experimental groups

A pacing-induced preconditioning protocol was applied to hearts obtained from (i) control [PC, Preconditioned Control], (ii) control, farnesol-treated [PC + F, Preconditioned Control + Farnesol] (iii) cholesterol-fed [PCh, Preconditioned Cholesterol-fed], and (iv) cholesterol-fed, farnesol-treated [PCh + F, Preconditioned Cholesterol-fed + Farnesol] groups. A time-matched non-preconditioning protocol was also applied to hearts from (v) control [NPC, Non-preconditioned Control], (vi) control, farnesol-treated [NPC + F, Non-preconditioned Control + Farnesol] (vii) cholesterol-fed [NPCh, Non-preconditioned Cholesterol-fed], and (viii) cholesterol-fed, farnesol-treated [NPCh + F, Preconditioned Cholesterol-fed + Farnesol] groups (n =7 in each group).

Preconditioning with rapid ventricular pacing

After a 10 min aerobic working perfusion, all hearts were subjected to 3 intermittent periods of either a high-demand preconditioning or a non-preconditioning protocol for 5 min and 15 sec each followed by 4 min and 45 sec of aerobic working perfusion as described previously [10, 20] (Fig. 2). In the non-preconditioned groups, periods of aerobic Langendorff perfusion were applied without pacing. In preconditioned groups hearts were paced at 10 Hz (600 bpm) under Langendorff perfusion for 4 min and 45 sec followed by 30 sec of non-paced aerobic Langendorff perfusion. The last 30 sec allowed spontaneous restoration of sinus rhythm before switching to working conditions. Langendorff perfusion was used during the periods of pacing, and therefore during corresponding control periods as well, since the heart paced at 10 Hz cannot maintain enough aortic flow to perfuse the coronary circulation [20]. Cardiac functional parameters were recorded at the end of each period of working perfusion and at the 10th min of test ischemia.

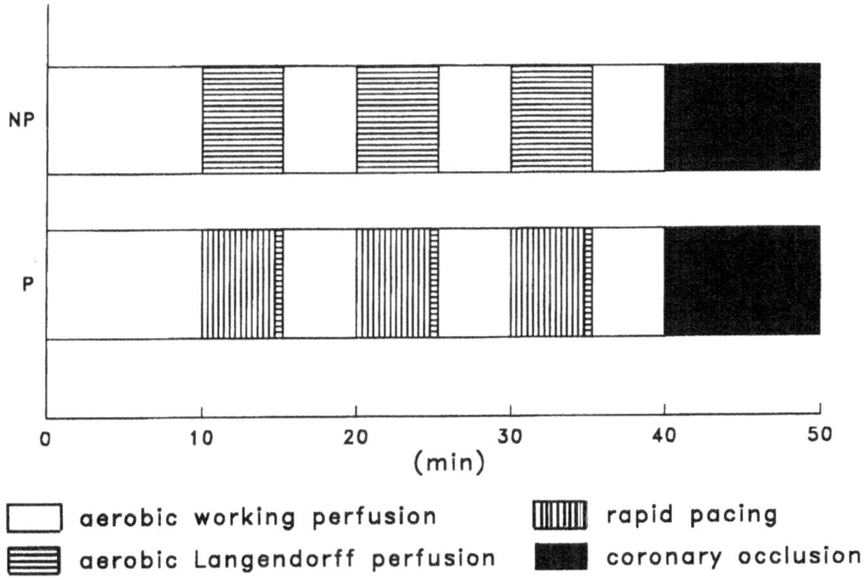

Fig. 2. Bar graph showing the perfusion protocol of isolated rat hearts subjected to pacing-induced preconditioning (P) or non-preconditioning (NP) protocols followed by test ischemia produced by coronary occlusion. Rapid pacing at 10 Hz (600 bpm) was carried out under Langendorff perfusion.

Test ischemia with coronary occlusion

After preconditioning/non-preconditioning protocol test ischemia of 10 min duration was produced by occlusion of the main coronary artery. A 10 min period of regional ischemia was chosen for assessing myocardial function, because this short-term regional ischemia induced considerable deterioration of myocardial function, but it did not result in ischemia-induced arrhythmias that might have disturbed measures of myocardial function [21]. Reperfusion was not applied after coronary occlusion in this study, since after a 10 min coronary occlusion, a high incidence of ventricular fibrillation (VF) occurs, and pacing-induced preconditioning does not protect against VF in this experimental protocol as we described [10, 19, 20]. Ischemic lactate dehydrogenase (LDH) release was measured from coronary effluents by an automatic analyzer (Hitachi 911) using Boehringer-Mannheim (Mannheim, Germany) kits as described [22].

Measurement of cardiac NO

NO was assayed by an *in vivo* spin-trapping method followed by electron spin resonance analysis of left ventricular tissue samples as described [10, 23, 24]. Briefly, the spin-trap diethyl-dithio-carbamate (DETC, 200 mg/kg), 50 mg/kg $FeSO_4$ and 200 mg/kg sodium-citrate were slowly administered intravenously into the femoral vein under ether anesthesia to all groups (n = 4 in each groups). Five minutes after DETC, $FeSO_4$, and citrate treatment, hearts were isolated and perfused in Langendorff mode for 30 sec to eliminate blood, then 100 mg tissue samples of the left ventricles were placed into quartz tubes, and frozen in liquid nitrogen until assayed for electron spin resonance spectra of $NO\text{-}Fe^{2+}\text{-}(DETC)_2$ complex. To obtain the background spectra of $Cu^{2+}(DETC)_2$ complex, animals were given 200 mg/kg DETC only. As a positive control, 3 mg/kg Na-nitroprusside was given intravenously followed by the spin-trapping procedure. Electron spin resonance spectra were recorded with Bruker ECS106 (Rheinstetten, Germany) spectrometer operating at X band with 100 kHz modulation frequency at a temperature of 160 K, using 10 mW microwave power to avoid saturation. Scans were traced with 2.85 G modulation amplitude, 340 G sweep width, and 3356 G central field. After subtraction of the background signal of $Cu^{2+}\text{-}(DETC)_2$, analysis of NO content was performed with double integration of all spectra.

Statistics

Data were expressed as means ± S.E. of the mean [S.E.M.] and analyzed with one way analysis of variance (ANOVA). If a difference was established, each group was compared to the control non-preconditioned group using a modified *t*-test corrected for simultaneous multiple comparisons according to the Bonferroni method [25].

Results

Preconditioning

When test ischemia was preceded by preconditioning with rapid ventricular pacing in hearts obtained from control animals (PC group, Fig. 3),AF, LVEDP, and LDH release were significantly improved as compared to non-preconditioned controls (NPC group, Fig. 3). HR and CF were not influenced by preconditioning; and preconditioning and non-preconditioning protocols did not significantly alter the parameters before test ischemia (data not shown). Farnesol treatment did not significantly influence the response to a test ischemia after a non-preconditioning protocol in control hearts (NPC + F group, Fig. 3), nor did it alter the protective effect of preconditioning (PC + F group, Fig. 3) in control hearts.

When hearts were isolated from cholesterol-fed, untreated rats, the protective effect of pacing-induced preconditioning was not seen (PCh group, Fig. 3), i.e. during test ischemia, AF, LVEDP, and LDH release were not improved by preconditioning. The protective effect of preconditioning reappeared when a preconditioning protocol was applied in hearts obtained from cholesterol-fed, farnesol-treated group (PCh + F group, Fig. 3); however, farnesol treatment did not significantly affect the severity of ischemia after a non-preconditioning protocol in hearts of cholesterol-fed rats (NPCh + F group, Fig. 3).

Cardiac NO content

In tissue samples of the left ventricle, high-cholesterol diet markedly decreased the intensity of the specific spectra of $NO\text{-}Fe^{2+}$-diethyl-dithio-carbamate complex assessed by electron spin resonance as compared to the control group. Farnesol-treatment did not affect cardiac NO content in either control or cholesterol-fed groups (Figs 4 and 5).

Discussion

The present results show that cholesterol-enriched diet for 24 weeks inhibits the cardioprotective effect of pacing-induced preconditioning in isolated working rat hearts with coronary occlusion, and eliminates the specific electron spin resonance spectra of $NO\text{-}Fe^{2+}\text{-}(DETC)_2$ in tissue samples of the left ventricle. Treatment with a key polyprenyl derivative, farnesol, does not mimic preconditioning but recaptures the

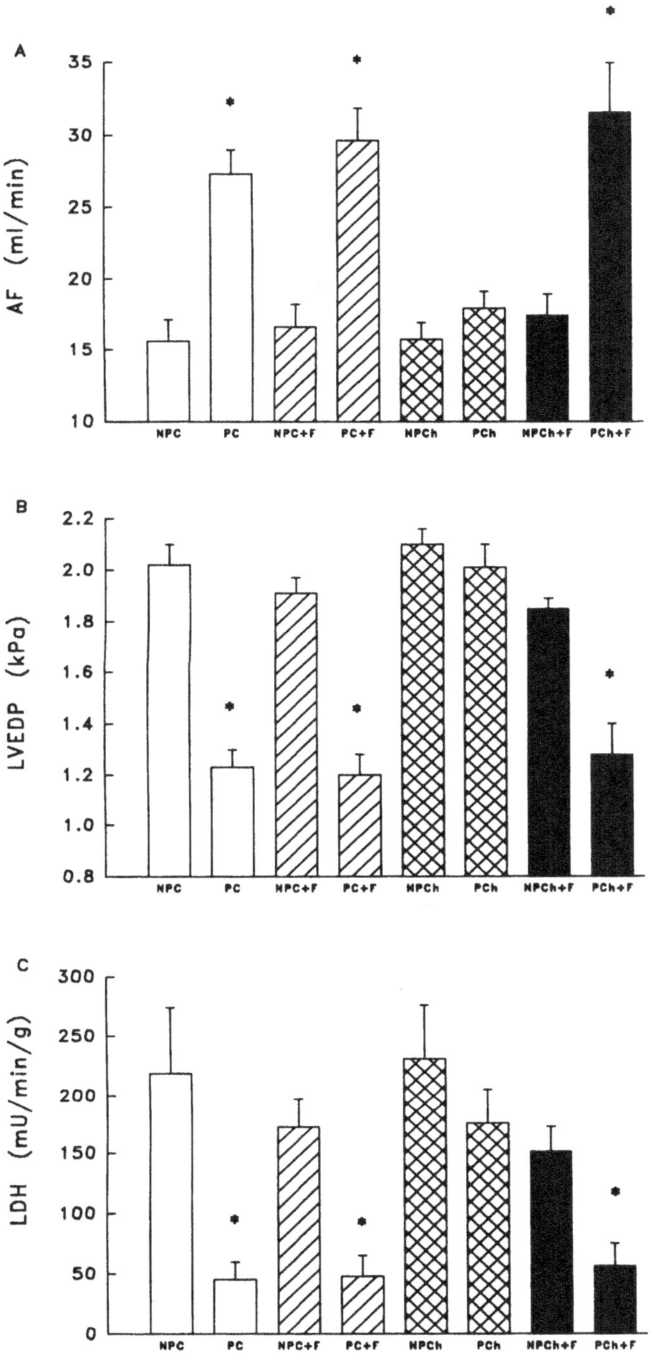

Fig. 3. Effects of pacing-induced preconditioning on coronary occlusion-induced changes in aortic flow (AF, panel A), left ventricular end-diastolic pressure (LVEDP, panel B), and lactate dehydrogenase release (LDH, panel C) in isolated working rat hearts. Measurements were made in the 10th min of regional ischemia. NPC, NPC + F, NPCh, NPCh + F, non-preconditioned hearts isolated from control, farnesol-treated control, cholesterol-fed, and farnesol-treated cholesterol-fed rats, respectively; PC, PC + F, PCh, PCh + F, preconditioned hearts isolated from control, farnesol-treated control, cholesterol-fed, and farnesol-treated cholesterol-fed rats, respectively. Data are means ± S.E.M., n = 7 in each groups, *($p < 0.05$) shows significant difference from NPC group. (1 mmHg = 13.3 Pa).

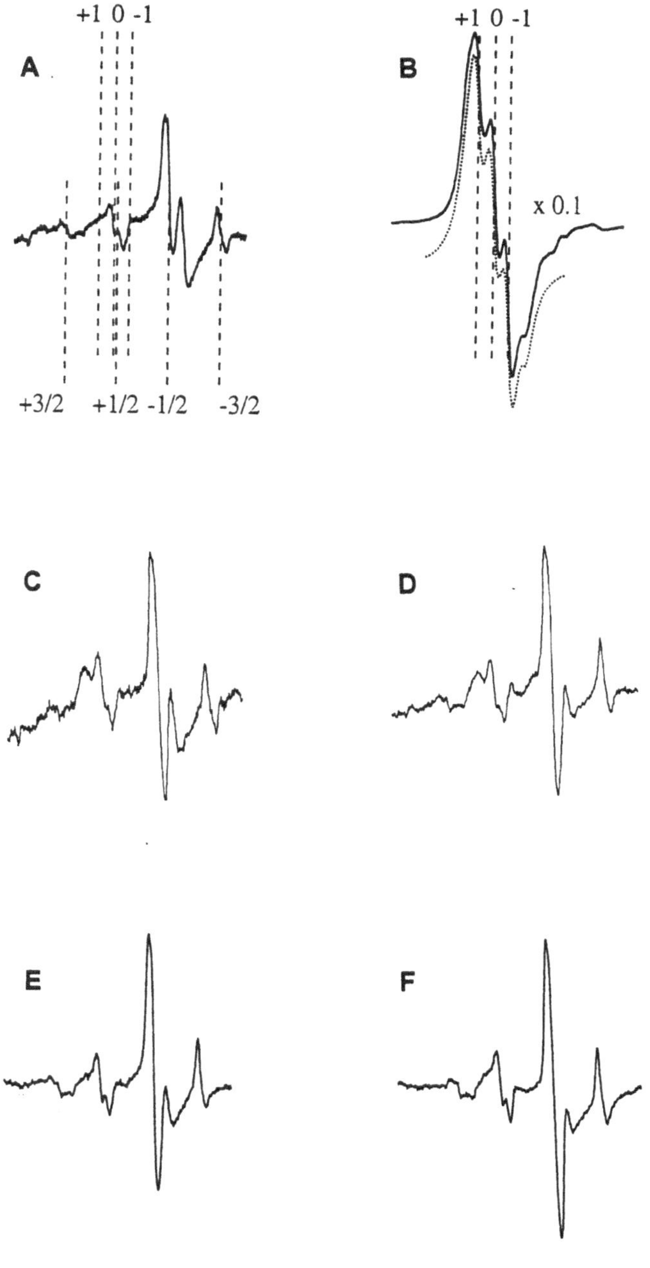

Fig. 4. Measurement of nitric oxide (NO) by electron spin resonance (ESR) after in vivo spin-trapping with Fe^{2+}-diethyl-dithio-carbamate (DETC) complex, representative spectra from all groups. ESR parameters: X band, 100 kHz modulation frequency, 160 K, 10 mW microwave power, 2.85 G modulation amplitude, 340 G sweep width, and 3356 G central field. *Curve A:* background spectrum of $Cu(DETC)_2$ complex; +1 0 –1: hyperfine splitting of NO-Fe^{2+}-$(DETC)_2$ triplet; +3/2 +1/2–1/2–3/2: hyperfine splitting of $Cu(DETC)_2$ complex. *Curve B:* solid line shows positive control (intravenous bolus dose of 3 mg/kg Na-nitroprusside); dotted line shows a fitted curve; x0.1 shows reduced gain compared to other spectra; +1 0 –1: hyperfine splitting of NO-Fe^{2+}-$(DETC)_2$ triplet. *Curve C:* control. *Curve D:* farnesol-treated control. *Curve E:* cholesterol-fed. *Curve E:* farnesol-treated cholesterol-fed.

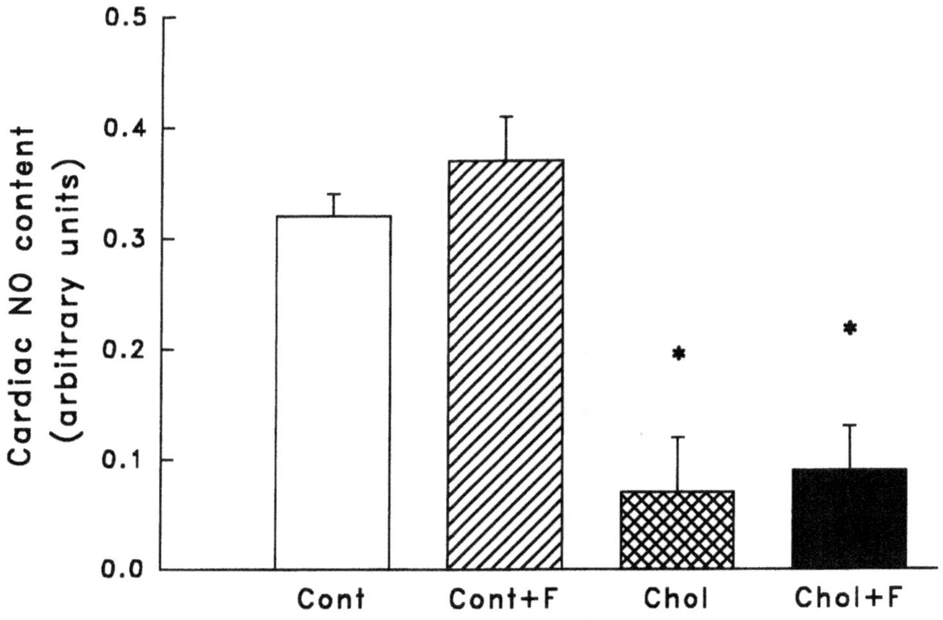

Fig. 5. Cardiac NO content in control (Cont), farnesol-treated control (Cont + F), cholesterol-fed (Chol), and farnesol-treated cholesterol-fed (Chol+F) rats. Data are means ± S.E.M., n = 4 in each groups, *($p < 0.05$) shows significant difference from control.

effect of preconditioning in cholesterol-fed rats, however, it does not restore cardiac NO synthesis. These findings strongly suggest that hypercholesterolemia may inhibit preconditioning through inhibition of synthesis of polyprenyl derivatives. The role of NO in the preconditioning signal cascade in hypercholesterolemia remains controversial.

Cholesterol-diet, polyprenyl derivatives, and preconditioning

Hyperlipidemia, especially hypercholesterolemia is a major risk factor for the development of ischemic heart disease. There is a linear relationship between elevation of serum total cholesterol concentration and the incidence of myocardial infarction, which is mostly attributed to the development of atherosclerosis (see for review [11]). However, we have recently demonstrated that hypercholesterolemia blocks the early phase of protection from preconditioning probably irrespective of the development of atherosclerosis in conscious rabbits [7, 8], which shows that hypercholestero-lemia may exert a direct effect on the myocardium. In accordance, a few studies demonstrated structural and functional alterations in the myocardium due to hyper-cholesterolemia [26–28].

The rate limiting enzyme of cholesterol biosynthesis, HMG-CoA reductase, is regulated by a feed-back inhibition by cholesterol. Thus, the presence of excess cholesterol leads to a blockade of the formation of farnesyl-PP, thereby leading to a diminished endogenous synthesis of several polyprenyl products like farnesylated proteins (G-proteins, ras-family, lamin B, etc.), ubiquinone, dolichol, and cholesterol-derivatives (see for review [13], Fig. 1). The family of trimeric G proteins (G_s, G_i, and G_t) are membrane associated. The membrane association requires polyisoprenylation of their γ-subunit [14]. Therefore, inhibition of protein-polyisoprenylation may affect several signal transduction pathways [29] (see for reviews: [12, 30]). Although the biochemical mechanisms of preconditioning is not completely known, several G protein-dependent pathways were shown to be involved in preconditioning [31–33]. Our present study showing that farnesol treatment does not mimic but, rather, restores preconditioning in hypercholesterolemic rats suggests that hypercholesterolemia may inhibit pre-conditioning via an inhibition of polyprenyl biosynthesis, thereby leading to a disruption of signal-transduction pathways.

NO and preconditioning

The role of NO in the conventional no-flow ischemia-induced preconditioning is controversial [34–36], however, our previous studies support the role of NO-cGMP system in pacing-induced preconditioning in rabbits and rats [10, 24, 37, 38]. Several studies demonstrate that hypercholesterolemia leads to an NO-deficient state [39–41]. We have recently shown that NO is involved in the trigger mechanism of pacing-induced preconditioning, and that the lack of preconditioning in hypercholesterolemia may be due to an inhibition of basal cardiac NO synthesis [9]. Our present study confirms that a high-cholesterol diet decreases cardiac NO content and blocks the effect of preconditioning; however, it shows that while farnesol treatment recaptures preconditioning, it does not restore cardiac NO level in cholesterol-fed rats. These findings suggest that polyprenyl biosynthesis does not affect cardiac NO metabolism, and that NO may be involved in a less important, 'replaceable' triggering step of the signal transduction cascade of preconditioning. To elucidate the controversial role of NO in preconditioning further studies are needed which should look at NO synthesis during the evolution of the preconditioning effect.

Conclusions

In summary, the results presented here show that farnesol treatment recaptures preconditioning, but it does not restore basal cardiac NO synthesis in experimental hypercholesterolemia. This suggests that cholesterol diet blocks preconditioning via a polyprenyl-dependent mechanism.

Acknowledgement

This work was supported by grants from National Scientific Research Found (OTKA F19946, D23736), National Technical Development Committee (OMFB 05201950616), Ministry of Education (FKFP 1284/1997), and the Hungarian Space Research Office, Budapest, Hungary.

References

1. Murry CE, Jennings RB, Reimer KA: Preconditioning with ischemia: A delay of lethal cell injury in ischemic myocardium. Circulation 74: 1124–1136, 1986
2. Parratt JR: Possibilities for the pharmacological exploitation of ischaemic preconditioning. J Mol Cell Cardiol 27: 991–1000, 1995
3. Baxter GF, Yellon DM: Ischaemic preconditioning of myocardium: A new paradigm for clinical cardioprotection. Br J Clin Pharmacol 38: 381–387, 1994
4. Baxter GF, Marber MS, Yellon DM: Myocardial stress response, cytoprotective proteins and the second window of protection against infarction. In: CL Wainwright, JR Parratt, (eds). Myocardial Preconditioning. R.G. Landes Company, 1996, pp 233–250
5. Walker DM, Yellon DM: Ischaemic preconditioning: From mechanisms to exploitation. Cardiovasc Res 26: 734–739, 1992
6. Downey JM: Ischemic preconditioning: Nature's own cardioprotective intervention. Trends Cardiovasc Med 2: 170–176, 1992
7. Szilvassy Z, Ferdinandy P, Szilvassy J, *et al*: The loss of pacing-induced preconditioning in atherosclerotic rabbits: role of hypercholesterolemia. J Mol Cell Cardiol 27: 2559–2569, 1995
8. Szekeres L, Szilvassy Z, Ferdinandy P, Nagy I, Karcsu S, Csati S: Delayed cardiac protection against harmful consequences of stress can be induced in experimental atherosclerosis in rabbits. J Mol Cell Cardiol 29: 1977–1983, 1997
9. Ferdinandy P, Szilvassy Z, Horvath LI, *et al.*: Loss of pacing-induced preconditioning in rat hearts: role of nitric oxide and cholesterol-enriched diet. J Mol Cell Cardiol 29: 3321–3333, 1997
10. Ferdinandy P, Csont T, Csonka C, *et al.*: Capsaicin-sensitive local sensory innervation is involved in pacing-induced preconditioning in rat hearts: Role of nitric oxide and CGRP? Naunyn Schmiedebergs Arch Pharmacol 356: 356–363, 1997
11. Roberts WC: Preventing and arresting coronary atherosclerosis. Am Heart J 130: 580–600, 1995
12. Goldstein JL, Brown MS: Regulation of the mevalonate pathway. Nature 343: 425–430, 1990
13. Casey PJ: Biochemistry of protein prenylation. J Lipid Res 33: 1731–1740, 1992
14. Finegold AA, Schafer WR, Rine J, Whiteway M, Tamanoi F: Common modifications of trimeric G proteins and ras protein: involvement of polyisoprenylation. Science 249: 165–169, 1990
15. Hano O, Thompson-Gorman SL, Zweier JL, Lakatta EG: Coenzyme Q_{10} enhances cardiac functional and metabolic recovery and reduces Ca^{2+} overload during postischemic reperfusion. Am J Physiol 266: H2174–H2181, 1994
16. Roach PD, Balasubramaniam S, Hirata F, *et al.*: The low-density lipoprotein receptor and cholesterol synthesis are affected differently by dietary cholesterol in the rat. Biochem Biophys Acta 1170: 165–172, 1993
17. Horton JD, Cuthbert JA, Spady DK: Regulation of hepatic 7α-hydroxylase expression and response to dietary cholesterol in the rat hamster. J Biol Chem 270: 5381–5387, 1995
18. Ferdinandy P, Das DK, Tosaki A: Pacing-induced ventricular fibrillation leading to oxygen free radical formation in aerobically perfused rat heart. J Mol Cell Cardiol 25: 683–692, 1993
19. Ferdinandy P, Szilvassy Z, Droy-Lefaix MT, Tarrade T, Koltai M: K_{ATP} channel modulation in working rat hearts with coronary occlusion: Effects of cromakalim, cicletanine, and glibenclamide. Cardiovasc Res 30: 781–787, 1995
20. Ferdinandy P, Szilvassy Z, Koltai M, Dux L: Ventricular overdrive pacing-induced preconditioning and no-flow ischemia-induced preconditioning in isolated working rat hearts. J Cardiovasc Pharmacol 25: 97–104, 1995
21. Ferdinandy P, Koltai M, Tosaki A, *et al.*: Cicletanine improves myocardial function deteriorated by ischemia/reperfusion in isolated working rat hearts. J Cardiovasc Pharmacol 19: 181–189, 1992
22. Ferdinandy P, Szilvássy Z, Csont T, *et al.*: Nitroglycerin-induced direct protection of the ischaemic myocardium in isolated working hearts of rats with vascular tolerance to nitroglycerin. Br J Pharmacol 115: 1129–1131, 1995
23. Zweier JL, Wang P, Kuppusamy P: Direct measurement of nitric oxide generation in the ischemic heart using electron paramagnetic resonance spectroscopy. J Biol Chem 270: 304–307, 1995
24. Ferdinandy P, Szilvassy Z, Balogh N, *et al.*: Nitric oxide is involved in active preconditioning in isolated working rat hearts. Ann NY Acad Sci 793: 489–493, 1996
25. Wallenstein S, Zucker CL, Fleiss JL: Some statistical methods useful in circulation research. Circ Res 47: 1–9, 1980
26. Hexeberg S, Willumsen N, Rotevatn S, Hexeberg E, Berge RK: Cholesterol induced lipid accumulation in myocardial cells of rats. Cardiovasc Res 27: 442–446, 1993
27. Haukur M, Leeson TS: Comparative electron microscope studies of the myocardium in adult rats fed normal and cholesterol diets. J Mol Cell Cardiol 7: 195–202, 1975

28. Senges J, Weihe E, Brachmann J: Effects of hypercholesterolaemia without ischemia on some electrophysiological and ultrastructural properties of the rabbit heart. J Mol Cell Cardiol 13: 253–264, 1981
29. Bradfute DL, Simoni RD: Non-sterol compounds that regulate cholesterogenesis. J Biol Chem 269: 6645–6650, 1994
30. Omura S, Takeshima H: Farnesyl-protein transferase inhibitors. Drugs Fut 19: 751–755, 1994
31. Downey JM, Cohen MV: Signal transduction in ischemic preconditioning. Z Kardiol 84: 77–86, 1995
32. Reimer KA, Jennings RB: Ischemic preconditioning: A brief review. Basic Res Cardiol 91: 1–4, 1996
33. Ravingerová T, Pyne NJ, Parratt JR: Ischaemic preconditioning in the rat heart: The role of G-proteins and adrenergic stimulation. Mol Cell Biochem 147: 123–128, 1995
34. Vegh A, Szekeres L, Parratt J: Preconditioning of the ischaemic myocardium; Involvement of the L-arginine nitric oxide pathway. Br J Pharmacol 107: 648–652, 1992
35. Weselcouch EO, Baird AJ, Sleph P, Grover GJ: Inhibition of nitric oxide synthesis does not affect ischemic preconditioning in isolated perfused rat hearts. Am J Physiol Heart Circ Physiol 268: H242–H249, 1995
36. Lu HR, Remeysen P, De Clerck F: Does the antiarrhythmic effect of ischemic preconditioning in rats involve the L-arginine nitric oxide pathway. J Cardiovasc Pharmacol 25: 524–530, 1995
37. Szilvassy Z, Ferdinandy P, Bor P, Jakab I, Lonovics J, Koltai M: Ventricular overdrive pacing-induced anti-ischemic effect: A conscious rabbit model of preconditioning. Am J Physiol 266: H2033–H2041, 1994
38. Szilvassy Z, Ferdinandy P, Bor P, *et al.*: Loss of preconditioning in rabbits with vascular tolerance to nitroglycerin. Br J Pharmacol 112: 999–1001, 1994
39. Lefer AM, Ma X: Decreased basal nitric oxide release in hypercholesterolemia increases neutrophil adherence to rabbit coronary artery endothelium. Arteriosc Thromb 13: 771–776, 1993
40. Deliconstatinos G, Villioyou V, Stavrides JC: Modulation of particulate nitric oxide synthase activity and peroxynitrite synthesis in cholesterol enriched endothelial cell membranes. Biochem Pharmacol 49: 1589–1600, 1995
41. Schmidt K, Graier WF, Kostner GM, Mayer B, Kukovetz WR: Activation of soluble guanilate cyclase by nitrovasodilators is inhibited by oxidized low-density lipoprotein. Biochem Biophys Res Commun 172: 614–619, 1990

Molecular and Cellular Biochemistry **186**: 35–41, 1998.
© 1998 *Kluwer Academic Publishers.*

Effect of ethanol on myocardial infarct size in a canine model of coronary artery occlusion-reperfusion

Masao Itoya, Jeffrey D. Morrison and H. Fred Downey

Department of Integrative Physiology, University of North Texas Health Science Center at Fort Worth, Fort Worth, TX, USA

Abstract

This study was conducted to determine if elevated blood alcohol prior to acute coronary artery occlusion affects myocardial infarct size in an *in vivo* canine model. Seven pentobarbital anesthetized open-chest dogs received 10 min iv infusion of ethanol (0.08 g/kg/min). Ten min after ethanol, the left anterior descending coronary artery (LAD) was occluded distal to its first major branch for 60 min. The LAD was then reperfused for 5 h. Following electrically induced ventricular fibrillation, the area at risk of infarction was delineated with dye. The area of infarction was identified by staining with triphenyl tetrazolium chloride. Eleven untreated control experiments were also conducted. Mean blood ethanol concentration was 155 ± 26 mg/dl just prior to LAD occlusion and 47 ± 3 mg/dl after 4 h reperfusion. Ethanol infusion had no effect on systemic hemodynamic variables during ischemia. In ethanol treated animals, the area at risk was 19.7 ± 3.0% of the left ventricle, and the infarct size was 20.9 ± 4.8% of the area at risk. In control experiments, the area at risk was 23.0 ± 4.1% of the left ventricle ($p > 0.05$), and the infarct size was 21.6 ± 3.8% of the area at risk ($p > 0.05$). Collateral blood flow to ischemic region did not differ between the two groups, and the relationships between infarct size and collateral flow were similar for control and untreated hearts. Acute ethanol exposure prior to coronary artery occlusion and subsequent reperfusion does not affect myocardial infarct size in the heart of the anesthetized dog. (Mol Cell Biochem **186**: 35–41, 1998)

Key words: ethanol, myocardial infarction, tetrazolium, collateral flow

Introduction

Epidemiological studies have suggested that moderate alcohol consumption is associated with a reduction in mortality from ischemic heart disease [1, 2]. However, acute effects of alcohol in the presence of coronary heart disease appear harmful rather than beneficial. Alcohol reduces cardiac output and left ventricular work [3], reduces exercise duration until angina [4], and exercise time until onset of ischemic ST segment depression [5] in patients with angina. In addition, alcohol has been known to induce coronary artery spasm in some patients with variant angina [6]. Contrarily, some studies in rats suggested that alcohol has a protective effect against myocardial ischemic injury. Bernauer [7] reported that ethanol infusion (2 g/kg iv) reduced myocardial infarct size due to coronary artery occlusion in rat heart, although a lower dose of ethanol (1 g/kg) did not have a clearly evident effect. Mallov *et al.* [8] demonstrated that oral administration of ethanol (0.5–6.0 g/kg) reduced the severity of myocardial necrosis produced by epinephrine in rat. However, the 60 min period of reperfusion in Bernauer's study might have resulted in an inaccurate estimation of infarct size by triphenyl tetrazolium chloride (TTC) staining [9, 10, 11]. Because the mechanism by which myocardial necrosis is caused by epinephrine differs from that of coronary artery occlusion [12], it is difficult to apply the results of the study of Mallov *et al.* [8] in defining the effects of ethanol on myocardial infarction.

Address for offprints: H. Fred Downey, Department of Integrative Physiology, University of North Texas Health Science Center at Fort Worth, 3500 Camp Bowie Boulevard, Fort Worth, TX 76107-2699, USA

Moreover, both studies were performed in rat heart, and to our knowledge no study has reported the effects of ethanol on myocardial infarct size in a large animal model, such as the dog.

Thus, the present study was conducted to determine if elevated blood ethanol affects the extent of myocardial infarction in a canine coronary occlusion-reperfusion model. Since the coronary artery was reperfused for 5 h following ischemia, this duration was sufficient to delineate infarcted tissue by the TTC method. We employed a moderate dose of ethanol to produce blood ethanol concentration of approximately 100 mg/dl, which is a usual legal definition of intoxication.

Materials and methods

The investigation was approved by the Institutional Animal Care and Use Committee of the University of North Texas Health Science Center at Fort Worth and was conducted in accordance with the *Guide to the Care and Use of Laboratory Animals* (NIH publication 85-23, revised 1985).

Surgical preparation

Thirty-five adult mongrel dogs of either sex (15–35 kg) were used in this study. Dogs were anesthetized with 30 mg/kg pentobarbital sodium, intubated, and mechanically ventilated with room air supplemented with oxygen. Positive end-expiratory pressure (2 cm H_20) was applied to minimize atelectasis. Supplemental pentobarbital sodium was given as needed to maintain stable anesthesia. Arterial blood gases were measured frequently and ventilation was adjusted, as needed, to maintain blood gases within physiological ranges. A vinyl catheter was inserted into a right femoral artery to measure aortic blood pressure and into a right femoral vein for administration of ethanol, supplemental pentobarbital, and fluids. Two vinyl catheters were inserted into the thoracic aorta via the left femoral artery to collect reference blood samples during microsphere injection (see below). A left thoracotomy was performed in the fifth intercostal space, and the heart was suspended in a pericardial cradle. Left intraventricular pressure was measured with a Millar catheter tip transducer inserted into the left atrial appendage and advanced across the mitral valve. The first derivative of left ventricular pressure (dP/dt) was obtained by differentiating the left ventricular pressure signal. Myocardial contractility was assessed from the maximum rate of pressure development (+dP/dt) and diastolic function from the maximum rate of relaxation (–dP/dt). A polyethylene catheter (PE100) was inserted into the left atrium for microsphere injection.

The left anterior descending coronary artery was isolated immediately distal to the first major diagonal branch and a 3-0 silk suture was passed around the coronary artery. The ends of the silk suture were threaded through a vinyl tube to form a coronary snare. Precordial electrocardiograms were monitored with bipolar leads.

Determination of collateral blood flow

Regional myocardial blood flow was measured using radioactive microspheres. Microspheres were agitated on a vortex mixer and in an ultrasonic bath for at least 15 min before use. Microspheres labeled with ^{113}Sn were injected through a left atrial catheter, followed by 5 ml saline flush. Beginning just before and continuing for 3 min after injection, duplicate reference arterial blood samples were withdrawn from the thoracic aorta at a rate of 4 ml/min. Adequacy of microsphere mixing in the blood perfusate was verified by comparing radioactivities in the duplicate reference blood samples.

After determining infarct size (see below), ventricular sections were divided into non-ischemic and central ischemic regions. Lateral border zones were excluded to avoid errors associated with measuring blood flow in samples of heterogeneous composition. Tissue samples were divided into subendocardial, mid-myocardial, and subepicardial thirds for blood flow measurement. Radioactivities of tissue and blood reference samples were measured in a gamma spectrometer (model 1185, Canberra Industries, Meriden, CT, USA). Blood flow to these tissue samples (ml/min/g) was calculated as previously described [13].

Collateral blood flow into the inner two thirds of the ischemic wall during coronary occlusion has been shown to be an excellent predictor of infarct size [14, 15]. Thus, it was necessary to ensure that collateral flow was similar in control and experimental hearts, and that infarct size was altered independently of differences in collateral flow between the groups. Data from dogs with collateral blood flow greater than 0.2 ml/min/g in the inner two thirds of the ischemic wall were excluded.

Determination of myocardial infarct size

After excision of the heart, the size of the anatomical area at risk of infarction was determined with a dual perfusion technique. The aorta was retrogradely perfused with saline containing 1.0% Evans blue dye while the left anterior descending coronary artery was perfused with saline alone. The solutions were infused simultaneously for 1–3 min at a constant pressure of 100 mmHg. After excision of the atria and right ventricle, the left ventricle was cut into 4–8 transverse sections approximately 5 mm thick. The sections were weighed, and outlines of their basal surfaces were traced onto the clear plastic overlays. The area at risk, as demarcated by the absence of dye, was then traced on the overlays. The slices were then incubated at 37°C for 20 min in phosphate buffer (0.1 M, pH 7.4) containing 1% triphenyl tetrazolium chloride to demarcate the infarct area [16]. The infarct size, as indicated by the absence of red formazan precipitate, was traced on the overlays. Area at risk and infarct size in each overlay were determined with Sigma Scan (Jandel Scientific Co.), a personal computer based area calculating program. These data and the weight of the slice were used to calculate the mass of the area at risk and the infarct area.

Determination blood ethanol

Arterial blood samples were collected via a catheter in the left femoral artery for determination of blood ethanol concentration. Whole blood ethanol concentration was analyzed fluorometrically with TDx REA Ethanol Assay (Abbott Laboratories, North Chicago, IL, USA) [17].

Experimental protocol

Dogs were divided into two groups. In the control group (n = 21), the LAD was occluded by tightening the snare for 60 min and then reperfused for 5 h. The ethanol treated group (n = 14) received a 10 min iv infusion of ethanol (0.08 g/kg/min). Ten min after the completion of ethanol infusion, the LAD was occluded and reperfused as in the control group. The existence of myocardial ischemia was confirmed by ST elevation on the ECG. In all dogs, lidocaine (1 mg/kg) was injected for prophylaxis of arrhythmias 2 min before LAD occlusion and before reperfusion. After completion of the experimental protocol, 500 U/kg heparin was injected iv to facilitate postmortem coronary perfusion. Ventricular fibrillation was induced electrically, and the heart was excised for analysis of area at risk, infarct size, and collateral blood flow to ischemic myocardium.

Statistical analyses

Values are expressed as mean ± S.E.M. Differences in between-group data were evaluated by unpaired *t*-test. Within group data measured over time were compared by a repeated measures analysis of variance, and the Student-Newman-Keuls test was used as a post hoc test. Analysis of covariance was used to test for between-group variation in the relationships of infarct size and collateral blood flow. Statistical significance was assumed at $p < 0.05$.

Results

Mortality and exclusions

Twenty-one dogs were enrolled in the control group. Six dogs died from ventricular fibrillation during coronary artery occlusion or reperfusion, and four were excluded due to high collateral flow. Data from the eleven remaining control dogs were used for the analyses. Fourteen dogs were enrolled initially in the ethanol treated group. One dog died from ventricular fibrillation during coronary occlusion and two during reperfusion. Four dogs were excluded because of high collateral flow during coronary artery occlusion. Thus, seven ethanol treated dogs contributed data for the following analyses. No significant difference in mortality between groups was detected by chi-square test.

Blood ethanol concentration

Figure 1 shows time course of blood ethanol concentration. Blood ethanol concentration rose to about 300 mg/dl during the ethanol infusion, and was 155 ± 26 and 76 ± 13 mg/dl just prior to and at the end of coronary artery occlusion,

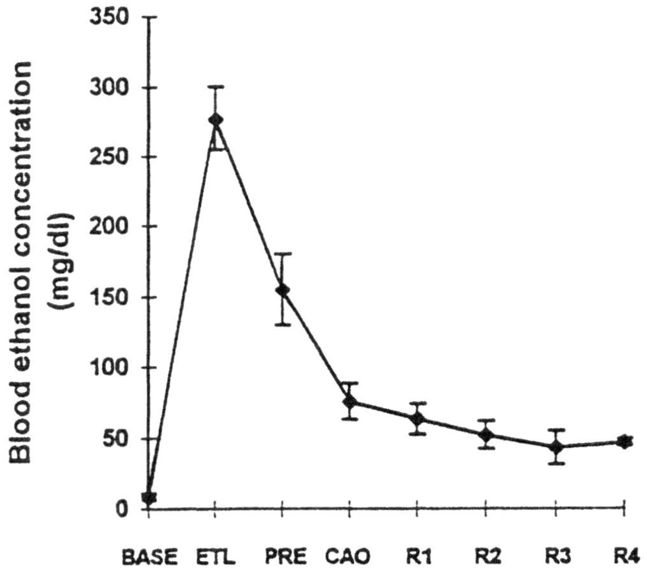

Fig. 1. Time course of blood ethanol concentration in the ethanol treated group. Data are means ± S.E.M. BASE – baseline; ETL – 9 min after ethanol infusion; PRE – 2 min before coronary occlusion; CAO – 58 min after coronary occlusion; Rn – n hour after reperfusion.

respectively. Ethanol level decreased slowly during reperfusion, and was 47 ± 3 mg/dl after 4 hr of reperfusion.

Hemodynamic variables

Hemodynamic measurements are summarized in Table 1. Left ventricular +dP/dt, an index of myocardial contractility, decreased during ethanol infusion, but gradually rose after cessation of ethanol infusion, although it did not fully recover to the baseline value. Mean arterial pressure, peak left ventricular pressure, and –dP/dt tended to decrease during ethanol infusion, but these changes were not statistically significant. HR was not affected by ethanol infusion. +dP/dt decreased during coronary occlusion, and gradually recovered during reperfusion in both groups. Mean aortic pressure also decreased progressively during reperfusion. No statistically significant differences were detected between two groups for any hemodynamic variables at any time.

Infarct size and collateral blood flow

The areas at risk as a percentage of the left ventricle were similar in the control and ethanol treated groups (23.0 ± 4.1% vs. 19.7 ± 3.0%, $p > 0.5$, Fig. 2.), and infarct sizes were also similar in the two groups (21.6 ± 3.8% vs. 20.1 ± 4.8%, $p < 0.9$, Fig. 2.). Average collateral blood flow into the ischemic region during LAD occlusion did not differ between the two groups ($p < 0.7$, Fig. 3.). Since infarct size varies with collateral flow, it was important to determine if

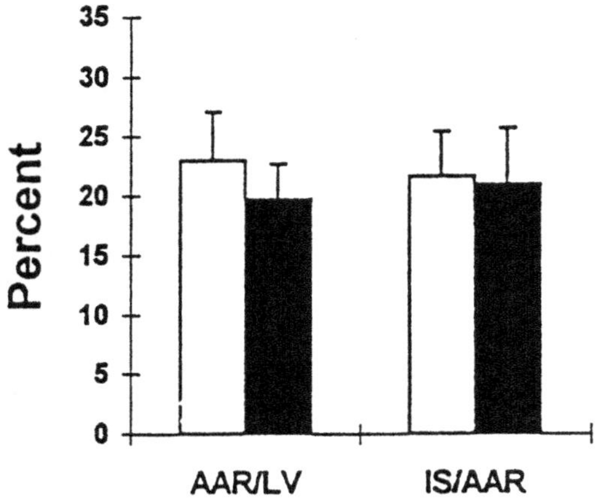

Fig. 2. The area at risk expressed as percent of left ventricle (AAR/LV) and infarct size expressed as percent of the area at risk (IS/AAR). Data are means ± S.E.M. Open bar – control group; closed bar – ethanol treated group.

Table 1. Hemodynamic data

VARIABLE	BASELINE	ETHANOL	PRE-CAO	CAO	REP 1h	REP 3h	REP 5h
MAP (mm Hg)							
Control			112 (3)	107 (6)	107 (4)	95 (4)	95 (7)
Ethanol	111 (4)	107 (4)	115 (3)	103 (3)	98 (5)*	97 (7)*	92 (5)*#
HR (beats/min)							
Control			142 (6)	149 (4)	148 (5)	145 (6)	151 (7)
Ethanol	152 (9)	154 (9)	154 (10)	149 (10)	154 (8)	162 (7)	161 (4)
LVP (mm Hg)							
Control			132 (5)	127 (5)	131 (6)	120 (6)	123 (8)
Ethanol	126 (5)	122 (5)	132 (5)	119 (5)	113 (7)	114 (9)	113 (7)
RPP (x100)							
Control			188 (11)	185 (11)	197 (15)	170 (18)	183 (19)
Ethanol	194 (20)	188 (20)	202 (21)	173 (17)	173 (18)	182 (21)	176 (12)
+dP/dt (mm Hg/s)							
Control			1780 (95)	1500 (59)*	1430 (50)*	1450 (77)*	1580 (65)
Ethanol	1960 (99)	1540 (107)#	1760 (94)	1520 (122)#	1420 (147)#	1560 (160)#	1640 (166)
–dP/dt (mm Hg/s)							
Control			2350 (170)	2370 (110)	2260 (189)	1990 (182)	2050 (285)
Ethanol	2480 (185)	2050 (123)	2520 (91)	2050 (141)	1910 (230)	2020 (244)	2000 (198)

Values are means (S.E.M.). MAP, mean arterial pressure; HR – heart rate; LVP – peak left ventricular pressure; RPP – rate-pressure product; +dP/dt and -dP/dt – maximum rate of left ventricular pressure development and relaxation; BASELINE – baseline condition; ETHANOL – 9 min after ethanol infusion; PRE-CAO – 2 min before coronary occlusion; CAO – 58 min after coronary occlusion; REP – reperfusion. #$p < 0.05$ vs BASELINE in the same group, *$p < 0.05$ vs PRE-CAO in the same group. No statistically significant differences were noted between two groups for any of the variables at any time.

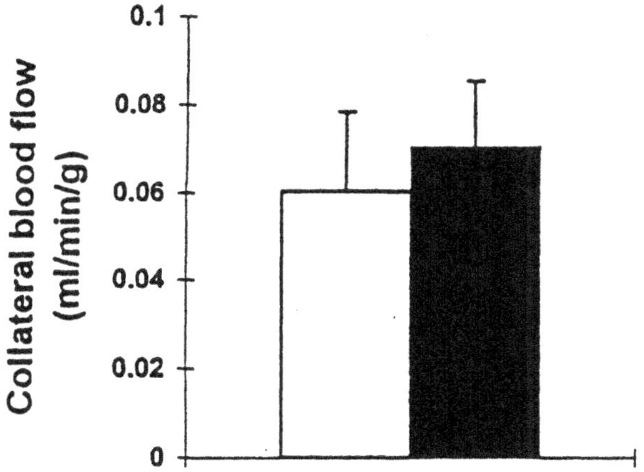

Fig. 3. Collateral blood flow to inner 2/3 of the ischemic wall during coronary occlusion. Data are means ± S.E.M. Open bar – control group; closed bar – ethanol treated group.

ethanol treatment impacted this relationship. Figure 4 shows this relationship was not altered by ethanol ($p > 0.7$).

Discussion

We found that moderate ethanol infusion did not affect the extent of infarct size in a canine coronary occlusion-reperfusion model. This finding is consistent with the study of Bernauer which found that ethanol had no clear effect on infarct size in rat. Blood ethanol concentrations in that study were 150 ± 6 mg/dl before coronary ligation and 100 ± 8 mg/dl before reperfusion, similar to values in our study. We estimated the dose of ethanol required to produce blood ethanol concentration of approximately 100 mg/dl. For an average person, consumption of approximately 80 ml of pure alcohol (equivalent to 160 ml of 100 proof whiskey) will produce a blood ethanol level in the range of 90–150 mg/dl [18]. The concentration of 100 mg/dl is also taken as the legal limit to operate an automobile in most states. In the present study, ethanol concentration before and during coronary artery occlusion were 155 ± 26 and 76 ± 13 mg/dl, respectively, close to the desired blood ethanol concentration.

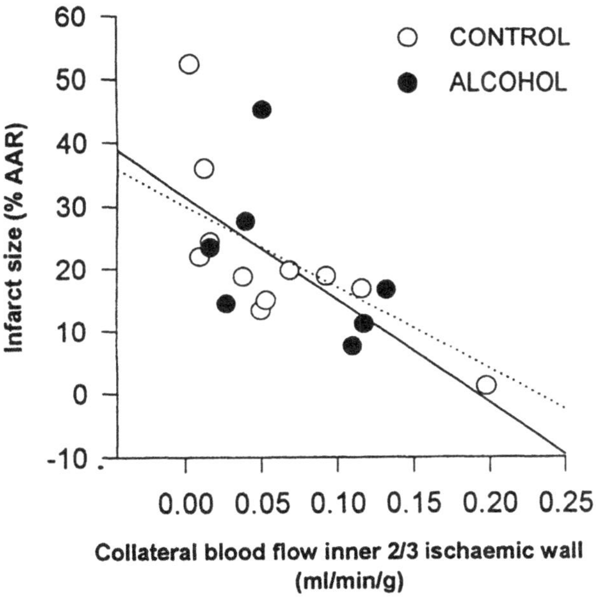

Fig. 4. Infarct size (IS/AAR) as a function of collateral flow to the inner 2/3 of the ischemic area for control (solid line) and ethanol treated hearts (dashed line).

Although effects of ethanol on cardiac function and hemodynamics have been extensively investigated, some findings are still controversial. Nakano *et al.* [19] found in anesthetized open-chest dogs that ethanol increased heart rate, blood pressure and contractility at 200 mg/dl blood ethanol concentration, but these variables decreased at 300 mg/dl. Friedman *et al.* [20] reported that ethanol decreased left ventricular dP/dt and ejection fraction at ethanol concentration of 199 ± 43 mg/dl in anesthetized closed-chest dogs. Mendoza *et al.* [21] did not find an obvious change in hemodynamics at 67 ± 13 mg/dl in anesthetized dogs, but they did find that coronary blood flow was significantly increased at 190 mg/dl. Cheng *et al.* [22] found that mildly elevated blood ethanol (67 ± 14 mg/dl) depressed myocardial contractility and decreased stroke volume in conscious dogs with unchanged heart rate.

From these studies, it appears that the relationship between hemodynamics and blood alcohol concentration was not always consistent. It seems to depend on differences in experimental model (conscious or unconscious, open-chest or closed-chest), and in the procedure of ethanol administration (po or iv, continuous infusion or bolus injection). Furthermore, acetaldehyde, a metabolite of ethanol, produces different hemodynamic effects [20]. In the present study, left ventricular contractility assessed from ±dP/dt decreased transiently during ethanol infusion, but recovered gradually after discontinuation of ethanol. In any case, no marked changes in hemodynamic variables which might have impacted infarct size were evident.

Limitations of the TTC method under conditions of short term reperfusion

In Bernauer's study, myocardial necrosis was greater in hearts reperfused for 60 min compared to those reperfused for shorter periods. This may reflect reperfusion damage [23, 24], as well as an insufficient reperfusion time to remove dehydrogenases stained by TTC. At least ninety min reperfusion in dog [27], 2 h reperfusion in pig [9], and 3 h reperfusion in rabbit [11] hearts are required for credible infarct size measurements. This problem has not been addressed specifically for the rat heart, but the 1 h reperfusion used by Bernauer *et al.* was probably insufficient for reliable infarct sizing.

Possibility of a cardioprotective effect of ethanol

As described in the introduction, Bernauer found that intravenous ethanol administration (2 g/kg) reduced the infarct size with reperfusion after 60 min of coronary occlusion, although he failed to find a clear effect with a lower dose of ethanol. Blood ethanol concentration in his study with 2 g/kg ethanol was 310 ± 7 mg/dl immediately before coronary occlusion, about two times higher than in our experiments. The study of Mallov *et al.* also demonstrated that the degree of protection against myocardial necrosis increased as the dose of ethanol was raised. These studies indicated that the cardioprotective effect of ethanol appeared to be dose-dependent. Accordingly, although we did not find any clear effect with a moderate dose of ethanol in the present study, we can not disregard the possibility that a higher dose of ethanol might have been effective in reducing infarct size. Bernauer suggested that negative inotropic effects of alcohol might have resulted in a reduction of myocardial oxygen consumption and that this was responsible for the infarct size-limiting effect of alcohol. On the contrary, Friedman *et al.* [20] demonstrated that alcohol increased myocardial oxygen consumption, even though cardiac function was reduced. Mallov *et al.* employed various agents to clarify the mechanism of the cardioprotective effect of alcohol, and concluded that platelet deaggregation or an osmotic effect of alcohol might be involved. They also concluded that the decrease in myocardial contractility was not responsible for the cardioprotective effect of alcohol. Further research on the cardioprotective effects of alcohol is warranted.

Conclusion

Alcohol infusion did not alter myocardial infarct size due to coronary artery occlusion in anesthetized dogs.

Acknowledgements

Arthur G. Williams, Jr., and Chong-Hong He provided expert technical assistance in conducting the experiments. Dennis P. Shingleton, Technical Manager of Laboratories, Department of Pathology, UNTHSC, provided facilities and advice for measuring blood alcohol. This study was supported in part by NIH grant HL 35027.

References

1. Ahlawat SK, Siwach SB: Alcohol and coronary artery disease. Int J Cardiol 44: 157–162, 1994
2. Rohan TE: Alcohol and ischemic heart disease: A review. Aust NZ J Med 14: 75–80, 1984
3. Conway N: Haemodynamic effects of ethyl alcohol in patients with coronary heart disease. Brit Heart J 30: 638–644, 1968
4. Orlando J, Aronow WS, Cassidy J, Prakash R: Effect of ethanol on angina pectoris. Ann Int Med 84: 652–655, 1976
5. Ahlawat S, Siwach S, Jagdish: Indirect assessment of acute effects of ethylalcohol on coronary circulation in patients with chronic stable angina. Int J Cardio 33: 385–392, 1991
6. Fernandez D, Rosenthal JE, Cohen LS, Hammond G, Wolfson S: Alcohol induced Prinzmetal variant angina. Am J Cardiol 32: 238–239, 1973
7. Bernauer W: The effect of ethanol on arrhythmias and myocardial necrosis in rats with coronary occlusion and reperfusion. Eur J Pharmacol 126: 179–187, 1986
8. Mallov S, Gilmour RF: Inhibition of epinephrine-induced myocardial necrosis in rats byadministration of single dose of ethanol. Drug Alcohol Depend 2: 397–407, 1977
9. Freeman I, GrunwaldAM, Robin B, Rao PS, Bodenheimer MM: Effect of reperfusion on use of triphenyltetrazorium chloride to differentiate viable from non-viable myocardium in area at risk. Cardiovasc Res 24: 109–114, 1990
10. Schaper W: Experimental infarcts and the microcirculation. In: DJ Hearse, DM Yellon (eds.) Therapeutic Approaches to Myocardial Infarct Size Limitation. Raven Press, New York, 1984, pp 79–90
11. Shirato C, Miura T, Ooiwa H, Toyofuku T, Wilborn WH, Downey JM: Tetrazorium artifactually indicates superoxide dismutase-induced salvage in reperfused rabbit heart. J Mol Cell Cardiol 30: 23–44, 1989
12. Rona G: Catecholamine cardiotoxicity. J Mol Cell Cardiol 17: 291–306, 1985
13. Heymann MA, Payne BD, Hoffman JIE, Rudolph, AM: Blood flow measurement withradionuclide-labeled particles. Prog Cardiovasc Dis 20: 55–79, 1977
14. Li GC, Vasquez JA, Gallagher KP, Lucchesi BR: Myocardial protection with preconditioning. Circulation 82: 609–619, 1990
15. Murry CE, Richard VJ, Jennings RB, Reimer KA: Myocardial protection is lost before contractile function recovers from ischemic preconditioning. Am J Physiol 260: H796–H804, 1991
16. Nachlas MM, Shnitka TK: Macroscopic identification of early myocardial infarcts by alterations in dehydrogenase activity. Am J Pathol 42: 379–405, 1963
17. Shaffar M, Stroupe SD: A general method for routine clinical chemistry on the Abbot TDx analyzer (abstract). Clin Chem 29: 1251,1983
18. Andrews LS, Snyder R: Toxic effects of solvents and vapors. In: CD Klaassen, MO Amdur, J Doull (eds.) Casarett and Doull's Toxicology. Macmillan Publishing Co Inc, New York, 1986, p 648
19. Nakano J, Kessinger JM: Cardiovascular effects of ethanol, its congeners and synthetic bourbon in dogs. Eur J Pharmacol 17: 195–201, 1972
20. Friedman H, Matsuzaki S, Choe SS, Fernando HA, Celis A, Zaman Q, Lieber CS: Demonstration of dissimilar acute hemodynamic effects of ethanol andacetaldehyde. Cardiovasc Res 13: 477–487, 1979
21. Mendoza LC, Hellberg K, Rickart A, Tillich G, Bing RJ: The effect of intravenous ethyl alcohol on the coronary circulation and myocardial contractility of the human and canine heart. J Clin Pharmacol & New Drugs 11: 165–176, 1971

22. Cheng CP, Shihabi Z, Little WC: Acute effects of mildly intoxication level of alcohol on left ventricular function in conscious dogs. J Clin Invest 85: 858–1865, 1990
23. Bienvenu K, Granger, DN: Leukocyte adhesion in ischemia/referfusion. Blood Cells 19: 279–289, 1993
24. Kirschner, RE, Fantini, GA: Role of iron and oxygen-derived free radicals in ischemia-reperfusion injury. J Am Col Surg 179: 103–117, 1994
25. Fishbein MC, Meerbaum S, Rit J, Lando U, Kanmatsuse K, Mercier JC, Corday E, Ganz W: Early phase acute myocardial infarct size quantification: Validation of the triphenyltetrazorium chloride tissue enzyme staining technique. Am Heart J 101: 593–600, 1981
26. Schaper W, Hoffmann M, Muller KD, Genth K, Carl M: Experimental occlusion of two small coronary arteries in the same heart: A new validation method for infarct sizemanipulation. Basic Res Cardiol 74: 224–229, 1979

Molecular and Cellular Biochemistry **186:** 43–51, 1998.
© 1998 *Kluwer Academic Publishers.*

Gene expression after short periods of coronary occlusion

Elisabeth Deindl and Wolfgang Schaper

Max-Planck-Institute for Physiological and Clinical Research, W. G. Kerckhoff-Institute, Department of Experimental Cardiology, Bad Nauheim, Germany

Abstract

Brief periods of coronary occlusion render the affected myocardium more tolerant to the otherwise devastating effects of long coronary occlusion. Besides this phenomena, called ischemic preconditioning, short periods of ischemia cause a regional dysfunction, namely myocardial stunning. The molecular mechanisms of both syndromes are not very well understood. We therefore investigated the expression of genes which may be involved in cardioprotection or repair processes. Using our porcine model of ischemia and reperfusion we were able to show an induction of genes coding for transcription factors (proto-oncogenes), for proteins involved in repair processes (heat shock genes), for proteins implicated in the calcium homeostasis (calcium-handling genes) and for growth factors. We could show that the increased mRNA levels are due to an enhanced transcriptional activity and not to a prolonged half-life of the transcripts. The angiogenic growth factor vascular endothelial growth factor (VEGF) represents an exception. It exhibits – in addition to a HIF-motif (Hypoxia Inducible Factor) in its promoter/enhancer – a protein binding region in its 3′ UTR which when occupied renders the mRNA more stable. However to what extent the expression of the distinct genes contributes to the cardioprotective effect of ischemic preconditioning or myocardial stunning can only be presumed. Increased mRNA stability can be confered via adenosine which is produced during ischemia by ATP-breakdown. The demasking of unknown genes – via differential display reverse transcription polymerase chain reaction (DDRT-PCR) – should provide a more comprehensive view of the mechanisms underlying both processes. (Mol Cell Biochem **186**: 43–51, 1998)

Key words: myocardial ischemia, gene expression, growth factors, phospholamban, calsequestrin heat shock proteins, preconditioning, stunning

Introduction

Despite the fact that long lasting coronary occlusions (more than 40 min) lead to an irreversible damage of the affected myocardium, short periods of repetitive coronary occlusions (between 4 and 15 min) have quite different effects: myocardial stunning and ischemic preconditioning. Stunning is defined as a long lasting (up to 48 h), but completely reversible contractile dysfunction in the absence of cell necrosis or visible damage of the tissue [1, 2]. Brief periods of ischemia render the affected myocardium also transiently more tolerant to the untoward effect of a subsequent long period of coronary occlusion. Infarction, usually transmural after 1 h occlusion, can be minimized or avoided when the reperfusion time between the short and the subsequent long occlusion does not last more than 30–45 min. This phenomenon is called ischemic preconditioning [3–5]. Abnormal intracellular calcium homeostasis, a target for free radicals, is implicated in both syndroms [6, 7]. Ischemic preconditioning is furthermore thought to be mediated via adenosine A_1-receptors or other receptors coupled to G_i proteins [8]. However the molecular mechanisms for both phenomena are not well understood.

Although cell death or visible morphological changes do not occur in both ischemic preconditioning and stunning, it is conceivable that there is not only the unmasking of a regulatory mechanism to protect the heart against further damage. The relatively long time necessary to restore contractile function implies the existence of molecular damage and subsequent repair processes. Changes in gene expression should give insight in mechanisms involved in

Address for offprints: E. Deindl, Max-Planck-Institute for Physiological and Clinical Research, W. G. Kerckhoff-Institute, Department of Experimental Cardiology, Benekestraße 2, D61231 Bad Nauheim, Germany

ischemic preconditioning and stunning and indicate where damage emerged.

To elucidate the molecular events we investigated the expression of proto-oncogenes, growth factors, heat shock proteins and calcium-binding proteins. To analyze whether increased transcript abundance was due to increased gene transcription or enhanced mRNA stability run-on assays were performed. Furthermore we tried to find new genes involved in both processes via DDRT-PCR

Material and methods

Animal model and experimental protocol

To study the expression of several groups of genes after short periods of ischemia and reperfusion, we developed a porcine model of ischemia in cooperation with Pieter Verdouw from the University of Rotterdam [9].

Crossbred Landrace X Yorkshire pigs of either sex were fasted overnight. After anesthesia with an intravenous injection of 20 mg/kg pentobarbital sodium, instrumentation and stabilization for at least 30 min, systemic hemodynamics and regional myocardial wall thickness were recorded. Stunning was achieved by 2 periods of 10 min of LAD (left anterior descending coronary artery) occlusion, separated by 30 min of reperfusion. After the second occlusion the myocardium was reperfused up to 210 min. Tissue samples from ischemic, ischemic-reperfused and non-ischemic regions of the heart were collected at the timepoints indicated (Fig. 1). As control we used sham-operated animals (80–440 min) and hearts from slaughterhouse pigs. If not indicated separately, the values for sham operated animals represented in graphs are combined values of 80 and 440 min.

Molecular biological techniques

Total RNA was isolated according to the method described by Chomzynski *et al.* [10]. Techniques and probes used to analyze gene expression (Northern blot, Southern blot, Western blot, slot blot, run-on assay, DDRT-PCR) are described elsewhere [9, 11–15]. Signal quantification was done via PhosphorImager or a densitometer.

Results and discussion

Regional myocardial wall thickening

Systemic hemodynamics and regional myocardial wall thickness were recorded throughout the whole experimental protocol as a parameter for contractile function.

Within seconds of the first occlusion, there was a complete loss of regional systolic wall thickening in the ischemic area (from 34 ± 7% to –3 ± 10%, Fig. 2). During the first 30 min of reperfusion, there was only a partial recovery of function 13 ± 9% ($p < 0.05$ versus preocclusion). After the second 10-min occlusion, recovery was slightly less, but a gradual increase to 19 ± 9% occurred during the subsequent 210 min. Systolic wall thickening of the nonischemic area and sham operated animals did not change significantely during the course of experiments.

Proto-oncogenes

The molecular response of eucaryotic cells to environmental changes like stress or membrane depolarization is complex and involves the rapid expression of a variety of genes [16,

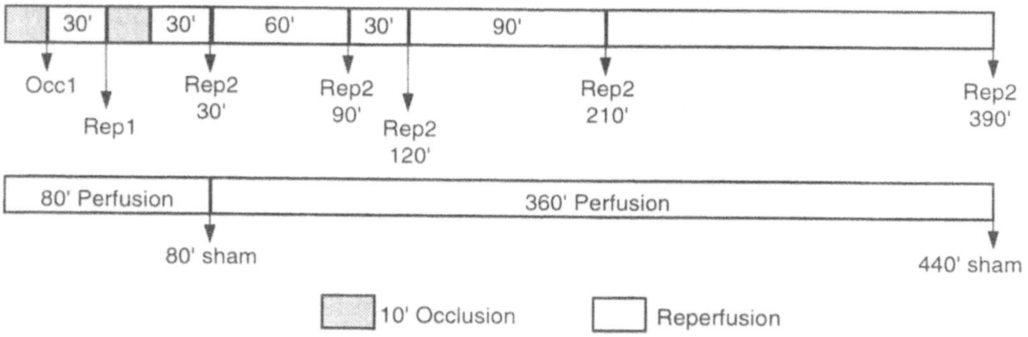

Fig. 1. Protocol for myocardial stunning in pigs. The experimental design is depicted as a temporal axis. The left anterior descending artery was occluded twice for 10 min separated by 30 min of reperfusion (Rep1). After a second reperfusion period (indicated by arrows) animals were sacrified and tissue was removed for further investigations. The sham operated animals were sacrified after 80 or 440 min of reperfusion. (By courtesy of R. Zimmermann)

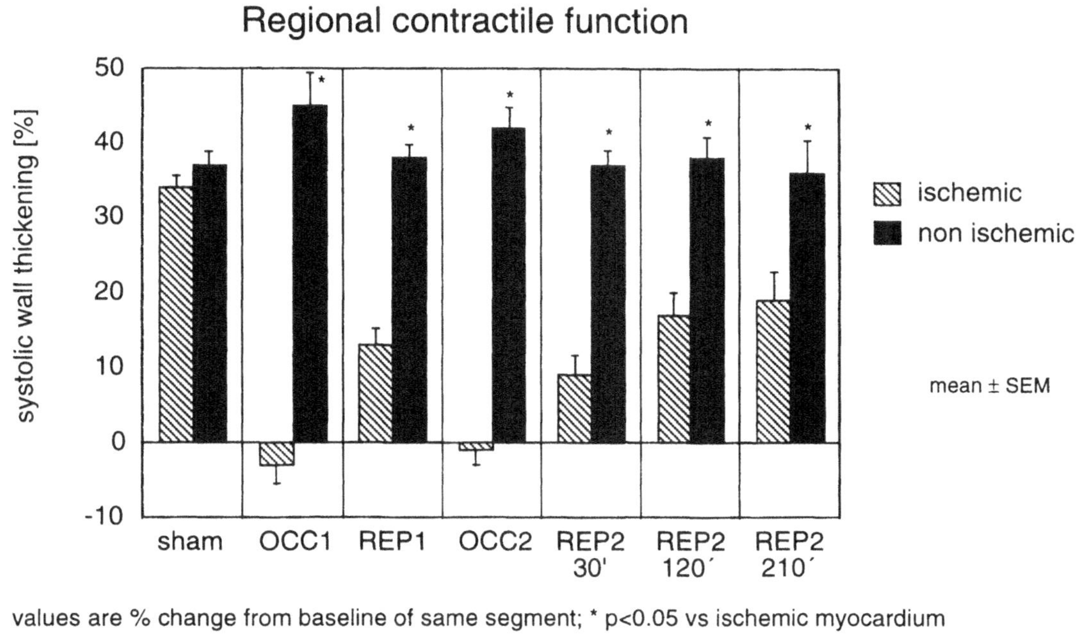

Fig. 2. Systolic wall thickening of ischemic and non-ischemic myocardium during the cycles described in Fig. 1. OCC = occlusion; REP = reperfusion. Values are percentage changes from baseline of the same segment. (By courtesy of R. Zimmermann) *p < 0.05 vs. ischemic myocardium.

17]. Most of these so-called immediate early genes code for DNA-binding proteins and are potent transription factors [18, 19]. We analyzed the expression of c-myc, c-fos, c-jun, junB and fosB which share several homologous domains, namely the basic DNA binding region as well as the leucine zipper, which help in dimerization [9, 20–24]. Egr-1 is another immediate early gene which codes for a structurally unrelated protein that has 3 zinc finger domains, by which it specifically interacts with DNA [25].

Northern blot results revealed an increased mRNA expression for c-fos and Egr-1 during the first period of occlusion which peaked in an about 7-fold induction during the first 30 min of the second reperfusion period (Fig. 3). For junB we found a similar strong induction already during the first reperfusion period (Fig. 3). c-jun did not show any increased mRNA amounts during the periods of occlusion and reperfusion which is not astonishing as junB has been shown to be a negative regulator for the expression of c-jun [25]. Although Northern blot results did not show altered mRNA levels for c-myc due to ischemia, run-on analysis displayed - similar as for the other proto-oncogens - an enhanced transcription rate [15]. These data reflect a decreased mRNA-stability for c-myc under conditions of ischemia. For fos-B we could not detect any transcript in the ischemic pig myocardium (data not shown), which is consistant with observations in the ischemic liver [26].

The induction of proto-oncogenes after brief periods of occlusion in ischemic myocardium points towards their role as transcription factors, exerting their function on genes involved in repair processes. The relatively small changes in expression of c-jun by ischemia and reperfusion might be due to differences in its promotor structure compared to c-fos, Egr-1 and junB [27]. As junB expression rises rapidly in the first reperfusion period, its protein product might inhibit the upregulation of c-jun. Egr-1 transactivates the α-MHC-gene in myocytes [28]. The skeletal α-actin promoter, which does not contain an AP-1 binding site, was shown to be transactivated by co-expressed fos and jun in ventricular myocytes [29]. These genes represent examples in which direct binding of proto-oncogenes was demonstrated; however it is not known whether these genes are affected by ischemia. However the promotor of calcium-handling genes like Ca^{2+}-ATPase, phospholamban and calsequestrin, which are expressed in the post ischemic myocardium contain binding domains for AP-1, the heterodimer formed by fos and jun [30].

Calcium-handling proteins

Increasing evidence points to a molecular damage of calcium homeostasis in stunned myocardium. Ito *et al.* found that intracoronary calcium infusion restores contractility in stunned myocardium [1]. Studies on the expression of calcium-handling genes could therefore contribute insight into the aetiological mechanisms. Beside Ca^{2+}-ATPase,

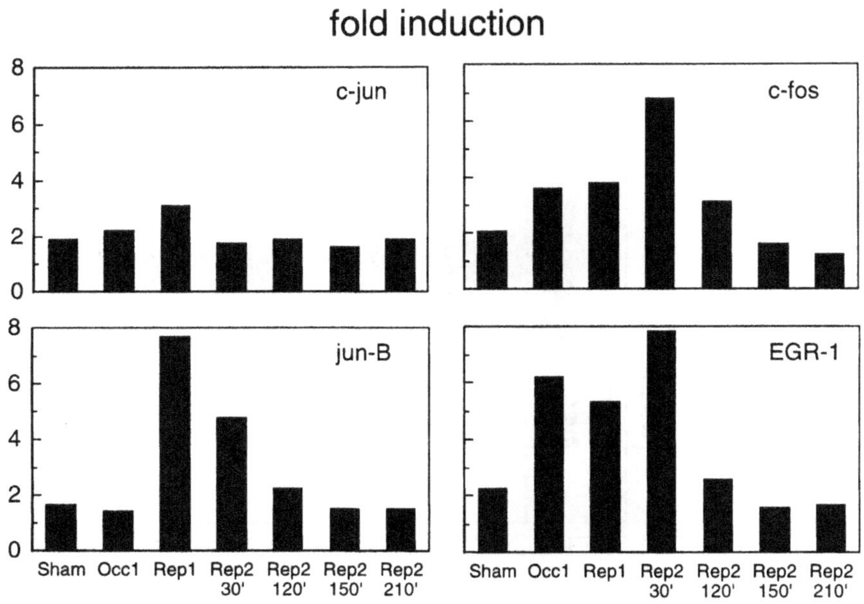

Fig. 3. Bar graph showing the mRNA expression of the proto-oncogenes analyzed via Northern blot (simplified representation). Densiometric values were corrected with GAPDH mRNA levels and expressed in columns (data are expressed as means) as fold induction of mRNA in ischemic myocardium compared with normal (nonischemic) perfused myocardium.

phospholamban, and calsequestrin we investigated the expression of calmodulin, the major intracellular calcium receptor [11].

Our results show a clear correlation between the transcript level of calcium-binding proteins and ischemia (Fig. 4). All analyzed transcripts showed a strong induction in particular during the second period of reperfusion; phospholamban which modulates the activity of Ca^{2+}-ATPase up to 2.5-fold, Ca^{2+}-ATPase itself up to 1.6-fold and Calsequestrin up to 2.3-fold. The increase in mRNA levels was due to enhanced

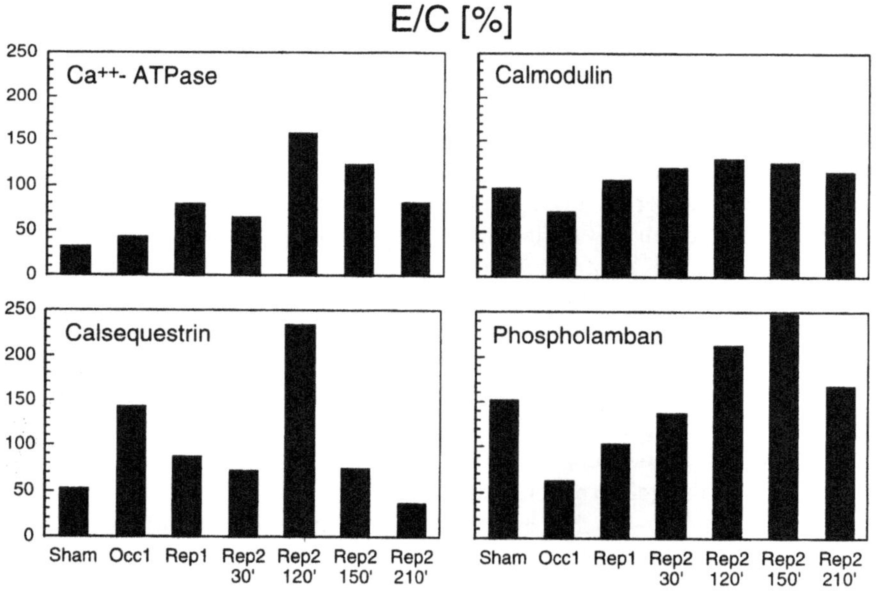

Fig. 4. Bar graph showing the mRNA expression of the calcium-handling proteins analyzed via Northern blot (simplified representation). Calcium regulatory protein gene expression was determined by densiometry and data of the experimental signals (E) were divided by data of their corresponding control signals (C) normalised by the matching GAPDH mRNA signal.

transcriptional activity as shown by run-on assays [15]. Calmodulin did not display any altered mRNA amounts but was transcriptionally upregulated as shown by run-on analysis which suggests an increased turnover of the transcript (unpublished results, R. Knöll).

In this context it is interesting to mention that the increased mRNA levels of the calcium-binding proteins do not correlate with an increased expression of β-myosin and actin, genes coding for contractile proteins (data not shown). The combination of enhanced transcription of calcium handling proteins and unchanged mRNA level of contractile proteins points to a disruption of electro-mechanical coupling in myocardial stunning.

By this line of reasoning we conclude that the calcium-handling proteins were damaged during the episodes of occlusion and reperfusion and require repair processes.

Heat shock proteins

Heat shock proteins are either constitutively expressed or induced via non-lethal stress. They exist in several families and are principally classified according to their molecular weight. They function in refolding damaged proteins or, like ubiquitin – so called because of its abundant presence in all eucariotic cells – in assisting protein degeneration [31, 32].

Our studies on the expression of heat shock proteins showed that brief ischemia is a powerful inducer of hsp mRNA. hsp 70, 60, 27 and ubiquitin were constitutively expressed in normal heart [12]. However short periods of ischemia induced a fast expression of hsp70 (data not shown) and 27 (Fig. 5), followed by ubiquitin (Fig. 6). The increased transcript levels of the heat shock proteins were associated with the temporary formation of new ubiquitin-protein conjugates, but are not followed by a rise in hsp27 concentration in the myocardium (data not shown). Uncoupling between increased transcription of hsp27 mRNA and tissue level of this protein might reflect a negative control of the translational regulation, or could also be caused by a high turnover of this protein under ischemic/reperfusion conditions as it is observed during brain ischemia in rats [33]. Consistent with this hypothesis is the known feature of the basally expressed hsp27, which within minutes of the exposure to stress undergoes a rapid post-translational modification by phos–phorylation generating unstable isoforms [34, 35]. The expression of hsp60 remained unchanged (data not shown). This protein, located in mitochondria, plays a role in the proper folding and assembly of proteins in the mitochondrial matrix [36]. Unchanged levels of hsp60 mRNA in our model are in keeping with the observation of intact mitochondrial function in stunned myocardium [1].

Analogous to the proto-oncogenes and the genes coding for calcium-handling proteins run-on assays showed that the elevated mRNA amounts of the heat shock proteins are due to an increased transcription activity and not the result of an enhanced mRNA stability [12].

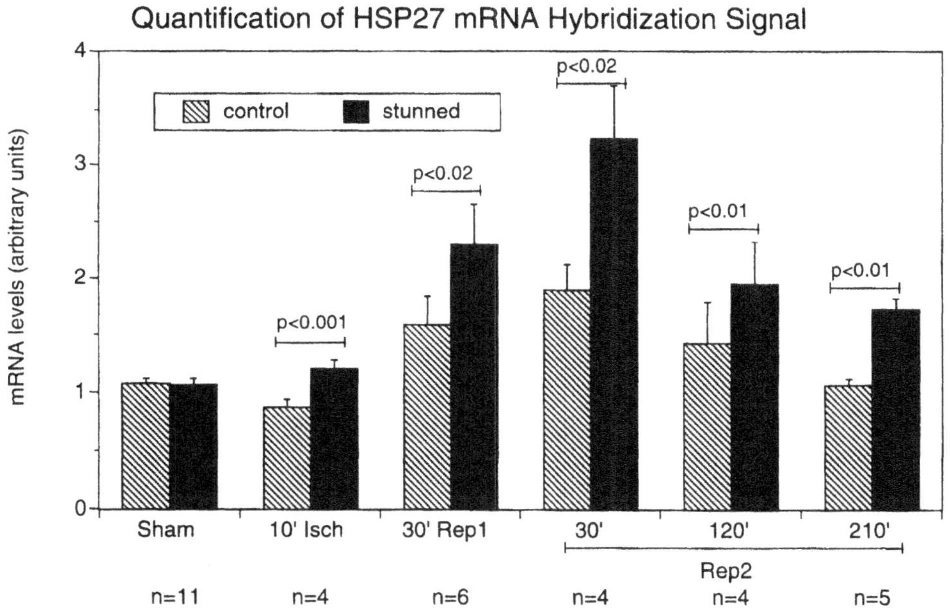

Fig. 5. Northern blot results of hsp27 expression represented in bar graphs. GAPDH hybridisation signals were used as a standard for densiometric quantification of mRNA levels. n indicates number of animals studied in each experimental group. Error bars = S.E.M.

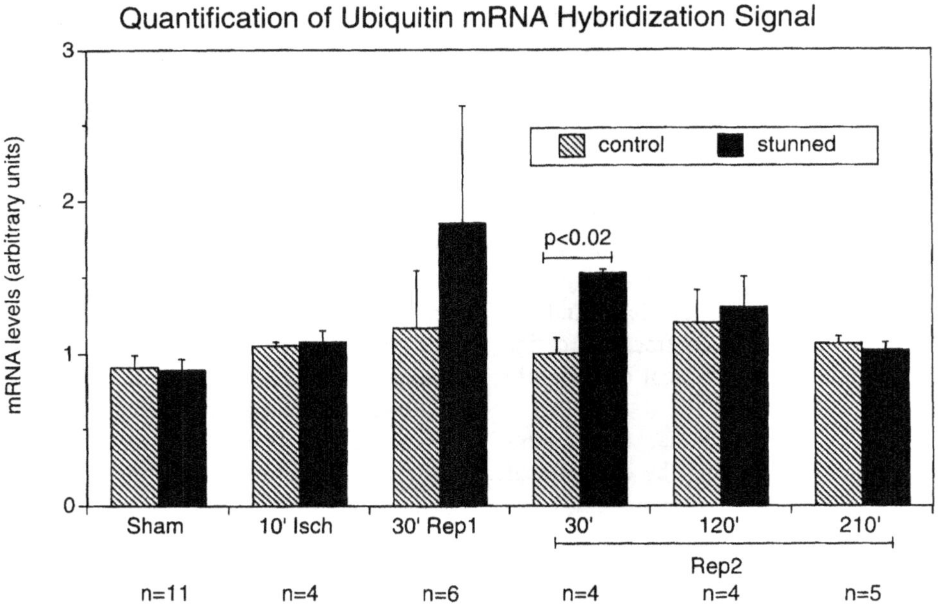

Fig. 6. Bar graph showing the expression of ubiquitin mRNA during the cycles of occlusion and reperfusion. For normalisation GAPDH mRNA was used. n indicates number of animals studied in each experimental group. Error bars = S.E.M.

Our results on the expression of heat shock proteins point to a molecular damage of the cardiomyocytes caused by the repetitive cycles of occlusion and reperfusion. In that context, heat shock proteins may exert their protective function either in refolding denatured proteins via hsp27 or by elimination irreversibly damaged proteins marked by ubiquitin for proteolysis by a proteosome complex [37].

Growth factors

Previous work of our laboratory showed that transcripts of fibroblast growth factor 1 (FGF-1) and VEGF, growth factors important for angiogenesis, are constitutively expressed in the heart where proliferative activity is almost undetectable. Short periods of occlusion of the coronary artery led to an upregulation of the VEGF mRNA but did not increase angiogenic activity [38]. We therefore hypothesised that growth factors may exert trophic functions necessary for the survival of cardiac myocytes and endothelial cells. IGFs, which play a central role in regulating growth of mesodermal cells, in differentiation and in activating metabolic processes are known to act as trophic factors in neurons [39, 40]. From IGF-I and IGF-II several spliceforms exist [41, 42]. They can bind to specific binding proteins (IGFBP) which are thought to regulate their bioavaibility and modulate the biological activity of the IGFs [43]. One function of the IGFs is the activation of the glucose transporters GLUT1 and GLUT4 which also belong to the IGF-system.

Our results on the expression of IGF-I, IGF-II and the IGFBP 2–6 showed a constitutive expression of all transcripts in the porcine myocardium [13]. Brief periods of ischemia and reperfusion resulted in an early induction only of IGF-II and IGFBP-5 compared to slaughterhouse pigs (Fig. 7). For IGF-II we found similar increased mRNA amounts in sham operated animals which suggests that IGF-II is, at least in our model, a stress inducible gene. IGFBP-5 was induced by ischemia/reperfusion but not by the stress of anesthesia and surgery and its peak expression was somewhat later than that of IGF-II. Differential expression of a trophic factor and its binding protein could be the basis of a model of time-limited protection as we can see in ischemic preconditioning. To elucidate whether IGF-II exhibits a cardioprotective function we injected IGF-II (0.25ug/ul) prior to 60 min of occlusion and 120 min of reperfusion in the porcine myocardium and measured infarct size [44]. There was a significant reduction of infarct size of about 20% of area at risk. Equimolar concentrations of IGFBP-5 or lavendustin, a tyrosine-kinase inhibitor, abolished the cardioprotective effect of IGF-II, indicating that IFGBP-5 may be one factor accounting for the limited temporal duration of preconditioning.

In view of these findings it was interesting to see that the glucose transporters GLUT1 and GLUT4, which are activated by the IGFs, showed an increased expression pattern in ischemic and control tissue like IGF-II but somehow delayed (A. Kluge, R. Zimmermann, W. Schaper, unpublished results). These results together with a recent publication of

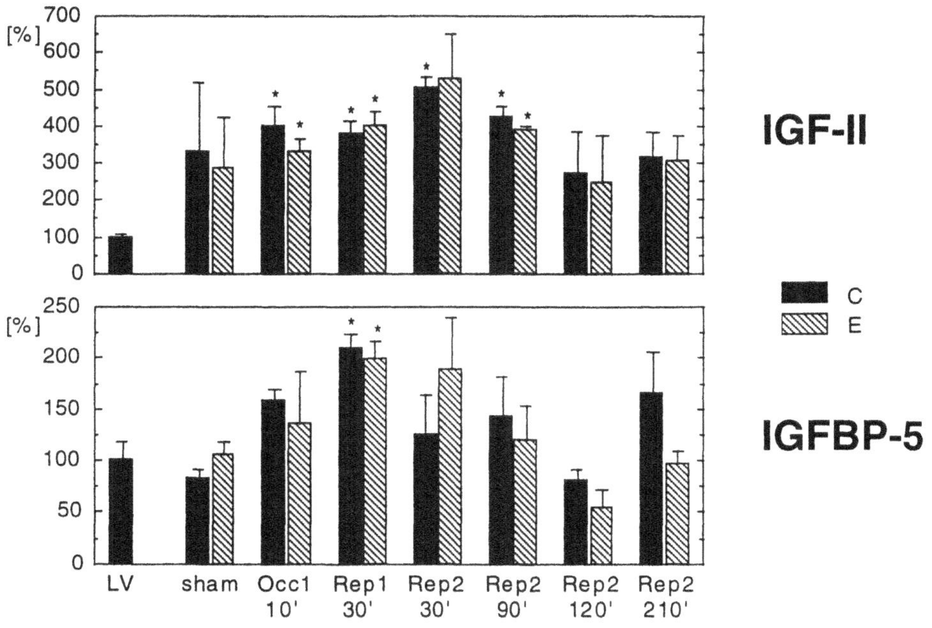

Fig. 7. Bar graphs representing the expression of IGF-II mRNA (upper part) and IGFBP-5% (lower part). Quantification of the mRNA levels was performed by means of slot blot analysis. Data are means, error bars = S.E.M. The average of the values obtained for the left ventricle of non-sham control pigs was standardised at 100%.

Lopaschuk *et al.* account for the role of GLUT1 and GLUT4 in cardioprotection [45].

Conclusion

Short periods of coronary occlusion and reperfusion lead to the enhanced expression of a variety of genes. To what extent the expression of the distinct genes contributes to ischemic preconditioning or myocardial stunning is not known. Protooncogenes, also called immediate early genes, are expressed very rapidly after coronary occlusion. They seem to contribute to both mechanisms, exerting their function as transcription factors on the expression of the other genes found to be upregulated. The early response was followed by an increased expression of the heat shock proteins. This response and the relatively long time necessary for restoration of contractile function (several hours to days) point to molecular damage caused by the repetitive cycles of occlusion and reperfusion. Molecular damage can vary in degree and it is probable that different molecules exhibit different resistance toward ischemic injury. We hypothesize that some denatured proteins may become refolded by hsp27 or hsp70. More severely injured molecules are covalently ubiquitinated and are thereby labeled for destruction via proteolysis. Both pathways are essential for cellular survival. Rapid elemination of nonfunctional proteins may be as essential as the repair of less intensively damaged proteins through refolding. Our results indicate that calcium-handling proteins belong to the type of proteins which have experienced damage and require repair processes or *de novo* synthesis. IGF-II is in our model a stress-inducible gene which also can act as a cardioprotectant via tyrosine kinase receptors and the GLUT-system. It was interesting for us to see that brief periods of coronary occlusion and reperfusion stimulate transcriptional activity of growth factors in two different pathways. Whereas the expression of VEGF mRNA is induced in a hypoxia specific pathway, presumably via the binding of HIF-1 at a specific recognition site in the promoter/enhancer of VEGF, IGF-II mRNA expression is induced in a non-hypoxia-specific but stress related pathway [46]. Furthermore the VEGF mRNA contains a motif at its polyA$^+$-tail which enables the stabilization of the transcript. This motif can be bound by a protein which is normally under the control of the von Hippel-Lindau suppressor gene but is also induced by hypoxia [47]. Specific binding of the protein at the 3′end of the VEGF mRNA leads to a prolonged half-time of the transcript. IGF-II functions to increase glucose uptake, to suppress protein degradation and to support protein synthesis and the uptake of amino acids [48]. IGF-II may exert a protective role in the myocardium by aiding the survival of myocytes after damage by stress, in analogy to its trophic functions in the brain, an organ that is terminally differentiated like the heart [39]. GLUT1 and GLUT4 are responsible for the glucose transport in the heart [49]. Young *et al.* demonstrated that if canine hearts are subjected to low-flow ischemia, there is a significant increase in the translocation of both GLUT1 and GLUT4 from the intracellular pool to the sarcolemmal membrane

[50]. Increased expression and translocation may therefore have important consequences for the myocardium to withstand short periodes of ischemia [45]. Accelerated glucose uptake and glycolysis may provide an important source of ATP to maintain membrane ion flux and afford some protection to the ischemic heart [51]. However it should be recognized that during severe ischemia high glycolytic rates may contribute to ischemic injury secondary to the production of glycolytic products such as lactate and protons [52]. To asses the role of the gene products shown to be upregulated in other mechanisms, it is necessary to obtain more information. We therefore started to define unknown genes involved in ischemic preconditioning and myocardial stunning. Using the technique of DDRT-PCR we found, beside the increased expression of known genes (ubiquitin and hsp27), 20 differentially expressed genes of unknown function which are currently being further characterized [14].

Acknowledgement

We thank the Steinkopff Verlag, Darmstadt, Germany for giving us the permission to reproduce diagrams published by Zimmermann *et al.* [53].

References

1. Ito B, Tate H, Kobayashi M, Schaper W: Reversibly injured, post-ischemic canine myocardium retains normal contractile reserve. Circ Res 61: 834–846, 1987
2. Heyndrickx GR, Millard RW, McRitchie RJ, Maroko PR, Vatner SF: Regional myocardial, functional, and electrophysiological alterations after brief coronary occlusion in conscious dogs. J Clin Invest 56: 978–985, 1975
3. Murry CE, Jennings RB, Reimer KA: Preconditioning with ischemia: A delay of lethal cell injury in ischemic myocardium. Circulation 74: 1124–1136, 1986
4. Sack S, Mohri M, Arras M, Schwarz ER, Schaper W: Ischaemic preconditioning - time course of renewal in the pig. Cardiovasc Res 27: 551–555, 1993
5. Schaper W, Sharma HS, Zimmermann R, Mohri M, Schaper J: J Mol Cell Cardiol, pp S-12-7 (abstr), 1992
6. Bolli R: Mechanism of myocardial 'stunning'. Circulation 82: 723–738, 1990
7. Vegh A, Szekeres L, Parrat J: Preconditioning of the ischaemic myocardium; involvement of the oxidative pathway. Br J Pharmacol 107: 648–652, 1992
8. Liu Y, Downey JM: Preconditioning against infarction in the rat heart does not involve a pertussis toxin sensitive G protein. Cardiovasc Res 27: 608–611, 1993
9. Brand T, Sharma HS, Fleischmann KE, Duncker DJ, McFalls EO, Verdouw PD, Schaper W: Proto-oncogene expression in porcine myocardium subjected to ischemia and reperfusion. Circ Res 71: 1351–1360, 1992
10. Chomczynski P, Sachhi N: Single-step method of RNA isolation by acid guanidinium thiocyanate-phenol- chloroform extraction. Anal Biochem 162: 156–159, 1987
11. Fraß O, Sharma HS, Knöll R, Duncker DJ, McFalls EO, Verdouw PD, Schaper W: Enhanced gene expression of calcium regulatory proteins in stunned porcine myocardium. Cardiovasc Res 27: 2037–2043, 1993
12. Andres J, Sharma HS, Knöll R, Stahl J, Sassen LMA, Verdouw PD, Schaper W: Expression of heat shock proteins in the normal and stunned porcine myocardium. Cardiovasc Res 27: 1421–1429, 1993
13. Kluge A, Zimmermann R, Münkel B, Verdouw PD, Schaper J, Schaper W: Insulin-like growth factor II is an experimental stress inducible gene in a porcine model of brief coronary occlusions. Cardiovasc Res 29: 708–716, 1995
14. Knöll R, Zimmermann R, Arras M, Schaper W: Characterization of differentially expressed genes following brief cardiac ischemia. Biochem Biophys Res Comm 221: 402–407, 1996
15. Knöll R, Arras M, Zimmermann R, Schaper J, Schaper W: Changes in gene expression following short coronary occlusions studied in porcine hearts with run-on assays. Cardiovasc Res 28: 1062–1069, 1994
16. Almendral JM, Sommer D, MacDonald-Bravo H, Buckhardt J, Perera J, Bravo R: Complexity of the early response to growth factors in mouse fibroblasts. Mol Cell Biol 8: 2140–2148, 1988
17. Greenberg ME, Ziff EB. Stimulation of 3T3 cells induces transcription of the c-fos proto-oncogene. Nature 311: 433–438, 1984
18. Morgan JI, Curran T: Stimulus-transcription coupling in the nervous system: Involvement of the inducible proto-oncogenes *fos* and *jun*. Annu Rev Neurosci 14: 421–451, 1991
19. Simpson PC: Proto-oncogenes and cardiac hypertrophy. Annu Rev Physiol 51: 189–202, 1988
20. Kelly K, Cochran BH, Stiles CD, Leder P: Cell specific regulation of the c-myc gene by lymphocyte mitogens and platelet-derived growth factor. Cell 35: 603–610, 1983
21. Ryder K, Nathans D: Induction of protooncogene *c-jun* by serum growth factor. Proc Natal Acad Sci USA 85: 8464–8467, 1988
22. Ryder K, Lau L, Nathans D: A gene activated by growth factors is related to the oncogene *v-jun*. Proc Natl Acad Sci USA 85: 1487–1491, 1988
23. Cohen D, Ferreira P, Gentz R, Franza R, Curran T: The product of a fos related gene *fra-1*, binds cooperatively to the AP-1 site with *jun*: Transcription factor AP-1 is comprised of multiple protein complexes. Gene Dev 3: 173–184, 1989
24. Zerial M, Toschi L, Ryseck RP, Schuermann M, Müller R, Bravo R: The product of a novel growth factor activated gene, *fosB*, interacts with JUN proteins enhancing their DNA binding activity. EMBO J 1989: 805–813, 1989
25. Sukhatme VP, Cao X, Chang LC, Tsai-Morris CH, Stamenkowich D, Ferreira PCP, Cohen DR, Edwards SA, Shows TB, Curran T, LeBeau MM, Adamson ED: A zinc finger-encoding gene coregulated with c-fos during growth and differentiation, and after cellular depolarization. Cell 53: 37–43, 1988
26. Schiaffonati L, Rappocciolo E, Tacchini L, Cairo G: A. B-Z. Reprogramming of gene expression in postischemic rat liver: Induction of proto-oncogenes and hsp-70. J Cell Physiol 143: 79–87, 1990
27. Auwerx J, Staels B, Sassone-Corsi P: Coupled and uncoupled induction of *fos* and *jun* transcription by different second messengers in cells of hematopoietic origin. Nucleic Acids Res 18: 221–228, 1990
28. Gupta MP, Gupta M, Zak R, Sukhatme V: Egr-1, a serum inducible zinc finger protein, regulates transcription of the rat cardiac a-myosin heavy chain gene. J Biol Chem 266: 12813–12816, 1991
29. Bishopric NH, Webster KA: Fos and Jun regulate preferential induction of the sceletal α-actin gene in neonatal rat heart cells. Circulation 84 (suppl II): II-87, 1991
30. Schaper W: Patterns of myocardial gene expression after ischemia

and reperfusion. In: S Haunso, J Aldershvile, JH Svendsen (eds). Coronary Microcirculation during Ischaemia and Reperfusion. Munksgaard, Copenhagen, 1997, pp 315–326

31. Rechsteiner M: Ubiquitin. Plenum Press, New York, 1988, pp 333
32. Hershko A, Ciechanover A: The ubiquitin system for protein degradation. Annu Rev Biochem 61: 761–807, 1991
33. Jacewicz M, Kiessling M, Pulsinelli WA: Selective gene expression in focal cerebral ischemia. J Cereb Blood Flow Metab 6: 263–272, 1986
34. Welch WJ: Phorbol ester, calcium ionophore, or serum added to quiescent rat embryo fibroblast cells all result in the elevated phosphorylation of two 28,000-dalton mammalian stress proteins. J Biol Chem 262: 3058–3062, 1985
35. Arrigo AP, Welch WJ: Characterization and purification of the small 28000-dalton mammilian heat shock protein. J Biol Chem 262: 15359–15369, 1987
36. Morimoto RI: Heat shock: the role of transient inducible responses in cell damage, transformation, and differentiation. Cancer Res 3: 295–301, 1991
37. Jentsch S: Ubiquitin-dependent protein degradation: A cellular perspective. Trends Cell Biol 2: 98–103, 1992
38. Sharma HS, Sassen L, Knöll R, Verdouw PD: Circulation, 1168 (abstr), 1992
39. Neff NT, Prevette D, Houenou LJ, *et al.* Insulin-like growth factors-putative muscle-derived agents that promote motoneuron survival. J Neurobiol 24: 1578–1588, 1990
40. Humbel RE. Insulin-like growth factor I and II. Eur J Biochem 190: 445–462, 1990
41. de-Pagter-Holthuizen P, Jansen M, van der Kammen RA, van Schaik FMA, Sussenbach JS: Differential expression of the human insulin-like growth factor II gene: Characterization of the IGF-II mRNAs and an mRNA encoding a putative IGF-II associated protein. Biochim Biophys Acta 950: 282–295, 1988
42. Steenbergh PH, A.M.B.C. K-R, Cleutjens CBJM, Sussenbach JS: Complete nucleotide sequence of the high molecular weight human IGF-I mRNA. Biochem Biophys Res Commun 175: 507–514, 1991
43. Clemmons DR: IGF binding proteins: Regulation of cellular actions. Growth Regul 2: 80–87, 1992
44. Vogt A, Htun P, Kluge A, Zimmermann R, Schaper W: Insulin-like growth factor-II delays myocardial infarction in experimental coronary artery occlusion. Cardiovasc Res 33: 469–477, 1997
45. Lopaschuk GD, Stanley WC: Glucose metabolism in the ischemic heart. Circulation 95: 313–315, 1997
46. Semenza GL, Jiang BH, Leung SW, Passantino R, Concordet JP, Giallonga A: Hypoxia response elements in aldolase A, enolase and lactate dehydrogenase A gene promotor contain essential binding sites for hypoxia-inducible factor 1. J Biol Chem 271: 32529–32537, 1996
47. Levy AP, Levy NS, Goldberg MA: Hypoxia-inducible protein binding to vascular endothelial growth factor mRNA and its modulation by the von Hippel-Lindau protein. J Biol Chem 271: 25492–25497, 1996
48. Ewton DZ, Falen SL, Florini JR: The type II insulin-like growth factor (IGF) receptor has low affinity for IGF-1 analogs: Pleiotropic actions of IGFs on myoblasts are apparenty mediated by the type I receptor. Endocrinology 120: 115–123, 1987
49. Stephens J, Pilch P: The metabolic regulation and ventricular transport of GLUT4, the major insulin-responsive glucose transporter. Endocr Rev 16: 529–546, 1995
50. Young LH, Renfu Y, Russel R, Hu X, Caplan M, Ren J, Shulman GI, Sinusas AJ: Low-flow ischemia leads to translocation of canine heart GLUT-4 and GLUT-1 glucose transporters to the sarcolemma *in vivo*. Circulation 95: 415–422, 1997
51. Weiss JN, Lamp ST: Glycolysis preferentially inhibits ATP-sensitive K^+ channels in isolated guinea pig cardiac myocytes. Science 238: 67–69 1987
52. Neely JR, Grotyohann LW: Role of glycolytic products in damage to ischemic myocardium. Dissociation of adenosine triphosphate levels and recovery of function of reperfused ischemic hearts. Circ Res 55: 816–824, 1984
53. Zimmermann R, Andres J, Brand T, Fraß O, Kluge A, Knöll R, Vogt A, Schaper W: Kardiale Genexpression nach kurzen Koronarverschlüssen. Z Kardiol 84 (Suppl 4): 159–165, 1995

Molecular and Cellular Biochemistry **186**: 53–60, 1998.
© 1998 *Kluwer Academic Publishers.*

Mechanism of hypoxic preconditioning in guinea pig papillary muscles

Tanya Ravingerova,[1] Jan Eirik Løkebø,[2] Jens Munch-Ellingsen,[2] Rune Sundset,[2] Pål Tande[2] and Kirsti Ytrehus[2]

[1]*Institute for Heart Research, Slovak Academy of Sciences, Bratislava, Slovakia;* [2]*Department of Medical Physiology, Institute of Medical Biology, University of Tromsø, Tromsø, Norway*

Abstract

Brief ischemia or hypoxia has been found to protect the heart against susbsequent long-lasting ischemia and to improve contractile dysfunction as well to reduce cell necrosis and the incidence of lethal arrhythmias. This phenomenon, termed preconditioning (PC) has been demonstrated in different species. However, little is known about PC in guinea pigs. Moreover, electrophysiological changes underlying protection have not been studied so far in conjuntion with force recovery in a setting of PC. The aim of the study was to study PC in a guinea pig papillary muscle, using recovery of contractility after long hypoxic challenge as the main end-point of protection, and to investigate concominant electrophysiological alterations. In guinea pig papillary muscle preparations contracting isometrically (paced at 2 Hz), transmembrane action potentials (AP) and developed force (DF) were recorded by conventional microelectrode technique and a force tranducer. In addition, effective refractory periods (ERP) were determined. Hypoxia was induced by superfusion with 100% N_2 ($pO_2 < 5$ kPa) and pacing at 3,3 Hz. In the control group, long hypoxia lasted for 45 min and was followed by 30 min reoxygenation. In the PC group, muscles were subjected to 5 min hypoxia followed by 10 min recovery prior to sustained hypoxia/reoxygenation. Results: Long hypoxia induced a similar depression of DF in both, PC and control groups. However, a loss of contractile activity occured earlier in the PC group. AP duration and ERP decreased faster and were significantly shorter after PC. Upon reoxygenation, preconditioned muscles showed significantly better recovery of function (DF 86% of prehypoxic value vs. 36% in controls; $p < 0,05$). AP and ERP were completely restored in both, PC and control groups. Guinea pig papillary muscle can be preconditioned with a brief hypoxic challenge against contractile dysfunction upon long-lasting hypoxia/reoxygenation. Shortening of AP and loss of contractility occured more quickly during hypoxia and may participate in the protective effect of preconditioning. Possible mechanisms might involve facilitated opening of K_{ATP}-dependent channels. (Mol Cell Biochem **186**: 53–60, 1998)

Key words: hypoxia, preconditioning, guinea pig papillary muscle, action potentials, developed force

Introduction

Brief episodes of ischemia (hypoxia) have been found to increase tolerance of the heart to a subsequent long-lasting ischemic challenge [1]. This adaptive phenomenon, termed preconditioning, has been found to occur in different animal species (dogs, rabbits, rats) and to include limitation of the infarct size [2, 3], improved postischemic contractile recovery [4] and reduction of severe ventricular arrhythmias both, in open-chest animals and in isolated heart preparations [5–7].

However, little is known about preconditioning in the guinea pig. This species has received little attention, partially because of its extensive collateralization it does not allow infarct size studies. Also, only few reports describe preconditioning in a papillary muscle model [8, 9]. Moreover, mechanisms underlying protection against arrhythmias afforded by preconditioning still remain unclear. Our interest in a papillary muscle model was enhanced by the fact that electrophysiological changes have not been yet studied in conjunction with force recovery in a setting of preconditioning. In addition, in the absence of arterial perfusion, this

Address for offprints: T. Ravingerova, MD, PhD, Institute for Heart Research, Slovak Academy of Sciences, 9 Dubravska cesta, 842 33 Bratislava, Slovakia

preparation allows one to evaluate whether coronary flow is an essential factor for preconditioning.

Accordingly, this study was designed to test whether short hypoxic episodes can induce protection against a long hypoxic challenge in a guinea pig papillary muscle model. Posthypoxic recovery of contractile function has been chosen as a main end-point of protection. A futher goal was to evaluate concomitant alterations of electrical activity of the preparation during hypoxia/reoxygenation.

Materials and methods

The studies conformed with the *Guide for the care and use of laboratory animals*, published by the National Institute of Health (NIH publication No 85–23, revised 1985).

Guinea pigs weighing 250–300 g were killed by cervical dislocation. The hearts were rapidly excised and mounted in a dissection chamber allowing perfusion in a Langendorff mode with Tyrode solution containing Hepes buffer at room temperature. The most suitable papillary muscles (length > 2.0 mm, diameter < 1.0 mm) were carefully dissected from the ventricles, placed horizontally in a 2 ml organ bath and superfused at a flow rate of 5ml/min. Solution was gassed with 100% O_2 to give pO_2 of ~50 kPa and pH of 7,4 in the bath. Temperature in the bath was maintained at 35°C, and pO_2 was analyzed intermittently using an automated blood gas analyzer (Radiometer, Copenhagen, Denmark). The base of the muscle was fixed to the floor of the chamber by a stimulus electrode, while the tip of the muscle was attached to a stainless steel hook connected to a miniature quartz strain gauge (AE 801 SensoNor, Horten, Norway), which was connected with a custom designed dc bridge amplifier (P. Körner, Lund, Sweden). Muscles were stretched close to their physiological resting length and field stimulated at a rate of 2Hz with 2ms wide rectangular pulses of 1,5 time threshold intensity. The stimuli, obtained from a stimulator linked with a stimulus isolator, were delivered through a pair of platinum electrodes, one placed in the base of the muscle and the other in the bath. Transmembrane action potentials were recorded with conventional glass microelectrodes filled with 3 M KCl with dc resistance in the range of 15-30 MΩ. The microelectrodes were coupled via Ag-AgCl half-cell to the input of a high impedance ($10^{12}\,\Omega$) capacitance neutralising amplifier. Signals were displayed on a storage oscilloscope and digitised and stored on diskettes using a digital transient recorder with a maximum sampling frequency of 0.5 MHz. The digitised signals were analysed on a personal computer using software written by K. Steinnes (Tromsoe, Norway) to obtain membrane diastolic potential, action potential amplitude, maximal rate of depolarisation, action potential duration at 90% repolarisation, developed force and resting force. Effective refractory periods were determined by a programmed stimulation with a premature pulse applied after a train of 8 basic stimuli at progressively shorter time intervals until the extra pulse failed to evoke an action potential. In this communication, only the following parametres will be dealt with: action potential duration at 90% repolarisation (APD_{90}), developed force (DF), resting force (RF) and effective refractory periods (ERP).

Hypoxia was induced by a superfusion with a substrate-free buffer gassed with 100% N_2 in the Ar-covered organ bath. These interventions lowered pO_2 to < 5 kPa. In addition, to increase the intensity of stress, pacing rate was increased to 3,3 Hz.

Experimental protocols

Stability of the preparations was tested in the experiments with superfusion lasting for > 3 h, which did not cause any substantial deterioration of the parameters investigated. The muscles were initially allowed to stabilise for 1 h at 35°C and pacing rate 2 Hz. After recording of the baseline parametres of action potentials and contractility, the preparations were divided into 2 groups (n = 6 per group). The preconditioned group was subjected to 5 min hypoxia and 10 min reoxygenation followed by a sustained 45 min hypoxia and 30 min reoxygenation, while in the control group normoxic superfusion was continued for an additional 15 min before sustained hypoxia/reoxygenation.

Statistics

All data are presented as means ± S.E.M. The statistical significance of differences in mean values was evaluated by analysis of variance (ANOVA) followed by Student's *t*-test. Differences were considered significant when p values were < 0.05.

Results

Representative recordings demonstrate changes in developed force (DF) and action potentials (AP) before the onset of hypoxia, at the end of hypoxic challenge and after reoxygenation (Figs 1–3). At the end of the stabilisation period, no differences in DF and AP between the control and preconditioned groups were observed (Fig. 1). During long hypoxia, a dramatic decrease in DF and hypoxic contracture occured in both groups (Fig. 2). A major difference was in the configuration of AP, which appeared to be much shorter in the preconditioned muscles at the end of hypoxic period.

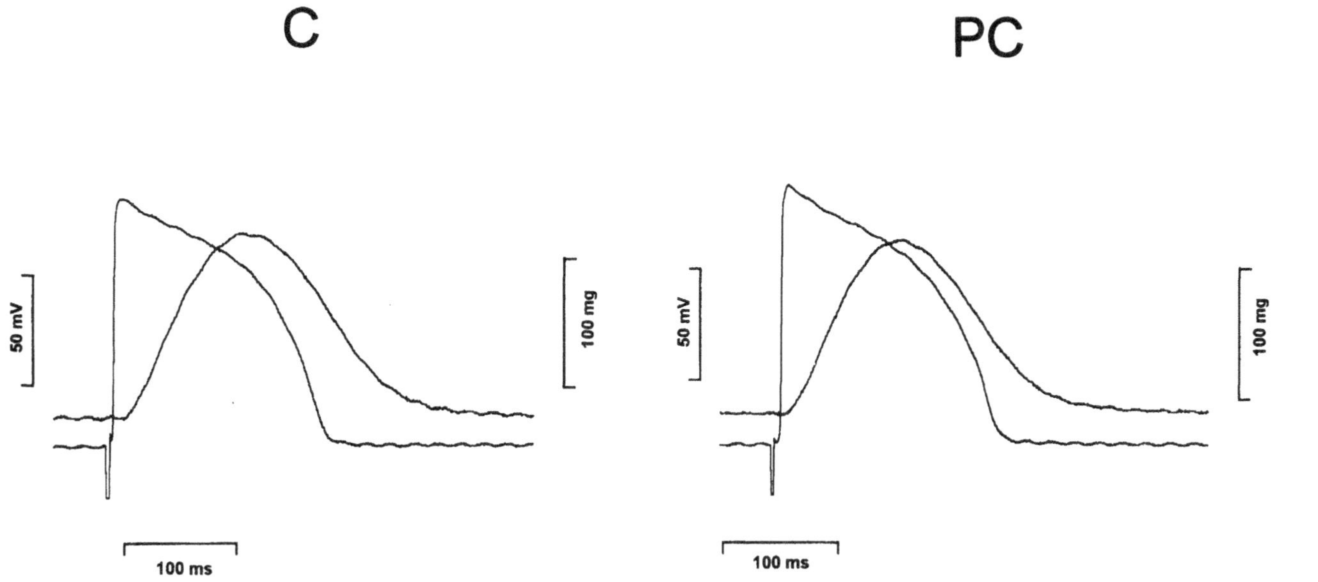

Fig. 1. Representative tracings of action potentials and developed force in the preconditioned (PC) and non-preconditioned (C) guinea pig papillary muscles during normoxia. Pacing rate 2Hz.

Reoxygenation resulted in a substantially better recovery of DF in the preconditioned group, whereas AP duration was completely restored in both groups to a same extent (Fig. 3).

Figure 4 shows timecourse of the changes in DF in both groups. *Short hypoxia* and reoxygenation (PC) induced a transient increase and fall in DF with partial (80%) recovery before the onset of long hypoxia. *Long hypoxia* induced a gradual decline of DF, the rate of which was faster in the PC group. At 5 min, DF in the PC group was significantly lower than in the control group. However, at the end of hypoxic

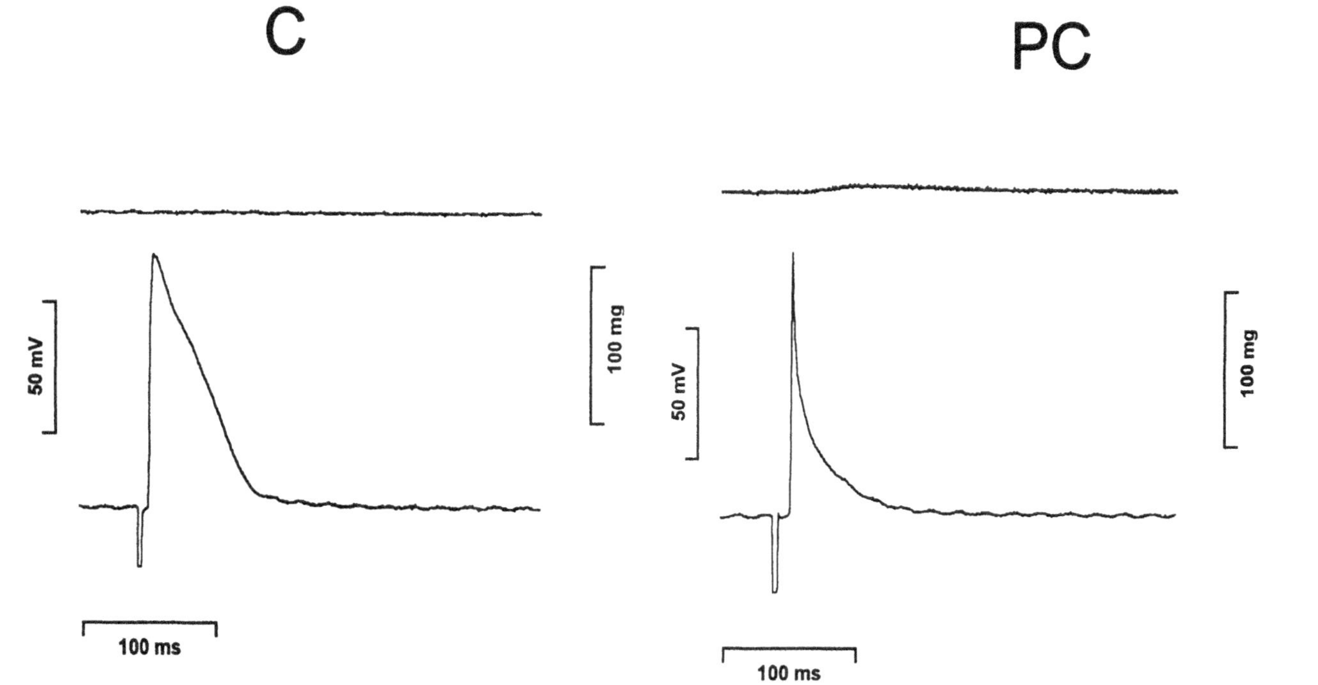

Fig. 2. Representative tracings of action potentials and developed force in the preconditioned (PC) and non-preconditioned (C) guinea pig papillary muscles during 45 min hypoxia. Pacing rate 3.3 Hz.

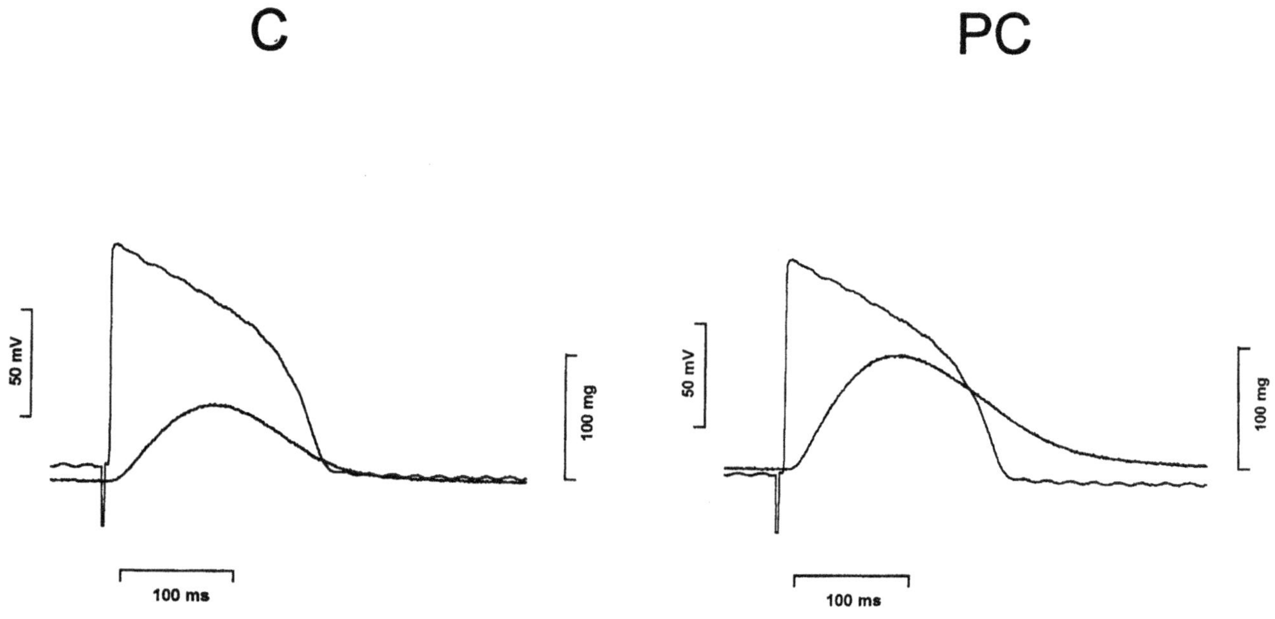

Fig. 3. Recovery of the control and preconditioned guinea pig papillary muscles after 45 min hypoxia and 30 min reoxygenation. Representative tracings of action potentials and developed force. Pacing rate 2Hz.

period, contractile activity was completely absent in both groups. *Reoxygenation* resulted in a significantly better recovery of contractility in the PC group within 30 min (DF was restored to 86 ± 9.5% of prehypoxic values as compared to 36 ± 4.4% in the control group, $p < 0.05$). Hypoxic contracture started to develop earlier in the preconditioned muscles, manifested by a shorter time to the initial increase in resting force (8.5 ± 2 vs. 15 ± 1.5 min in the control group,

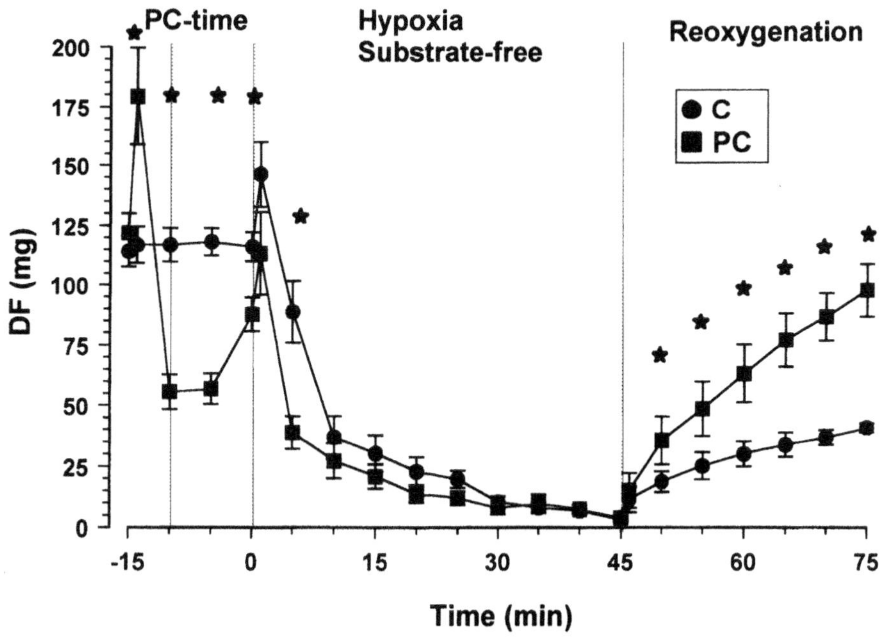

Fig. 4. Changes of developed force (DF) during hypoxia/reoxygenation in the guinea pig papillary muscle. Effect of preconditioning (PC). Data are means ± S.E.M.; n = 6 for each group; *$p < 0.05$; preconditioned vs. control group. Filled circles – control group; filled squares – preconditioned group.

p < 0.05) as shown on Fig. 5 (left panel; time to onset). However, there were no differences in the time to development of maximal contracture (38 ± 4 vs. 43 ± 1.5 min in the control group, $p > 0.05$) and in the magnitude of RF between both groups (Fig. 5, right panel). There were no differences in the restoration of RF upon reoxygenation, which was incomplete, so that contracture persisted to a similar degree in both groups (not shown).

Timecourse of the changes in electrical activity is shown on Figs. 6 and 7. Action potential duration (APD_{90}) and ERP during preconditioning were initially shortened with a subsequent complete recovery. During *long hypoxia*, APD_{90} and ERP decreased faster and were significantly shorter in the PC group (APD_{90} from 15 min and ERP from 5 min onwards until the end of hypoxia) as compared to the controls. Upon *reoxygenation*, APD_{90} and ERP were completely restored in both, PC and control groups.

Discussion

The effect of preconditioning depends on the balance between the intensity of the first stimulus and the duration and severity of the prolonged stress. As we have chosen the recovery of contractility as the main index of protection, our goal was to get a substantial difference in posthypoxic values of DF between the control and preconditioned groups. To achieve this, we have tested several protocols modifying the intensity of the first stress (by changing the duration of hypoxia, the degree of pO_2 lowering and the rate of pacing), as well as the duration of the second hypoxic challenge and reoxygenation. Finally, we have established a protocol in which, on the one hand, hypoxia lasting 45 min led to a relatively low (36%) recovery of DF after 30 min of reoxygenation without further improvement when reoxygenation was prolonged to 60 min. On the other hand, preconditioning consisting of a single hypoxic episode lasting 5 min followed by 10 min reoxygenation markedly improved posthypoxic recovery of DF reaching 86% during reoxygenation (Fig. 4). Since the preparation is lacking arterial perfusion, our results also confirm that the enhancement of collateral flow is not a prerequisite for the effective protection. These results are in agreement with the data of Walker *et al.* [8] who demonstrated better recovery of contractility after 45 min hypoxia in rabbit papillary muscles preconditioned by a single hypoxic episode. In the human myocardium, similar protocol of substrate-free hypoxic preconditioning afforded protection against contractile dysfunction after 45 min simulated ischemia and reoxygenation [10].

Similar to our results, Walker *et al.* [8] have also found that hypoxic contracture started to develop earlier in the preconditioned muscles. Earlier onset of contracture during sustained ischemia has been also observed in isolated rat hearts subjected to global ischemia after preconditioning, and it seems to be a characteristic feature of preconditioning, at least in this model [11–14]. The onset of ischemic contracture correlates with a loss of myocardial ATP, and particular role plays a reduction of glycolitic ATP production [15]. Therefore,

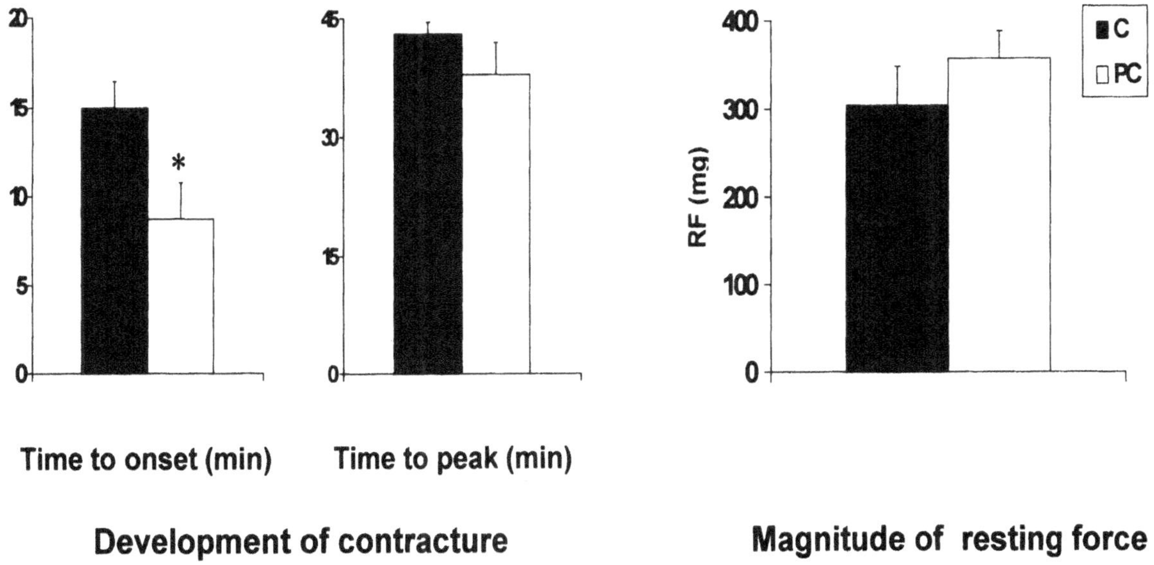

Fig. 5. Development of hypoxic contracture and changes of resting force (RF) during hypoxia in the guinea pig papillary muscle. Effect of preconditioning (PC). Time to onset of contracture (left) was measured as the duration of hypoxia which led to a 5 mm increase in prehypoxic RF. Magnitude of resting force (right) was measured at the peak of contracture. Data are means ± S.E.M.; n = 6 for each group; *p < 0.05; preconditioned vs control group. Filled bars – control group; open bars – preconditioned group.

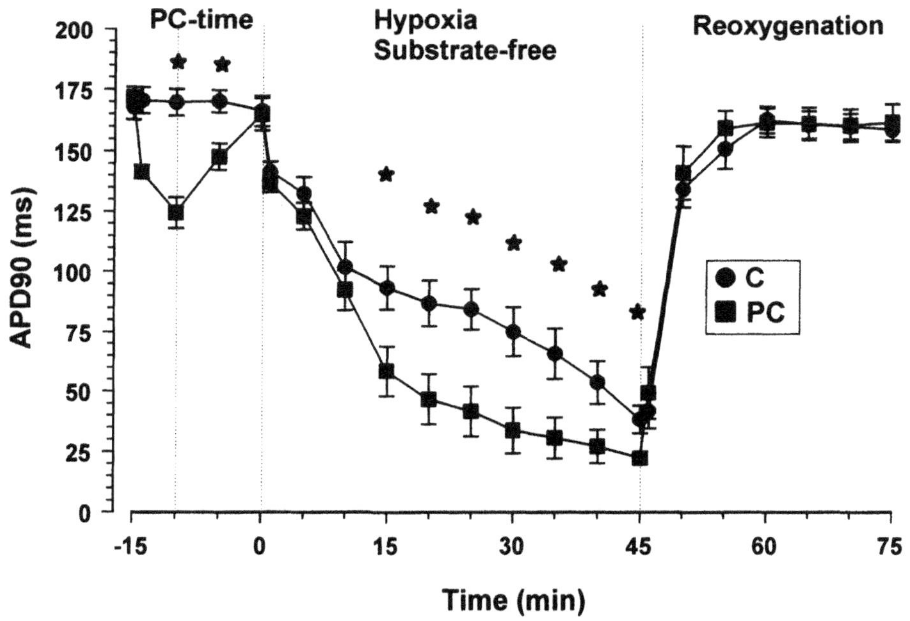

Fig. 6. Changes of action potential duration (APD_{90}) during hypoxia/reoxygenation in the guinea pig papillary muscle. Effect of preconditioning (PC). Data are means ± S.E.M.; n = 6 for each group; *$p < 0.05$; preconditioned vs. control group. Filled circles – control group; filled squares – preconditioned group.

acceleration of ischemic contracture after preconditioning could likely be attributed to a lower level of ATP at the onset of long ischemia in the preconditioned hearts, which has been demonstrated in rats and dogs [13, 16, 17].

The extent of ischemic contracture has been generally considered as an index of damage and its magnitude and onset inversely related to postischemic recovery [18]. On the other hand, despite the earlier onset of contracture during ischemia,

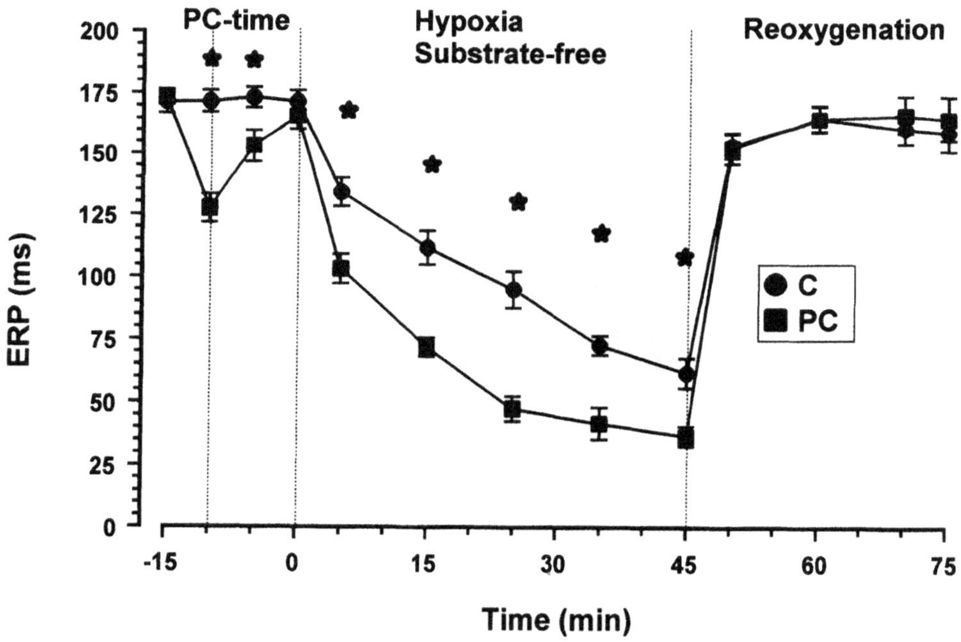

Fig. 7. Changes of effective refractoriness (ERP) during hypoxia/reoxygenation in the guinea pig papillary muscle. Effect of preconditioning (PC). Data are means ± S.E.M.; n = 6 for each group; *$p < 0.05$; preconditioned vs. control group. Filled circles – control group; filled squares – preconditioned group.

preconditioning results in the attenuation of ATP loss during prolonged ischemic episode and subsequent improved postischemic contractile recovery. It has been suggested that slower rate of high energy phosphates depletion in the preconditioned canine heart could account for this protective effect [16, 19]. On the other hand, in the study on rats, the rate of ATP depletion was similar in control and preconditioned hearts [17]. Thus, changes in energy metabolism cannot provide a universal explanation of the protective effect of preconditioning.

Possible mechanism of protection might involve a faster collapse of AP also reflected as a shortening of refractoriness during hypoxia after preconditioning. It has been found [9] that protection against electrophysiological derangements leading to arrhythmias (cellular uncoupling) is associated with a shortened AP in the rabbit myocardium after preconditioning due to activation of ATP sensitive potassium channels. K_{ATP}-channels are closed under normal physiological conditions. However, they become activated by lowering of the ATP level during ischemia or hypoxia, and as a consequence of their opening, IK_{ATP} develops resulting in shortening of the action potential duration [20]. Recent data indicate that K_{ATP}-channel activation by different factors, such as adenosine, H^+, or protein kinase C [21], may participate in the protective mechanisms of preconditioning, since K_{ATP} blockade attenuates protective effect of PC [22]. Mimicking preconditioning with K_{ATP}-channel opener cromakalim resulted in the improved postishemic contractile recovery in the isolated rat heart, and this effect was associated with the shortening of APD and ERP in guinea pig heart [23]. Shortened action potential could lead to a faster loss of contractility on the one hand, and attenuate cell death on the other hand, due to a reduced time of calcium influx into the cells and prevention of calcium overload [24]. Results of this study show that short hypoxia induced a marked elevation of developed force which could increase energy demands of the muscles and under conditions of constant flow result in a decrease of high energy phosphates before the onset of long hypoxia. Lower level of ATP might subsequently lead to a faster opening of the K_{ATP} channels. The increased rate of shortening of APD and ERP in the PC group may also indicate possible involvement of the K_{ATP} channels. Thus, it is conceivable to speculate that their facilitated opening could be a mechanism underlying the effect of preconditioning in a guinea pig model. However, this hypothesis requires further investigation.

Acknowledgement

This study was supported by a grant from the Norwegian Research Council.

References

1. Murry CE, Jennings RB, Reimer KA: Preconditioning with ischemia: A delay of lethal cell injury in ischemic myocardium. Circulation 74: 1124–1136, 1986
2. Liu GS, Thornton J, Van Winkle DM, Stanley AWH, Olsson RA, Downey JM: Protection against infarction afforded by preconditioning is mediated by A_1 adenosine receptors in rabbit heart. Circulation 84: 350–356, 1991
3. Ytrehus K, Liu Y, Downey JM: Preconditioning protects ischemic rabbit heart by protein kinase C activation. Am J Physiol 266: H1145–H1152, 1994
4. Cohen MV, Liu GS, Downey JM: Preconditioning causes improved wall motion as well as smaller infarcts after transient coronary occlusion in rabbits. Circulation 84: 341–349, 1991
5. Shiki K, Hearse DJ: Preconditioning of ischemic myocardium: Reperfusion-inducedarrhythmias. Am J Physiol 253: H1470–H1476, 1987
6. Vegh A, Komori S, Szekeres L, Parratt JR: Antiarrhythmic effects of preconditioning in anaesthetised dogs and rats. Cardiovasc Res 26: 487–495, 1992
7. Ravingerova T, Pyne NJ, Parratt JR: Ischaemic preconditioning in the rat heart: The role of G-proteins and adrenergic stimulation. Mol Cell Biochem 147: 123–128, 1995
8. Walker DM, Marber MS, Walker JM, Yellon DM: Preconditioning in isolated superfused rabbit papillary muscles. Am J Physiol 266: H1534–H1540, 1994
9. Tan HL, Mazon P, Verberne HJ, Sleeswijk ME, Coronel R, Opthof T, Janse MJ: Ischaemic preconditioning delays ischaemia induced cellular electrical uncoupling in rabbit myocardium by activation of ATP sensitive potassium channels. Cardiovasc Res 27: 644–651, 1993
10. Cleveland JC, Meldrum DR, Rowland RT, Banerjee A, Harken AH: Adenosine preconditioning of human myocardium is dependent upon the ATP-sensitive K^+ channel. J Mol Cell Cardiol 29: 175–182, 1997
11. Barakat O, Van Wylen DGL, Mentzer RM, Lasley RD: Ischemic preconditioning improves postischemic recovery of function but shortens time to onset of ischemic contracture in isolated rat hearts (Abstr.) Circulation 84, Suppl II: 433, 1991
12. Cave AC, Horowitz GL, Apstein CS: Can ischemic preconditioning protect against hypoxia-induced damage?: Studies on contractile function in isolated perfused rat hearts. J Mol Cell Cardiol 26: 1471–1486, 1994
13. Kolocassides KG, Galinanes M, Hearse DJ: Dichotomy of ischemic pre-conditioning: Improved postischemic contractile function despite intensification of ischemic contracture. Circulation 93: 1725–1733, 1996
14. Nasa Y, Yabe K, Takeo S: Beta-adrenoceptor stimulation-mediated preconditioning-like cardioprotection in perfused rat hearts. J Cardiovasc Pharmacol 29: 436–443, 1997
15. Kingsley PB, Sako EY, Yang MQ *et al.*: Ischemic contracture begins when anaerobic glycolysis stops: A ^{31}P-NMR study of isolated rat hearts. Am J Physiol 261: H469–H478, 1991
16. Reimer KA, Murry CE, Yamasawa I, Hill ML, Jennings RB: Four brief periods of myocardial ischemia cause no cumulative ATP loss or necrosis. Am J Physiol 251: H1306–H1315, 1986
17. Kolocassides KG, Seymour AL, Galinanes M *et al.*: Detrimental effects of ischemic preconditioning on ischemic contracture? NMR studies on energy metabolism and intracellular pH in the rat heart. (Abstr.) Circulation 90(4,2): I–476, 1994
18. Hearse DJ, Garlick PB, Humphrey SM: Ischemic contracture of the myocardium; mechanisms and prevention. Am J Cardiol 39: 986–993, 1977

19. Murry CE, Richard VJ, Reimer KA, Jennings RB: Ischemic preconditioning slows energy metabolism and delays ultrastructural damage during a sustained ischemic episode. Circ Res 66: 913–931, 1990
20. Noma A: ATP-regulated K^+ channels in cardiac muscle. Nature 305: 147–148, 1983
21. Light PE, Sabir A, Allen BG, Walsh MP, French AJ: Protein kinase C-induced changes in the stoichiometry of ATP binding activate cardiac ATP-sensitive K^+ channels. A possible mechanistic link to ischemic preconditioning. Circ Res79: 399–406, 1996
22. Gross GJ, Auchampach JA: Blockade of ATP-sensitive potassium channel prevents myocardial preconditioning in dogs. Circ Res 70: 223–233, 1992
23. Grover GJ, D'Alonzo AJ, Sleph PG, Dzwonczyk S, Hess T, Darbenzio RB: The cardioprotective and electrophysiological effects of cromakalim are attenuated by meclofenamate through a cyclooxygenase-independent mechanism. J Pharmacol Exptl Therapeut 269: 536–540, 1994
24. Behling RW, Malone HJ: K_{ATP} channel openers protect against increased cytosolic calcium during ischemia and reperfusion. J Mol Cell Cardiol 27: 1809–1817, 1995

Molecular and Cellular Biochemistry **186:** 61–68, 1998.
© 1998 *Kluwer Academic Publishers.*

Characterisation and validation of a murine model of global ischaemia-reperfusion injury

Mark S. Sumeray and Derek M. Yellon
The Hatter Institute, University College Hospitals, London, UK

Abstract

A specially designed Langendorff apparatus was constructed to allow perfusion of the isolated mouse heart. Hearts were randomised into groups to receive differing periods of global (zero flow) ischaemia or continuous perfusion (controls). During reperfusion, recovery of baseline force was recorded and perfusate collected for LDH assay (U/L/g wet weight). After 30 min reperfusion, hearts were stained with tetrazolium and planimetry performed to measure infarct size. Dose-response relationships were demonstrated for all 3 end-points against duration of ischaemic insult. Functional recovery and enzyme leakage correlated well with infarct size ($r = 0.77$, $p < 0.001$ and $r = 0.73$, $p < 0.001$ respectively). Transgenic mice may now be used to study the effect of specific phenotypic changes on the pathogenesis of ischaemia-reperfusion injury using a reliable and reproducible technique. (Mol Cell Biochem **186**: 61–68, 1998)

Key words: mouse, ischemia, ischaemia-reperfusion, myocardial infaction, murine, Langendorff, lactate dehydrogenase

Introduction

Ischaemia-reperfusion injury is responsible for many pathological syndromes in man. The importance of carefully conducted research to elucidate the cellular mechanisms underlying the resultant injury is reinforced by epidemiological data which confirm the place of coronary artery disease as the leading cause of death in the developed world. Traditionally there has been a considerable gulf between molecular biologists and physiologists in terms of the philosophy underlying their investigative techniques. However, recent advances in understanding the mouse genome and genetic manipulative techniques have created a unique opportunity to explore the influence of single gene alterations on complex physiological systems [1]. It is essential that physiologists respond to this challenge by careful design of appropriate, miniturised systems to allow phenotypic changes to be explored. Thorough characterisation and validation of such models is an essential but frequently neglected step before meaningful conclusions can be drawn from the data they generate. It is with these important considerations in mind that a new murine model of global ischaemia-reperfusion injury has been developed *in vitro*.

Materials and methods

The experiments were performed in accordance with the United Kingdom Home Office 'Guidance on the Operation of the Animals (Scientific Proceedures) Act 1986', published by Her Majesty's Stationary Office, London. All animals were obtained from the same source and housed under identical conditions.

Preparation of hearts for perfusion

Male Swiss White mice (25–35 g body wt) were used for all experiments. Animals were deeply anaesthetised with sodium pentobarbitone (60 mg/kg ip). As soon as an adequate depth of anaesthesia was confirmed the chest was opened excising the sternum and attached costal cartilages to give adequate access to the mediastinum. The heart was rapidly removed following transection of the pulmonary veins and great vessels, and placed in iced Krebs-Henseleit buffer (at 4°C). Thymic and fatty tissue was carefully trimmed to reveal the ascending aorta, which was then cannulated with a 20 G phalanged stainless steel cannula under the Krebs solution

Address for offprints: D.M. Yellon, The Hatter Institute for Cardiovascular Studies, Division of Cardiology, University of London Hospitals and Medical School, Grafton Way, London, WC1E 6DB, UK

to avoid air embolisation of the coronaries (Fig. 1). The aorta was secured on the cannula using a 4-0 silk suture and transferred to the perfusion apparatus. During transfer the cannula was continuously perfused through a side arm with Krebs solution under low pressure to prevent air entry. Once attached to the perfusion apparatus retrograde perfusion at constant pressure (80 mmHg) was commenced. The time from excision of the heart to the commencement of perfusion was recorded and never exceeded 2.5 min.

Buffer perfusion

The perfusion buffer consisted of modified Krebs-Henseleit buffer of the following composition: NaCl 118mM, $NaHCO_3$ 24.0mM, KCl 4.0mM, NaH_2PO_4 1.0mM, $CaCl_2$ 2.5 mM, $MgCl_2$ 1.2mM, di-sodium EDTA 0.5mM, glucose 10mM. The buffer was pumped via an in-line 2 μm glass microfibre filter to a reservoir at a fixed height above the aortic cannula (104 cmH_2O; 80 mmHg). A small incision was made in the pulmonary outflow tract to allow free drainage of coronary effluent. A fine micro-thermocouple wire was passed retrogradely into the right ventricle and attached to a digital thermometer (Physitemp Bat-12; Sensortek, Clifton, NJ). The apparatus was water-jacketed to maintain a constant myocardial temperature of 37°C. During ischaemia temperature was maintained by immersing the heart in an organ bath containing anoxic modified Krebs solution at 37°C.

A 4-0 silk suture (on a round-bodied needle) was placed through the apex of the left ventricle (Fig. 2), and threaded through a light weight rigid coupling rod which was in direct continuity with a force transducer (Statham, Gould Inc, Cleveland OH). The vertical position of the coupling rod and transducer was adjusted by fine manipulation using a hydraulic micromanipulator (Clark Electromedical instuments, Reading, UK). The frequency response characteristics of the force transducer, coupling rod and mounting gantry had been assessed previously and found to be flat to at least 50 Hz [2].The diastolic tension of the suture was adjusted to 1g during stabilisation of the heart.

At regular intervals during each experiment developed force was recorded at high paper speed (100 mm/s Gould Window Graf Recorder), and coronary effluent collected over 30 sec. The coronary flow rate was recorded by timed collection and the effluent frozen at –80°C for later assay of LDH activity.

Experimental protocol

After commencing retrograde aortic perfusion hearts were allowed to stabilise for 30 min. Any hearts in which coronary flow rate was excessive (>6 ml/min, usually due to an aortic tear), demonstrating bradycardia (<300 bpm), poor contractility (<1.0 g developed force) or persistent arrhythmia

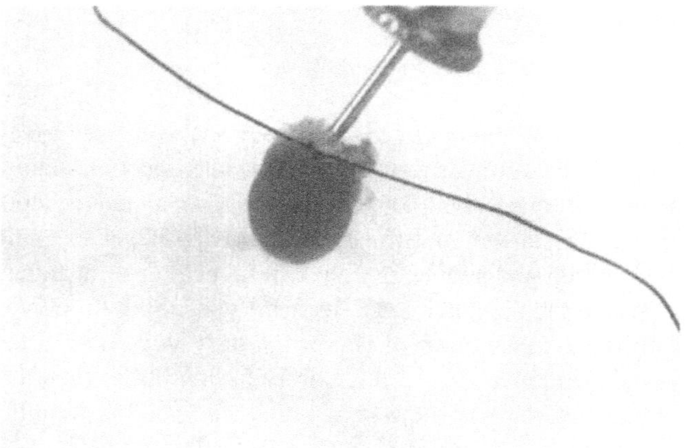

Fig. 1. Photograph showing the excised mouse heart mounted on a 20G phalanged cannula under iced buffer. The heart is secured with a 4-0 silk suture.

Fig. 2. Photograph showing the isolated mouse heart perfused on the Langendorff apparatus. A 4-0 silk suture attaches the apex of the left ventricle to a light weight rigid coupling rod which is attached to a force transducer. Perfusate is seen running off the end of the coupling rod.

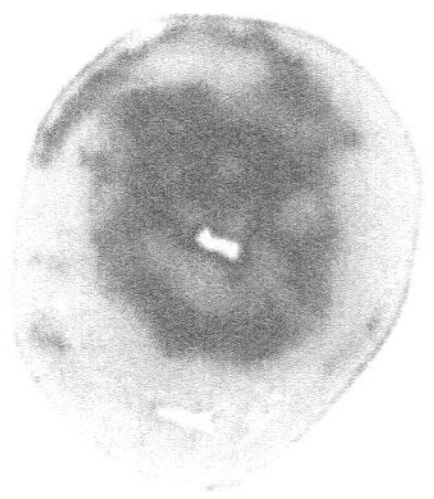

Fig. 3. A magnified camera image of a typical slice through a mouse heart which has been stained with tetrazolium to delineate viable myocardium. This slice is from a heart subjected to 30 min global ischaemia and shows extensive myocardial necrosis.

(SVT, VT or VF persisting after 20 min stabilisation) were excluded from further study. Baseline contractility and coronary flow measurements were made. Hearts were then lowered into the organ bath. The flow of Krebs-Henseleit buffer was stopped and global ischaemia was maintained for differing periods. Temperature was carefully controlled at 37°C.

Thirty-six hearts were randomised into 6 groups to receive differing periods of global ischaemia (10, 15, 20, 25 or 30 min) or continuous perfusion (controls). Reperfusion was continued for 30 min following the ischaemic insult. Recovery of contractile force, coronary flow rate, and heart rate were recorded.

Measurement of infarct size

Following 30 min reperfusion hearts were lowered into the organ bath. Triphenyltetrazolium chloride (1% solution) was injected down the side arm of the aortic cannula and infused into the coronary circulation. Once the hearts had stained dark red they were removed, blotted dry and weighed. They were then frozen and stored at –80°C. The following day they were defrosted and sliced into 1 mm sections parallel to the atrioventricular groove and then fixed in 2% formaldehyde overnight. Slices were then compressed between plexiglas plates 0.57 mm apart and a magnified video image generated by a specially modified camera (Fig. 3). The image was digitised (Apple Macintosh 7500) and the area of infarction and ventricular area (the entire LV and RV were at risk) planimetred using image analysis software (NIH Image v1.61). Graph paper (1mm squares) was used to calibrate the image areas, which were

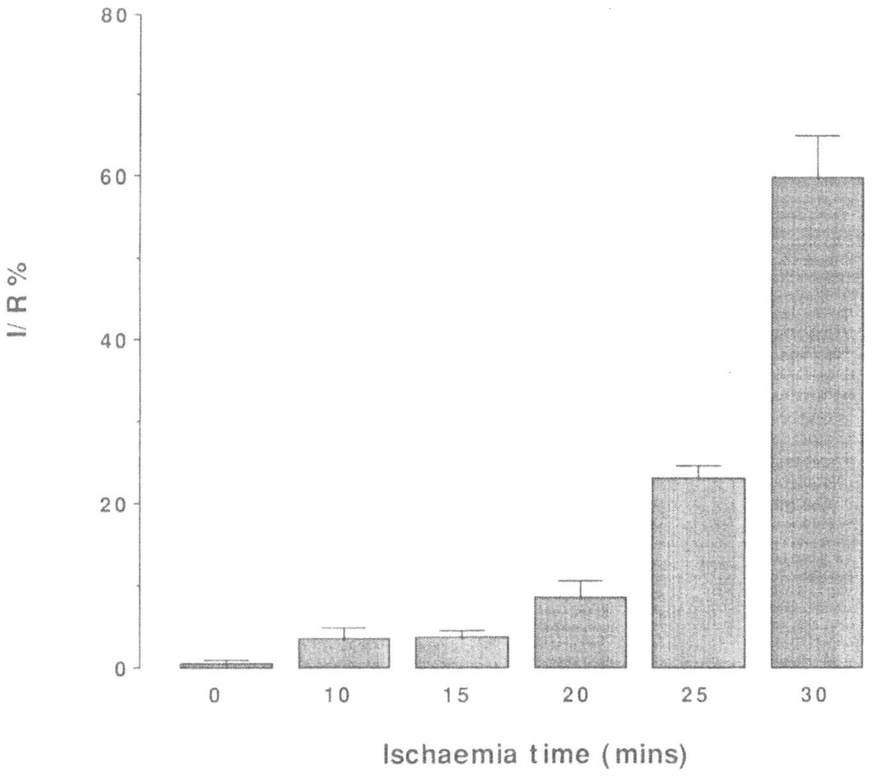

Fig. 4. Graph showing the relationship between infarct size (expressed as a percentage of total volume at risk) and duration of global ischaemia.

then multiplied by the slice thickness (0.57 mm) to generate the volume planimetered.

Measurement of LDH activity

Samples of coronary perfusate were collected and frozen at –80°C. Samples were later allowed to defrost and analysed for LDH activity. The assay was based on a spectrophotometric technique in which absorbance at 340 nm decreases as NADH is utilised during reduction of pyruvate to lactate catalysed by LDH (Sigma, Dorset UK.). LDH activity was then expressed as activity leaked per litre of perfusate per gram wet heart weight.

Satistical analysis

Results are expressed as mean values ± S.E. Comparisons between groups for variables at different time points were made using one-way analysis of variance (ANOVA). Correlation between end-points was assessed by Spearman rank coefficient. Statistical significance was defined by a p value < 0.05.

Results

In total 44 hearts were entered into the study but 8 were excluded during the stabilisation period and before any data had been recorded. This was due to aortic damage in 5 cases and persistent tachyarrhythmia in 2 cases. One heart was excluded due to severe bradycardia. The baseline characteristics are shown for all 6 groups (Table 1), and did not differ significantly between groups.

Infarct size

Volume of infarcted myocardium was expressed as a percentage of ventricular volume. Myocardium was stained with tetrazolium after 30 min of reperfusion. The adequacy of this reperfusion period has previously been established in this model (2) and is discussed below. The results are shown in Fig. 4, and demonstrate a graded injury in response to increasing periods of global ischaemia. Figure 5 shows that ventricular volume did not differ significantly between groups. It can be seen that there was minimal infarction until a 25 min global ischaemic insult had been administered (23.3 ± 1.5%). There was a significant increase in infarct size

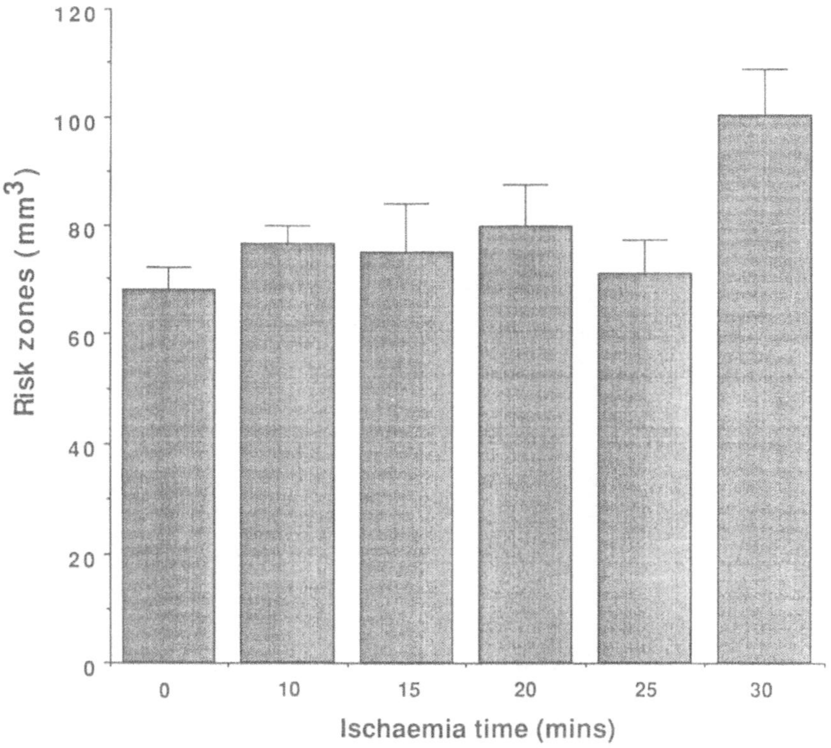

Fig. 5. Graph demonstrating that the total risk volumes between groups did not differ significantly.

Table 1. Table showing baseline characteristics of hearts in all groups at the end of 20 min stabilisation

Group	Body wt (g)	Heart wt (g)	Coronary flow rate (ml/min)	Heart rate (bpm)	Developed force (g)
Controls	28.4 ± 0.8	0.15 ± 0.01	3.4 ± 0.2	336 ± 8.2	2.5 ± 0.4
10 min	26.9 ± 0.3	0.15 ± 0.01	3.0 ± 0.4	355 ± 11.8	3.0 ± 0.2
15 min	27.7 ± 0.8	0.17 ± 0.01	3.5 ± 0.6	357 ± 15.6	3.0 ± 0.6
20 min	27.0 ± 0.9	0.19 ± 0.03	3.7 ± 0.8	343 ± 16.7	3.6 ± 0.5
25 min	26.1 ± 1.3	0.15 ± 0.01	3.8 ± 0.9	367 ± 16.7	2.8 ± 0.2
30 min	30.9 ± 0.9	0.22 ± 0.02	3.0 ± 0.4	372 ± 16.8	3.6 ± 0.3

compared with controls at 25 and 30 min (23.3 ± 1.5%, 59.8 ± 5.1%) indicating that tissue necrosis did not increase linearly with duration of ischaemia.

Contractile recovery

Developed force during the reperfusion period was recorded at regular intervals and, following an initial hypercontractile phase during the first 2 min, maximum recovery had occured by 30 min reperfusion. Contractile force was expressed as a percentage of baseline and is shown in Fig. 6. In contrast to tissue necrosis, contractile function was depressed considerably following a shorter duration of global ischaemia. All groups exhibited significantly reduced contractile function compared to controls, but hearts subjected to 25 and 30 min ischaemia did not differ significantly from those subjected to 20 min ischaemia. The component of contractile dysfunction that is clearly unrelated to myocardial necrosis probably reflects a significant 'stunning' effect (see discussion).

LDH release

Samples of coronary perfusate were collected at regular intervals during reperfusion and analysed for LDH activity.

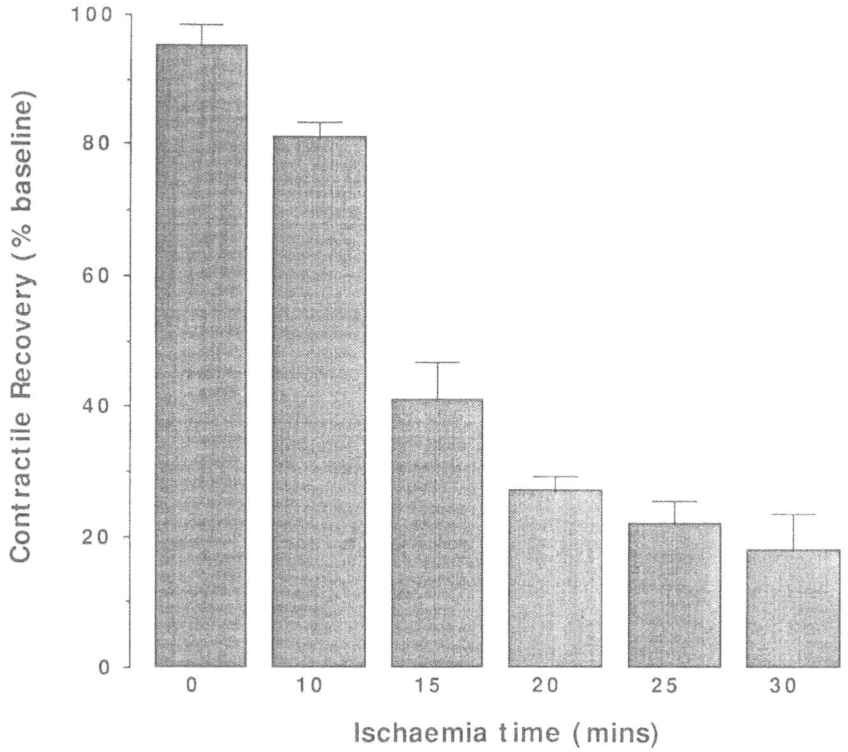

Fig. 6. Graph showing the relationship between recovery of contractile function after 30 min reperfusion and duration of global ischaemia.

Release had peaked at 30 min reperfusion and Fig. 7 shows values for each of the groups. Considerable variability was present and standard errors were large. Hearts subjected to periods of ischaemia in excess of 15 min released significant quantities of LDH compared to controls, but did not differ significantly from each other.

Comparison of infarct volume with contractile recovery and LDH release

Figures 8 and 9 show how contractile recovery and LDH release correlate with infarct size. As expected from Figs 2–4 hearts in the group subjected to 30 min ischaemia displayed significantly greater necrosis without a corresponding effect on either contractile recovery or release of LDH (both of which have begun to plateau after 20 min ischaemia). Different mechanisms of injury are likely to be responsible (see discussion).

Discussion

This study demonstrates that global ischaemic injury may be assessed in the isolated mouse heart, allowing the mechanisms underlying ischaemia-reperfusion injury to be further elucidated.

Infarct size determination

The feasibility of measuring infarct size by tetrazolium staining in such a small heart has now been established, despite the potential problems with this technique. A magnified video image can be digitised by computer video card allowing accurate planimetry using appropriate software (Fig. 3). The time allowed for reperfusion before staining with tetrazolium was 30 min. It was possible that this period might have been inadequate for effective washout of dehydrogenase enzyme from infarcted tissue. However, the

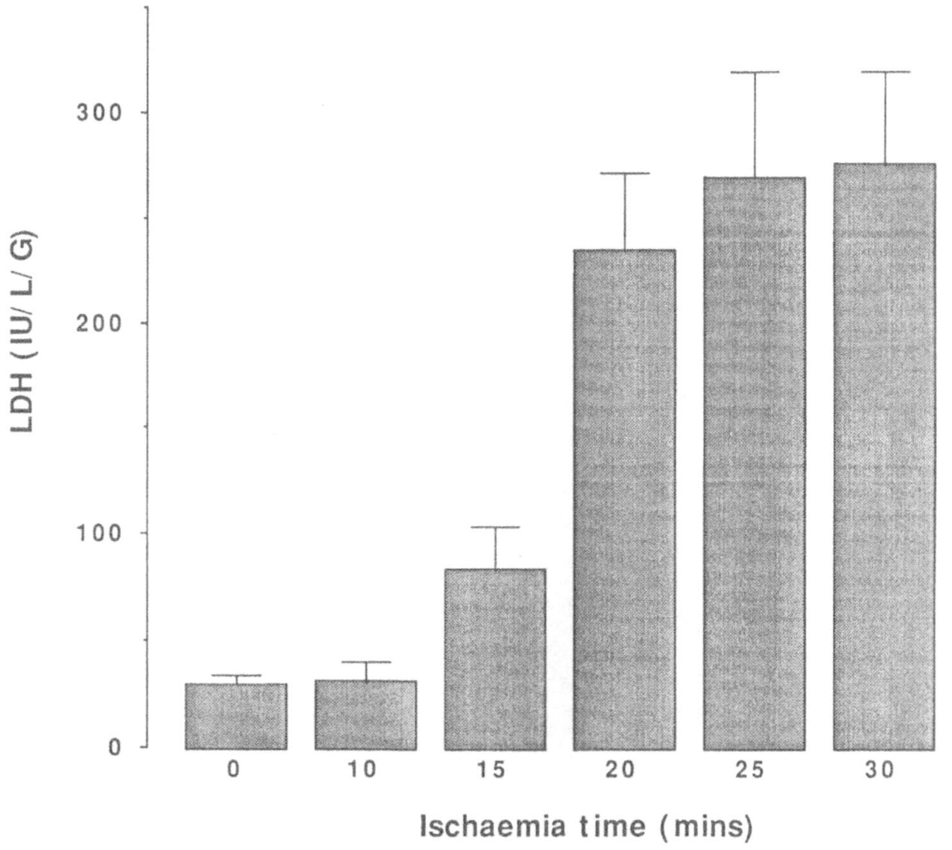

Fig. 7. Graph showing the relationship between the activtiy of LDH in the coronary perfusate collected at 30 min reperfusion and duration of global ischaemia.

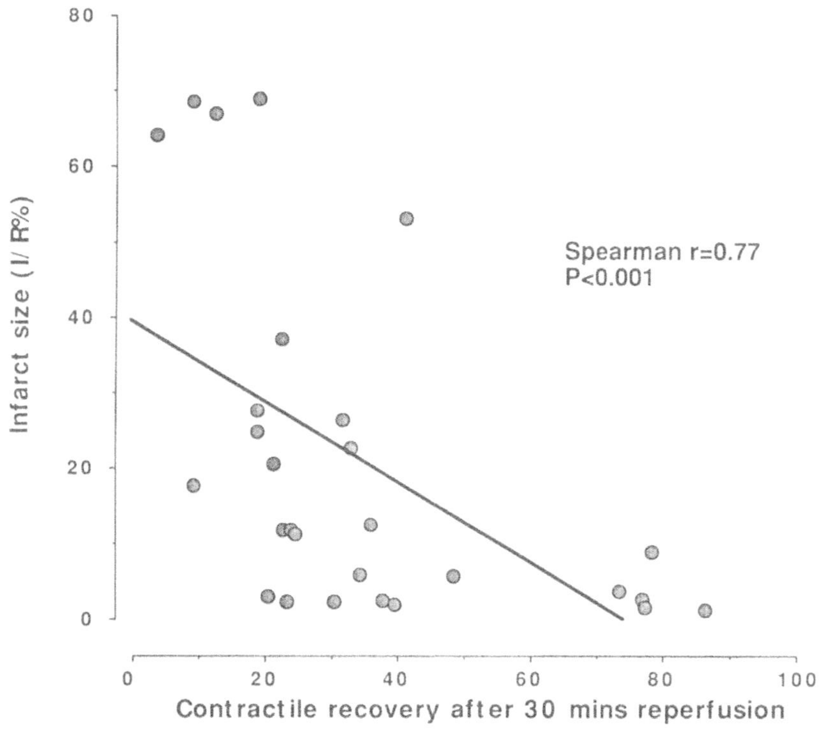

Fig. 8. Graph to show the correlation between contractile recovery and LDH release with infarct size at 30 min reperfusion.

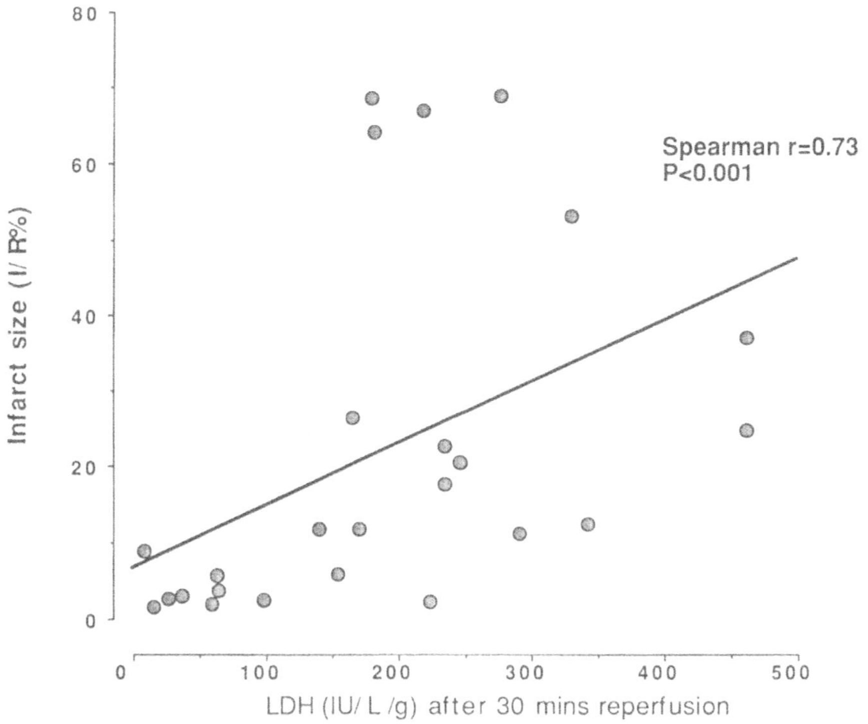

Fig. 9. Graph to show the correlation between contractile recovery and LDH release with infarct size at 30 min reperfusion.

mouse heart has a high coronary flow rate per gramme heart weight, and adequate washout is likely in this short time period. In order to verify this assumption preliminary experiments were carried out allowing a 2 h reperfusion time before staining in hearts subjected to 25 min of global ischaemia (n = 6). Infarct size in this group did not differ from those subsequently reperfused for 30 min only. This is in agreement with a previous study [2].

Arrhythmias

This model was not designed to assess the incidence and nature of arrhythmias occuring during the reperfusion period, and electrocardiographic recordings were not made. However, certain observations were of interest. In general the incidence of arrhythmia increased with more prolonged periods of ischaemia. 80% of hearts subjected to 10 or 15 min of global ischaemia suffered extrasystoles (which may have been of ventricular or supraventricular in origin) occurring approximately 10 min into the reperfusion period but disappearing after 20 min. One half of the hearts subjected to 20, 25 or 30 min of global ischaemia suffered ventricular fibrillation which began 15 min into the reperfusion period, and resolved by 30 min in approximately 80% of hearts.

Several other interesting areas have been highlighted. It would appear that there is a probably a significant component of myocardial stunning in this model, although this hypothesis has not been tested directly. Following 15 min of ischaemia, contractile recovery is reduced to 40% of baseline, but infarct size is minimal (4%). It is only after 25 min of ischaemia that there is appreciable necrosis (I/R 23%), and there is a disproportionate increase to 60% after 30 min ischaemia. However, there is very little further increase in contractile dysfuntion from 20–30 min ischaemia, indicating that contractility is likely to have been depressed by a mechanism independent of necrosis. It should be noted, however, that even after short periods of ischemia, some necrosis has occurred and may be responsible for the impaired contractility observed at these time points. It is also interesting to note that release of LDH rises rapidly between 15 and 20 min ischaemia and then reaches a plateau. This is in contrast with the time period of necrotic injury and indicates that LDH might be released from cells that are still viable. Alternatively, wash-out of LDH may be impaired following larger infarction due to lack of effective subendocardial reperfusion. It is possible that both these factors may have contributed to the effect observed. Taken together all three end-points provide valuabe information about the nature of the myocardial injury, but their differences emphasize the importance of avoiding the use of single end-points in models of ischaemia-reperfusion injury.

The isolated mouse heart has been perfused in Langendorff [2–4] and working mode [5–7] in previous studies. *In vivo* models, utilising regional ischaemia, have also been developed [8, 9] The potential for well characterised physiological models to be used with new transgenic technology in the investigation in the invesigation of a number of myocardial pathological conditions has been recognised [1]. It is essential that development of such models with their various end-points of injury should incorporate a thorough process of chararacterisation before meaningful conclusions can be drawn about the results they generate.

References

1. Chien K: Cardiac muscle diseases in genetically engineered mice: Evolution of molecular physiology. Am J Physiol 38: H755–H766, 1995
2. Marber M, Mestril R, Chi S, Sayen M, Yellon D, Dillmann W: Overexpression of the rat inducible 70-kD heat stress protein in a transgenic mouse increases the resistance of the heart to ischemic injury. J Clin Invest 95: 1446–56, 1995
3. Yoshida T, Watanabe M, Engelman D, *et al.*: Transgenic mice overexpressing GSHPx-1 are resistant to myocardial ischemia-reperfusion injury. J Mol Cell Cardiol 28: A169 (abstract), 1996
4. Gabel S, Steenbergen C, Murphy E: Effects of preconditioning in perfused mouse heart: 31P NMR and left ventricular pressure. J Mol Cell Cardiol 28: A184 (abstract), 1996
5. Ng W, Grupp I, Subramaniam A, Robbins J: Cardiac myosin heavy chain mRNA expression and myocardial function in the mouse heart. Circ Res 68: 1742–1750, 1991
6. Bittner H, Chen E, Peterseim D, Van-Trigt P: A work-performing heart preparation for myocardial performance analysis in murine hearts. J Surg Res 64: 57–62, 1996
7. Chen E, Bittner H, Davis R, Folz R, Van-Trigt P: Extracellular superoxide dismutase transgene overexpression preserves postischemic myocardial function in isolated murine hearts. Circulation 94(9 suppl): 11412–11417, 1996
8. Hutter J, Mestril R, Tam E, Sievers R, Dillmann W, Wolfe C: Overexpression of heat shock protein 72 in transgenic mice decreases infarct size *in vivo*. Circulation 94: 1408–1411, 1996
9. Michael L, Entman M, Hartley C, *et al.*: Myocardial ischemia and reperfusion: A murine model. Am J Physiol 269(6 Pt 2): H2147–H2154, 1995

Molecular and Cellular Biochemistry **186:** 69–77, 1998.
© 1998 *Kluwer Academic Publishers.*

Ischemic preconditioning in isolated perfused mouse heart: Reduction in infarct size without improvement of post-ischemic ventricular function

Lei Xi, Michael L. Hess and Rakesh C. Kukreja
Division of Cardiology, Medical College of Virginia, Virginia Commonwealth University, Richmond, USA

Abstract

Genetically engineered mice provide an excellent tool to study the role of a particular gene in biological systems and will be increasingly used as models to understand the signal transduction mechanisms involved in ischemic preconditioning (IP). However, the phenomenon of IP has not been well characterized in this species. We therefore attempted to examine whether IP could protect isolated mouse heart against global ischemia/reperfusion (GI/R) injury. Thirty adult mice hearts were perfused at constant pressure of 55 mmHg in Langendorff mode. Following 20 min equilibration, the hearts were randomized into three groups (n = 10/each): (1) Control Group; (2) IP2.5 Group: IP with two cycles of 2.5 min GI + 2.5 min R; (3) IP5 Group: IP with 5 min GI + 5 min R. All hearts were then subjected to 20 min of GI and 30 min R (37°C). Ventricular developed force was measured by a force transducer attached to the apex. Leakage of CK and LDH was measured in coronary efflux. Infarct size was determined by tetrazolium staining. Following sustained GI/R, infarct size was significantly reduced in IP2.5 (13.8 ± 2.3%), but not in IP5 (20.1 ± 4.0%), when compared with non-preconditioned control (23.6 ± 3.8%) hearts. CK & LDH release was also reduced in both IP2.5 and IP5 groups. No significant improvement in post-ischemic ventricular contractile function was observed in either IP groups. We conclude that IP with repetitive cycles of brief GI/R is able to reduce myocardial infarct size and intracellular enzyme leakage caused by a sustained GI/R in the isolated perfused mouse heart. This anti-necrosis cardioprotection induced by IP was not associated with the amelioration of post-ischemic ventricular dysfunction. (Mol Cell Biochem **186**: 69–77, 1998)

Key words: isolated mouse heart, ischemia/reperfusion injury, myocardial infarction, preconditioning, cardioprotection, ventricular function, signal transduction

Introduction

Since Murry *et al.* [1] first demonstrated the phenomenon in 1986 in a canine model, ischemic preconditioning (IP) has been extensively studied by a number of investigators. The potent endogenous cardioprotective effects afforded by IP, especially in enhancing myocardial cellular survival against irreversible damages caused by ischemia/reperfusion, have been consistently found in different experimental models including dogs [2–5], pigs [6], rabbits [7–12] and rats [13–21] both *in vitro* and *in vivo*. However, the signal transduction mechanisms involved in the early phase of IP remain unclear. There has been controversy in such mechanisms particularly across different animal species.

In recent years, the genetically engineered mice are being used as unique tools to investigate the role of genes in the biological systems. Transgenic mouse models have been reported to study the role of heat shock protein 70 and adenosine A1 receptor in myocardial protection [22–24]. These animals will be increasingly used to understand the mechanisms of ischemic preconditioning at the molecular level. However, the results from transgenic mice studies can not be entirely extrapolated to other species unless we have a thorough understanding of the cardiovascular physiology in normal mouse. In this context, mouse differs from other species in many ways. For example, the total body oxygen consumption and myocardial oxidative capacity seem to be very high in this species [25], which may have a major impact

Address for offprints: R.C. Kukreja, Division of Cardiology, Box 282, Medical College of Virginia, Richmond, VA 23298, USA

on the myocardial response to ischemia/reperfusion injury. The activity of xanthine oxidoreductase enzyme, a major source of free radicals, is much higher in mouse heart as compared to other species [26]. Combined with its lack of inhibition of mitochondrial adenosine triphosphatase (ATPase) [27], the susceptibility of mouse heart to ischemia/ reperfusion injury may be substantially increased. A recent study from our laboratory [28, 29] demonstrated that, in contrast to some previous reports in other species (rats or rabbits), whole body heat shock failed to precondition the isolated mouse heart despite induction of 72 kDa heat shock protein at 6 or 24 h after the heat stress. These results suggest that mouse may differ from other species in its response to stress. Moreover, IP, which has been proved in virtually all studied mammalian species, has not been well characterized in the mouse heart. Therefore, the purpose of the present investigation was to demonstrate that the classic IP could also afford an early phase of cardioprotection in the isolated perfused mouse heart against ischemia/reperfusion injury. We measured multiple end-points such as myocardial infarct size, ventricular contractile function, and intracellular enzyme release to ensure a more comprehensive assessment of the potential cardioprotection in the globally ischemic isolated mouse hearts.

Materials and methods

Animals

Adult male outbred ICR mice (body weight 26–42 g) were supplied by Harlan Sprague Dawley Co. (Indianapolis, IN). The animals were allowed to readjust the new housing environment for at least 3 days before any experiment. Standard rodent food and water were freely accessible to the animals. All animal experiments were conducted under the guidelines on humane use and care of laboratory animals for biomedical research published by the U.S. National Institutes of Health (NIH Publication No.85-23, revised 1985) and the experimental protocols were approved by the Animal Welfare Committee of Medical College of Virginia/ Virginia Commonwealth University.

Langendorff isolated perfused heart preparation

The animals were anesthetized with sodium pentobartital (100 mg/kg with 33 IU heparin, i.p.) and hearts were quickly removed from the thorax and placed into a small dish containing ice-cold modified Krebs-Henseleit (K-H) solution containing heparin. Under an illuminated magnifier, the aortic opening of mouse heart was immediately cannulated and tied on a 20 gauge stainless steel blunt needle. The heart was retrogradely perfused with the K-H solution containing (in mM) NaCl 118, $NaHCO_3$ 24, $CaCl_2$ 2.5, KCl 4.7, KH_2PO_4 1.2, $MgSO_4$ 1.2, Glucose 11, EDTA 0.5 at a constant pressure of 55 mmHg in the non-recirculating Langendorff mode. The K-H solution was pre-filtered by a micro-filter (0.45 μm diameter, Millipore Corp.) and constantly gassed with 95% O_2–5% CO_2 (pH 7.35–7.43). The perfusion solution was warmed through a water-jacketed glass cylinder/heat exchanger system with a warming/cooling bath (Brinkmann) and was constantly circulated by a water pump. The heart temperature was monitored continuously by a thermo-couple thermometer (Cole-Palmer, Model 8112-10) with a Type K micro-probe and maintained at 37 ± 0.2°C. The ambient temperature around the heart was also kept at 37°C throughout the experiment using a heating lamp and monitored by a digital thermometer (Fisher Scientific, Model 15-077-8D). During the no-flow ischemia, K-H buffer was periodically applied on the heart surface in order to keep it moisturized.

Assessment of ventricular function

Soon after the start of perfusion, a force-displacement transducer (Grass, Model FT03) was attached to the apex of mouse heart through a rigid, light weight stainless metal hook with No. 5 surgical thread and a low-resistance pulley. The resting tension of the heart was initially adjusted to 0.3 g as used by Plumier *et al.* [23] and kept without readjustment thereafter. The ventricular force was recorded by a polygraph (Beckman, R-511A) connected to the force transducer. The preamplifier and coupler were calibrated and balanced. The force transducer was calibrated prior to each experiment with four known amount of weights i.e., 0.2, 0.3, 0.5, 1.0 g. The coronary flow rate was calculated by collecting the efflux perfusate within certain time period. The hearts were not electrically paced.

Ischemia/reperfusion protocol

A detailed illustration of the ischemia/reperfusion protocol is shown in Fig. 1. In brief, after a 20 min equilibration period, heart rate, ventricular force and coronary flow were measured and recorded. The global ischemia (GI) was achieved by clamping the aortic inflow line whereas reperfusion (R) was accomplished by re-opening the aortic line. Thirty mouse hearts were randomly subjected to one of the following three experimental protocols (n = 10/each): (1) Control Group: no ischemic preconditioning (IP); (2) IP2.5 Group: IP with two cycles of 2.5 min GI + 2.5 min R; (3) IP5 Group: IP with 5 min GI + 5 min R. All of the hearts were then challenged by a sustained 20 min GI followed by 30 min R.

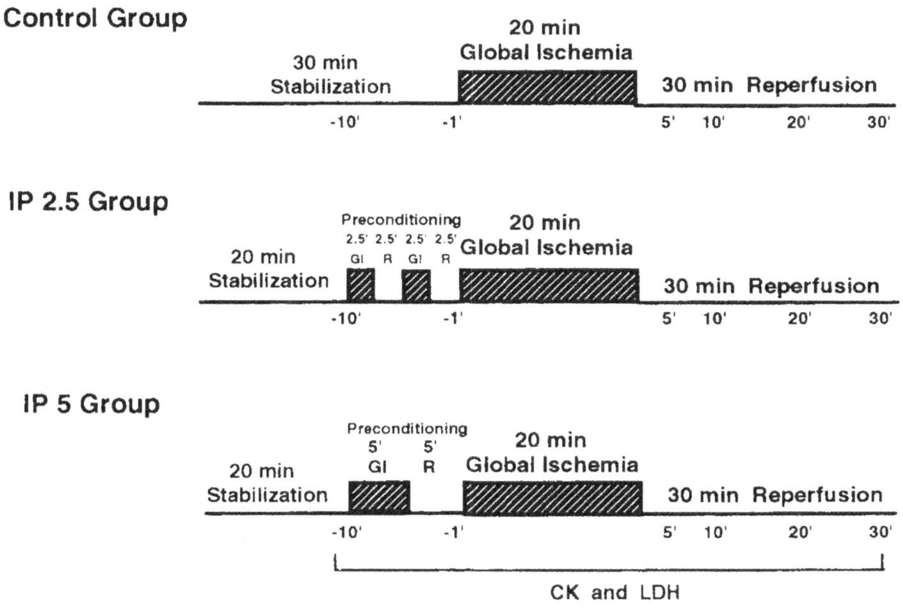

Fig. 1. Experimental protocol of ischemia/reperfusion in isolated perfused mouse hearts. Hearts were randomized into three groups (n = 10/each group): (1) Control: no preconditioning; (2) IP2.5: preconditioning with two cycles of 2.5 min global ischemia and 2.5 min reperfusion; (3) IP5: preconditioning with one cycle of 5 min global ischemia and 5 min reperfusion. All hearts were subjected to 20 min global ischemia and 30 min reperfusion. Cardiac function and leackage of intracellular enzymes (CK and LDH) were determined at the indicated time points. Myocardial infarct size was measured at the end of 30 min reperfusion.

Three hearts (i.e. 9% of the total 33 perfused mouse hearts) were excluded from further data analysis due to the presence of any of the following undesirable situations: (1) a significant time delay in aortic cannulation; or (2) damage of aorta during the process of cannulation; and (3) sustained arrhythmia during the 20 min stabilization period.

Estimation of creatine kinase (CK) and lactate dehydrogenase (LDH) release

Coronary effluent was collected from the isolated perfused mouse hearts at the end of a 20 min stabilization period; 1 min prior to the 20 min sustained global ischemia as well as 5, 10, 20, 30 min during the reperfusion. The LDH and CK measurements were done spectrophotometrically (Shimadzu, Model UV160U) using kits supplied by Sigma Chemical Co. (St. Louis, MO). The activities of enzymes were expressed as units per liter and also normalized against coronary flow rate and heart weight (U/min/g wet weight).

Measurement of myocardial infarct size

At the end of ischemia/reperfusion, the heart was removed from the Langendorff perfusion apparatus and immediately frozen and stored in a freezer overnight. Next morning, the frozen hearts were cut from apex to base into 8 transverse slices of approximately equal thickness (~0.8 mm). The slices were placed into a small cell culture dish and then incubated in 10% triphenyl-tetrazolium chloride (TTC) in phosphate buffer (Na_2HPO_4 88 mM, NaH_2PO_4 1.8 mM, pH 7.8) at room temperature for 30 min. The development of the red formazan pigment in living tissues relies on the presence of lactate dehydrogenase or NADH while failure to stain red indicates a loss of these constituents from necrotic tissue. After staining, the TTC buffer was replaced by 10% formaldehyde, the slices were fixed for the next 4–6 h before the areas of infarct tissue were determined by computer morphometry (Bioquant System IV). The risk area was the sum of total ventricular area minus cavities. The infarct size was calculated and presented as percentage of risk area.

Data analysis and statistics

All measurements are expressed as mean ± S.E.M. Data from the three experimental groups were analyzed by one-way ANOVA. If a significant value of F was obtained, the Student-Newman-Keuls test was used for pair-wise comparisons. Paired *t*-test was used to compare any pair of pre- and post-treatment values. A probability value of < 0.05 was considered significant.

Results

Morphometric and the pre-ischemia baseline values of cardiac hemodynamic and contractile parameters from the three experimental groups are summarized in Table 1. There was no significant difference among the groups for any of the parameters.

Ventricular contractile function

The time course of ventricular developed force and rate-force product during pre-ischemic as well as reperfusion period are shown in Fig. 2A and B. The repeated brief ischemia/reperfusion episodes (IP2.5) caused a consistent overshoot in cardiac contractile function as indicated by a significant increase in ventricular developed force ($p < 0.05$). On the other hand, a single cycle of IP (IP5) did not significantly alter the ventricular function. Following the 20 min sustained global ischemia, the ventricular developed force significantly decreased in the control and IP groups at 5 min after reperfusion (0.13 ± 0.05, 0.17 ± 0.08, 0.12 ± 0.05 g in control, IP2.5, and IP5, respectively, Fig. 2A). The contractile function was not significantly improved at any time point during the rest of the reperfusion period. A similar trend in the changes in the rate-force product was also observed (Fig. 2B). No significant differences in both resting tension (Fig. 3A) and heart rate (Fig. 3B) were found between control and IP groups at any time point throughout the experimental protocol ($p > 0.05$, n = 10/each group).

Table 1. Morphometric characters of the mice and the baseline values of hemodynamic and contractile parameters of the isolated perfused mouse hearts at the end of 20 min stabilizing period.

Experimental Groups	Control	IP 2.5	IP 5
Body Weight (g)	31.0 ± 1.1	33.3 ± 1.3	32.7 ± 0.9
Heart Wet Weight (mg)	248 ± 10	264 ± 11	265 ± 7
Heart Rate (bpm)	387 ± 14	351 ± 14	387 ± 10
Coronary Flow (ml/min)	1.93 ± 0.21	1.56 ± 0.16	1.65 ± 0.24
Resting Tension (g)	0.21 ± 0.03	0.17 ± 0.02	0.18 ± 0.02
Developed Force (g)	0.48 ± 0.10	0.51 ± 0.04	0.60 ± 0.06
Rate-Force Product (g x bpm)	186 ± 39	179 ± 18	231 ± 22

Values are mean ± S.E.M. (n = 10 / each group). IP 2.5 = mouse heart subjected to two cycles of 2.5 min ischemia and 2.5 min reperfusion; IP 5 = mouse heart subjected to 5 min ischemia and 5 min reperfusion. No significant difference was found among the experimental groups for any of the parameters.

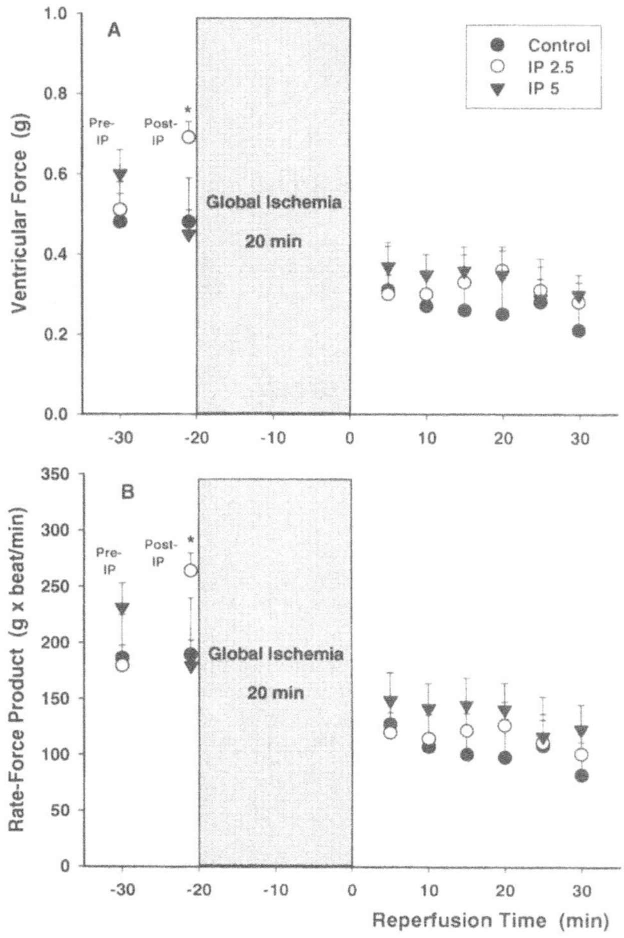

Fig. 2. Effect of ischemic preconditioning (IP) on ventricular contractile function of the isolated perfused mouse hearts before and after IP as well as the subsequent 20 min no-flow global ischemia and reperfusion. (A) time course of ventricular developed force, and (B) time course of rate-force product. Despite a consistent functional overshoot observed after IP in IP2.5 group (* means $p < 0.05$), no significant difference for both functional parameters was observed between control and IP groups at any time point during the 30 min reperfusion period ($p > 0.05$, n = 10/each group).

Intracellular enzymes (CK and LDH) leakage

Figure 4 shows the time course of CK (A) and LDH (B) release in the coronary effluent of isolated hearts during pre-IP, post-IP and following 20 min ischemia and 30 min of reperfusion period. Pre-IP CK values (U/L) were 4.9 ± 1.2, 5.1 ± 0.8, 8.8 ± 2.2 in control, IP2.5, and IP5 respectively which were not significantly different from the post-IP values (3.5 ± 0.6, 4.6 ± 1.6, 4.7 ± 1.2, respectively, $p > 0.05$). At 5 min of reperfusion, CK efflux increased significantly in the control group to 129.6 ± 50.2, but not in both IP groups (36.5 ± 15.7 for IP2.5; 16.3 ± 4.4 for IP5). Although the mean values of the enzyme efflux were remarkably higher in the control group (Fig. 4A) as

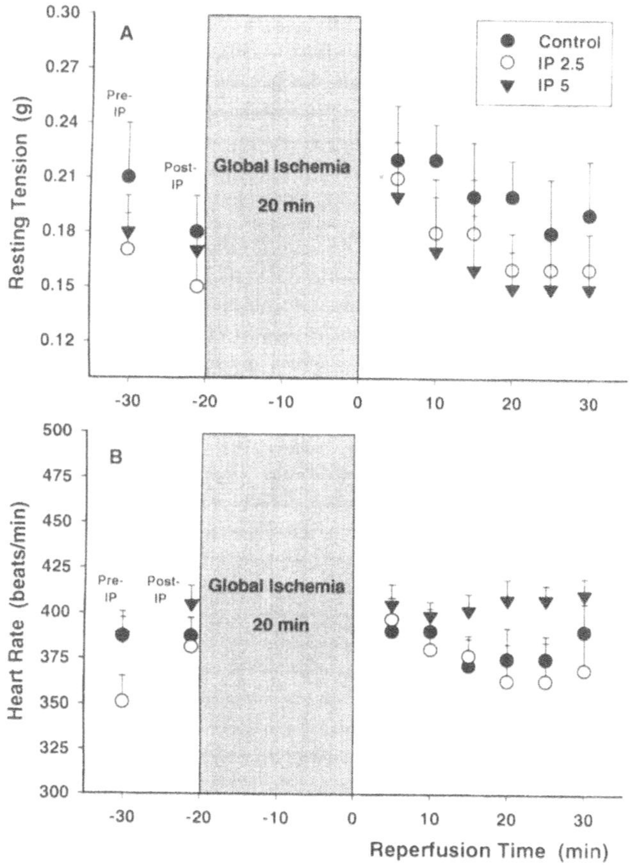

Fig. 3. Effects of ischemic preconditioning (IP) on resting tension (A) and heart rate (B) before and after 20 min no-flow global ischemia (IS). No significant difference for both parameters was found between Control and IP groups at any time point ($p > 0.05$, n = 10/each group).

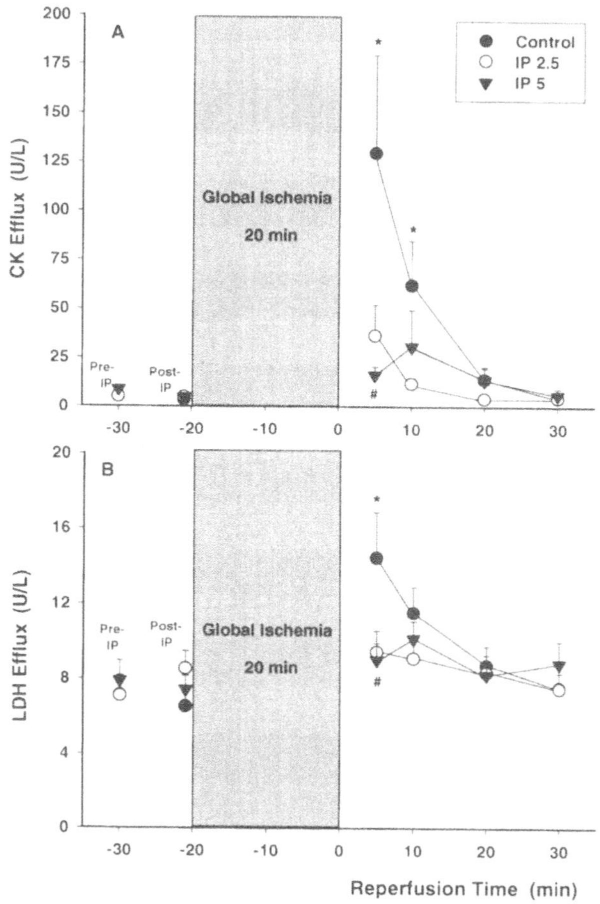

Fig. 4. Leakage of intracellular enzymes CK (A) and LDH (B) in coronary efflux from the isolated perfused mouse hearts subject to 20 min global ischemia and 30 min reperfusion. * indicates a significantly higher enzyme release as compared to pre-IP basal value and # indicates a significantly lower enzyme release as compared to control group (mean ± S.E.M., $p < 0.05$, n = 10/each group) at the time point of 5 min reperfusion.

compared with the IP groups, the difference failed to reach the statistical significance between IP2.5 and control groups due to the high variability within the group. Such a difference between control and IP groups was also observed in the release of LDH (Fig. 4B). A similar trend in the release of CK (A) and LDH (B) release was found when the values were expressed in terms of wet heart weight (Fig. 5).

The pre-ischemic baseline of coronary flow was 1.93 ± 0.21, 1.56 ± 0.16, 1.65 ± 0.24 (ml/min) for control, IP2.5, IP5 respectively. At the end of reperfusion, the level of coronary flow decreased to 1.62 ± 0.16 in control group, but increased to 1.93 ± 0.21 in IP2.5 and 1.98 ± 0.24 in IP5, although these changes were not statistically significant ($p > 0.05$).

Myocardial infarct size

Figure 6 presents the average infarct sizes for the three experimental groups. The mean values of the area at risk were not significantly different between the three groups. After 20 min global ischemia and 30 min reperfusion, a substantial amount of the ventricular muscle was irreversibly damaged in the control and IP5 hearts, but much less in IP2.5 hearts. The pale color infarct zone was predominantly seen in endocardial area. The infarct size calculated as percentage of the risk area was 23.6 ± 3.8% in control, reduced significantly to 13.8 ± 2.3% in IP2.5 group ($p < 0.05$). In contrast, the infarct size in IP5 was 20.1 ± 4.0% which was not significantly different from the control hearts.

Discussion

The major findings of the present study are summarized as follows: (1) the phenomenon of 'classic' or the early phase

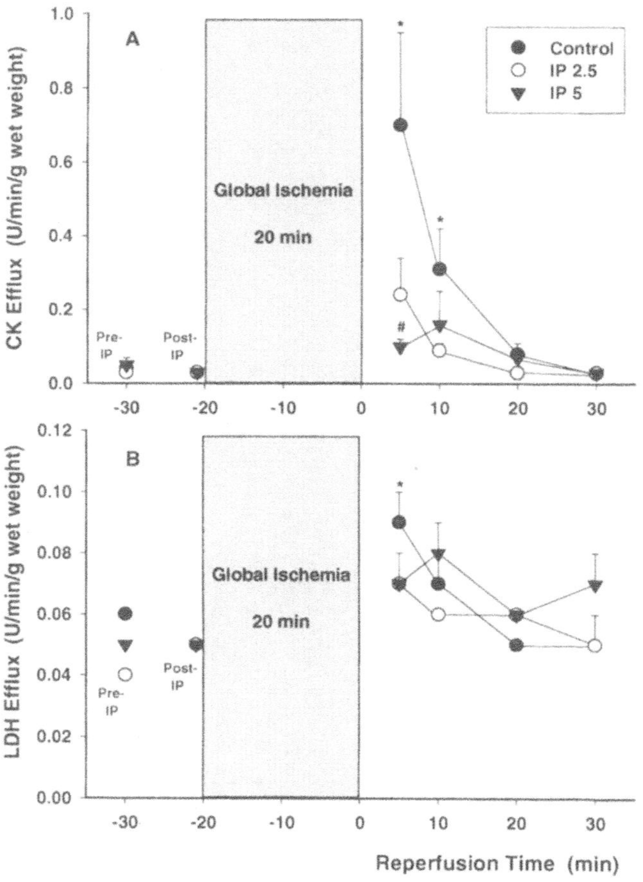

Fig. 5. Leakage of intracellular enzymes CK (A) and LDH (B) in coronary efflux. Values are normalized as unit per min per gram of heart wet weight. * indicates a significantly higher enzyme release as compared to pre-IP basal value and # indicates a significantly lower enzyme release as compared to control group (mean ± S.E.M., $p < 0.05$, n = 10/each group) at the time point of 5 min reperfusion.

ischemic preconditioning does exist in the isolated mouse heart despite its species-specific energetic such as high heart rate and lack of inhibition of mitochondria ATPase; (2) two cycles of IP with 2.5 min ischemia and 2.5 min reperfusion are necessary to induce the anti-infarct cardioprotection in the mouse hearts; (3) the IP-induced myocardial stunning was not a key element in the induction of cardioprotection in the isolated mouse hearts; (4) The anti-necrosis cardioprotection induced by IP was not associated with the amelioration of post-ischemic ventricular dysfunction.

Post-ischemic ventricular contractile function

It remains controversial whether the early phase of ischemic preconditioning affords significant beneficial effects against myocardial dysfunction. In *in situ* model, some investigators demonstrated IP-induced improvement of myocardial wall motion associated with a smaller infarct size following transient coronary occlusion in rabbits [3], whereas Ovize *et al.* [30] showed that regional myocardial contractility was not preserved in the preconditioned canine hearts subjected to a 60 min regional ischemia. On the other hand, in isolated perfused heart model, a number of investigators [14, 15, 17, 20, 31] have shown a dramatic improvement of the ventricular contractile function afforded by IP in rat or rabbit hearts following a sustained global ischemia. However, many other researchers [10, 11, 16, 19, 32] have failed to observe the similar functional improvement in a comparable experimental model and protocol. In the present study, we did not observe a significant difference in ventricular developed force, heart rate, rate-force product, and resting tension between control and preconditioned hearts (Figs 2 and 3) throughout the 30 min reperfusion period. The lack of IP-induced improvement of post-ischemic ventricular function is in accordance with the preliminary reports in globally ischemic isolated mouse hearts which also showed no or very limited improvement in the ventricular function afforded by IP [33, 34].

Infarct size and intracellular enzymes leakage

The powerful anti-infarct effects afforded by ischemic preconditoning have been consistently found by numerous investigators using different experimental models in several species (see Introduction for details). The present study confirms that the phenomenon of 'classic' or the early phase ischemic preconditioning (IP) does exist in the isolated mouse heart. The percentage of infarct size reduction (41.5%) observed in the present study is very similar to the preliminary reports by Sumeray and Yellon [34] who demonstrated a 42.1% reduction in infarct size in the preconditioned mouse hearts after 30 min global ischemia and 30 min reperfusion. A similar preliminary report by Gabel *et al.* [33] demonstrated that IP significantly attenuated the drop of pH in the isolated perfused mouse myocardium. We believe that the attenuation of tissue acidosis may be accountable at least partially to the infarct size reduction observed in the present study. Similarly, leakage of intracellular enzymes (CK and LDH) was remarkably decreased in both IP groups compared with control group (Figs 4 and 5), providing additional evidence of the cardioprotective effects induced by IP. The increase in post-ischemic coronary flow in the preconditioned hearts may also play a beneficial role in maintaining an adequate perfusion condition to the myocardium. In the present study, IP with two cycles of I/R (IP2.5) was able to reduce myocardial infarct size significantly, whereas a single cycle of I/R (IP5) failed to induce a significant anti-infarct cardioprotection. It is possible that shorter ischemic episodes of preconditioning

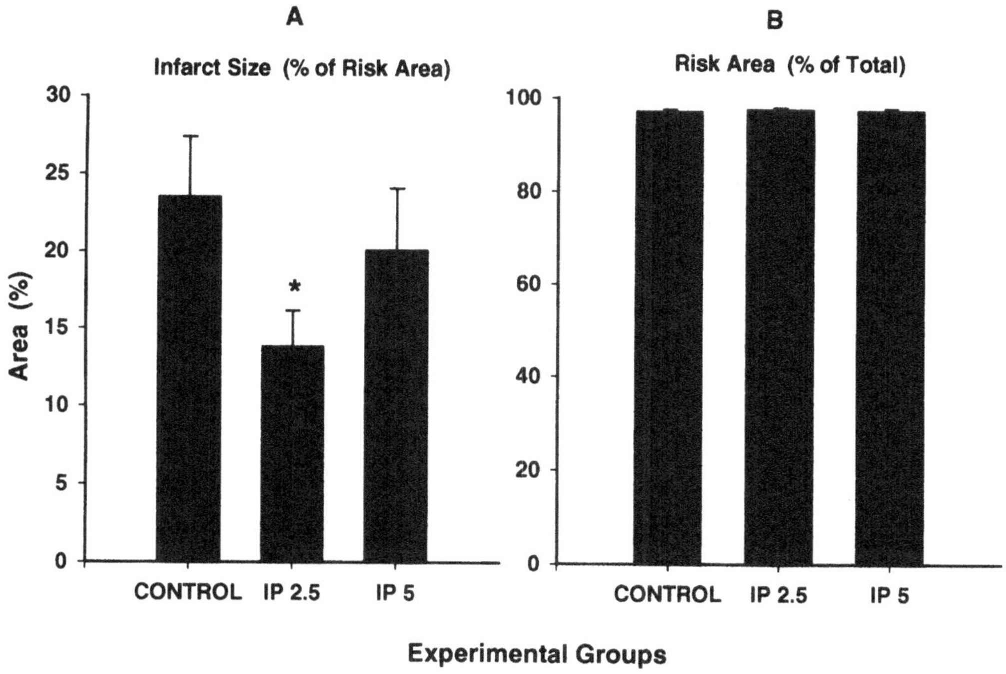

Fig. 6. Effects of ischemic preconditioning on infarct size (mean ± S.E.M.) in the isolated perfused mouse myocardium after 20 min no-flow global ischemia and 30 reperfusion. There is no difference between any pair of the experimental groups (n = 10/each group) for Risk Area (% of Total Area) (see B). However, the Infarct Size (% of Risk Area) was significantly reduced in IP2.5 (* means $p < 0.05$) but not in IP5 group as compared with control group (see A).

are more effective in triggering the signal transduction cascade in the mouse heart as compared to other species. Since brief episodes of ischemia and reperfusion may elicit bursts of oxygen-derived free radicals, it is reasonable to speculate that they may serve as important stimulus activating certain endogenous antioxidant cardioprotective mechanisms, although further investigations should be done in order to prove such a hypothesis. This finding is also in agreement with the previous reports demonstrating cardioprotective effects after 3-cycle of IP but not 1-cycle of IP in the *in situ* rat hearts [13]. However, this was in contrast with the previous studies from our laboratory where we were able to precondition rat heart after a single bout of 5 min ischemia and 10 min reperfusion *in vivo* [21, 35].

Myocardial stunning and preconditioning

Brief episodes of ischemia/reperfusion used in IP could cause myocardial stunning in some experimental models. However, it is controversial whether such an IP-induced myocardial stunning is a key element responsible for the cardioprotection afforded by IP against the subsequent sustained ischemia. Evidence has been reported to either prove [6, 36] or refute [37, 38] a role played by stunning in the phenomenon of ischemic preconditioning.

In the present study, we did not observe significant stunning following either one cycle of 5 min ischemia and 5 min reperfusion (IP5) or two cycles of 2.5 min ischemia and 2.5 min reperfusion (IP2.5). Therefore, it appears that myocardial stunning does not play a role in reducing infarct size in the preconditioned hearts. In contrast, to our surprise, IP2.5 caused a consistent overshoot in cardiac contractile function as indicated by the increase of ventricular developed force (Fig. 2). We do not know mechanisms inducing the up-regulation of ventricular function could also promote the anti-infarct effects observed in the IP2.5 group.

Critique of methodology

In the present study, we used a Langendorff mode isolated perfused mouse heart model which is similar to the documented studies in transgenic [22–24, 39, 40] or normal [34, 41] mice. We perfused the murine hearts at a constant pressure of 55 mmHg which was within the physiological range of 50–55 mmHg suggested by Ng *et al.* [41] for perfusing the mouse heart. This perfusion pressure is similar to that used by Yoshida *et al.* [39] but different from Marber *et al.* [22] who perfused mouse hearts at 80 mmHg. Although most studies successfully demonstrated a reduction in infarct size and/or improved functional recovery following ischemia/reperfusion in transgenic animals [22, 39, 40], the magnitude of these parameters were highly inconsistent among various studies. For example, the infarct size in the globally ischemic mouse hearts ranged from ~8 to ~60% following a 20 or 30 min ischemia [22, 28, 34, 39, 40]. Clearly, there are substantial differences in the experimental approaches and conditions among various research groups and further studies are needed to resolve these issues.

Species-specific myocardial energetics

We further speculate that the increased susceptibility of mouse heart to ischemia may contribute in part to the following three species-specific myocardial energetic factors: (1) a significantly elevated total body oxygen consumption (4 folds > human, dog, pig) and myocardial oxidative capacity (2 folds > human, dog, pig) [25]; (2) a lack of inhibition of mitochondria adenosine triphosphatase (ATPase), an enzyme which consumes 35–50% of ATP during myocardial ischemia [27, 42]; (3) a higher level of xanthine oxydoreductase, an enzyme which is the source for oxygen free radicals during ischemia/reperfusion injury in the cardiac muscle [26, 43, 44]. Liu and Downey [13] were the first group to report the IP phenomenon in rat - a species with a fast beating heart similar to mouse. They ruled out an obligatory role played by ATPase in IP and proposed an increased threshold of IP in rat hearts.

Conclusion

The present study has confirmed the existence of ischemic preconditioning in the normal mouse heart. This anti-necrosis cardioprotection induced by ischemic preconditioning was not associated with the amelioration of post-ischemic ventricular dysfunction. To date, the exact mechanism of the classic IP remains elusive, although a number of receptors, end-effectors, or mediating pathways have been suggested to be related to the signal transduction cascade including: (1) Adenosine receptors [7, 17, 20]; (2) α-adrenergic receptors [9, 18]; (3) Protein kinase C pathway [8, 9]; (4) ATP-sensitive potassium channels [4, 21]; and more recently (5) tyrosine phosphorylation [45–47]. Further studies are necessary to demonstrate if the mechanism(s) of preconditioning in the mouse heart is similar to those reported in other species. Studies on transgenic or knockout mice would eventually help in dissecting the specific target genes that lead to cardioprotective effects of preconditioning. Knowledge of the mechanisms involved in the preconditioning in mouse heart may help in developing some clinically relevant therapeutic and preventive approaches.

Acknowledgements

This research was supported in part by NHLBI grant HL-51045 (to R.C. Kukreja); L. Xi was supported by a postdoctoral fellowship from NIH training grants (HL-07537 & HL-07580). The authors are grateful to G.P. Matherne and J.P. Headrick at University of Virginia for their expert advice on the isolated perfused mouse heart model; to J.E. Levasseur and Y-Z Qian for their technical assistance.

References

1. Murry CE, Jennings RB, Reimer KA: Preconditioning with ischemia: A delay of lethal cell injury in ischemic myocardium. Circulation 74: 1124–1136, 1986
2. Li GC, Vasquez JA, Gallagher KP, Lucchesi BR: Myocardial protection with preconditioning. Circulation 82: 609–619, 1990
3. Cohen M, Liu G, Downey J: Preconditioning causes improved wall motion as well as smaller infarcts after transient coronary occlusion in rabbits. Circulation 84: 341–349, 1991
4. Gross GJ, Auchampach JA: Blockade of ATP-sensitive potassium channels prevents myocardial preconditioning in dogs. Circ Res 70: 223–233, 1992
5. Iwamoto T, Bai X-J, Downey HF: Preconditioning with supply-demand imbalance limits infarct size in dog heart. Cardiovasc Res 27: 2071–2076, 1993
6. Schott RJ, Rohmann S, Braun ER, Schaper W: Ischemic preconditioning reduces infarct size in swine myocardium. Circ Res 66: 1133–1142, 1990
7. Liu GS, Thornton J, Van Winkle DM, Stanley AWH, Olsson RA, Downey JM: Protection against infarction afforded by preconditioning is mediated by A_1 adenosine receptors in rabbit heart. Circulation 84: 350–356, 1991
8. Ytrehus K, Liu Y, Downey JM: Preconditioning protects ischemic rabbit heart by protein kinase C activation. Am J Physiol 266: H1145–1152, 1994
9. Tsuchida A, Liu Y, Liu GS, Cohen MV, Downey JM: α_1-adrenergic agonists precondition rabbit ischemic myocardium independent of adenosine by direct activation of protein kinase C. Circ Res 75: 576–585, 1994
10. Sandhu R, Diaz RJ, Wilson GJ: Comparison of ischemic preconditioning in blood perfused and buffer perfused isolated heart models. Cardiovasc Res 27: 602–607, 1993
11. Jenkins DP, Pugsley WB, Yellon DM: Ischaemic preconditioning in a model of global ischemia: infarct size limitation, but no reduction of stunning. J Mol Cell Cardiol 27: 1623–1632, 1995
12. Baines CP, Goto M, Downey JM: Oxygen radicals released during ischemic preconditioning contribute to cardioprotection in the rabbit myocardium. J Mol Cell Cardiol 29: 207–216, 1997
13. Liu Y, Downey JM: Ischemic preconditioning protects against infarction in rat heart. Am J Physiol 263: H1107–H1112, 1992
14. Cave AC, Hearse DJ: Ischemic preconditioning and contractile function: Studies with normothermic and hypothermic global ischemia. J Mol Cell Cardiol 24: 1113–1123, 1992
15. Volovsek A, Subramanian R, Rebousin D: Effects of duration of ischemia during preconditioning on mechanical function, enzyme release and energy production in the isolated working rat heart. J Mol Cell Cardiol 24: 1011–1019, 1992

16. Das DK, Engelman RM, Kimura Y: Molecular adaptation of cellular defences following preconditioning of the heart by repeated ischemia. Cardiovasc Res 27: 578–584, 1993
17. Lasley RD, Anderson GM, Mentzer RM Jr: Ischemic and hypoxic preconditioning enhance postischemic recovery of function in the rat heart. Cardiovasc Res 27: 565–570, 1993
18. Banerjee A, Locke-Winter C, Rogers KB, Mitchell MB, Brew EC, Cairns CB, Bensard DD, Harken AH: Preconditioning against myocardial dysfunction after ischemia and reperfusion by an α_1-adrenergic mechanism. Circ Res 73: 656–670, 1993
19. Arad M, de Jong JW, de Jong R, Huizer T, Rabinowitz B: Preconditioning in globally ischemic isolated rat hearts: Effect on function and metabolic indices of myocardial damage. J Mol Cell Cardiol 28: 2479–2490, 1996
20. Headrick JP: Ischemic preconditioning: bioenergetic and metabolic changes and the role of endogenous adenosine. J Mol Cell Cardiol 28: 1227–1240, 1996
21. Qian Y-Z, Levasseur J, Hess ML, Kukreja RC: K_{ATP} channel in the rat heart: blockade of ischemic and acetylcholine preconditioning by glibenclamide. Am J Physiol 271: H23–H28, 1996
22. Marber MS, Mestril R, Chi S-H, Sayen MR, Yellon DM, Dillmann WH: Overexpression of the rat inducible 70-kD heat stress protein in a transgenic mouse increases the resistance of the heart to ischemic injury. J Clin Invest 95: 1446–1456, 1995
23. Plumier J-CL, Ross BM, Currie RW, Angelidis CE, Kazlaris H, Pagoulatos GN: Transgenic mice expressing the human heat shock protein 70 have improved post-ischemic myocardial recovery. J Clin Invest 95: 1854–1860, 1995
24. Matherne GP, Linden J, Byford AM, Gauthier NS, Headrick JP: Transgenic A1 adenosine receptor overexpression increases myocardial resistance to ischemia. Proc Natl Acad Sci USA 94: 6541–6546, 1997
25. Barth E, Stammler G, Speiser B, Schaper J: Ultrastructural quantitation of mitochondria and myofilaments in cardiac muscle from 10 different animal species including man. J Mol Cell Cardiol 24: 669–681, 1992
26. de Jong JW, van der Meer P, Nieukoop AS, Huizer T, Stroeve RJ, Bos E: Xanthine oxidoreductase activity in perfused hearts of various species, including humans. Circ Res 67: 770–773, 1990
27. Rouslin W: The mitochondrial adenosine 5′-triphosphatase in slow and fast heart rate hearts. Am J Physiol 252: H622–H627, 1987
28. Xi L, Chelliah J, Nayeem MA, Levasseur JE, Kukreja RC: Whole body heat shock fails to protect mouse heart against ischemia/reperfusion injury in an isolated perfused heart model. J Mol Cell Cardiol 29: A213; 235 (abstract), 1997
29. Chelliah J, Xi L, Okubo S, Kukreja RC: Expression of HSP72 and antioxidant enzymes in mouse heart following whole body hyperthermia. J Mol Cell Cardiol 29: A194; 159 (abstract), 1997
30. Ovize M, Przylenk K, Hale SL, Kloner RA: Preconditioning does not attenuate myocardial stunning. Circulation 85: 2247–2254, 1992
31. Asimakis GK, Inners-McBride K, Medellin G, Conti VR: Ischemic preconditioning attenuates acidosis and postischemic dysfunction in isolated rat heart. Am J Physiol 263: H887–H894, 1992
32. Bolling SF, Olzanski DC, Childs KF, Gallagher KP: Does cardiac ‘preconditioning’ result in enhanced postischemic functional recovery? Surg Forum 42: 239–242, 1991
33. Gabel SA, Steenbergen C, London R, Murphy E: Preconditioning (PC) in perfused mouse hearts. J Mol Cell Cardiol 29: A229; 297 (abstract), 1997
34. Sumeray MS, Yellon DM: Ischaemic preconditioning reduces infarct size following global ischemia in the murine myocardium. J Mol Cell Cardiol 29: A72; Th129 (abstract), 1997
35. Schultz JEJ, Qian Y-Z, Gross GJ, Kukreja RC: The ischaemia-selective K-ATP channel antagonist, 5-hydroxydecanoate, blocks ischemic preconditioning in the rat heart. J Mol Cell Cardiol 29: 1055–1060, 1997
36. Cargnoni A, Ceconi C, Bernocchi P, Pasini E, Curello S, Ferrari R: Is stunning an important component of preconditioning? J Mol Cell Cardiol 28: 2323–2331, 1996
37. Miura T, Goto M, Urabe K, Endoh A, Shimamoto K, Iimura O: Does myocardial stunning contribute to infarct size limitation by ischemic preconditioning? Circulation 84: 2504–2512, 1991
38. Matsuda M, Catena TG, Vander Heide RS, Jennings RB, Reimer KA: Cardiac protection by ischemic preconditioning is not mediated by myocardial stunning. Cardiovasc Res 27: 585–592, 1993
39. Yoshida T, Watanabe M, Engelman DT, Engelman RM, Schley JA, Maulik N, Ho Y-S, Oberley TD, Das DK: Transgenic mice over-expressing glutathione peroxidase are resistant to myocardial ischemia reperfusion injury. J Mol Cell Cardiol 28: 1759–1767, 1996
40. Wang P, Wong PC, Sankarapandi S, Qin H, Chacko VP, Zweier JL: Elevated intracellular SOD expression in transgenic mice prevents myocardial reperfusion injury. Circulation 94(8): I-279,1625 (abstract), 1996
41. Ng WA, Grupp IL, Subramaniam A, Robbins J: Cardiac myosin heavy chain mRNA expression and myocardial function in the mouse heart. Circ Res 69: 1742–1750, 1991
42. Jennings RB, Reimer KA, Steenbergen C Jr: Effect of inhibtion of the mitochondrial ATPase on net myocardial ATP in total ischemia. J Mol Cell Cardiol 23: 1383–1395, 1991
43. Eddy LJ, Stewart JR, Jones HP, Engerson TD, McCord JM, Downey JM: Free radical-producing enzyme, xanthine oxidase, is undetectable in human hearts. Am J Physiol 253: H709–H711, 1987
44. Kukreja RC, Hess ML: The oxygen free radical system: From equations through membrane protein-interactions to cardiovascular injury and protection. Cardiovasc Res 26: 641–655, 1992
45. Maulik N, Watanabe M, Zu Y-L, Huang CK, Cordis GA, Schley JA, Das DK: Ischemic preconditioning triggers the activation of MAP kinase and MAPKAP kinase 2 in rat hearts. FEBS Lett 396: 233–237, 1996
46. Baines CP, Cohen MV, Downey JM: Protein tyrosine kinase inhibitor, genistein, blocks preconditioning in isolated rabbit hearts. Circulation 94: I-661; 3860 (abstract), 1996
47. Kukreja RC, Qian Y-Z: Tyrosine kinase pathway is involved in ischemic preconditioing in rat heart. J Mol Cell Cardiol 29: A230; 302 (abstract), 1997

Molecular and Cellular Biochemistry **186**: 79–86, 1998.
© 1998 *Kluwer Academic Publishers.*

Differential electrophysiologic effects of global and regional ischemia and reperfusion in perfused rat hearts. Effects of Mg^{2+} concentration

Amira Ponce Zumino,[1] Gustavo Baiardi,[1] Otto F. Schanne[2] and Elena Ruiz Petrich[2]

[1]*Department of Physiology, Faculty of Medical Sciences, University of Cuyo, Mendoza, Argentina;* [2]*Département de physiologie et biophysique, Faculté de médecine, Université de Sherbrooke, Sherbrooke, Québec, Canada*

Abstract

The effects of regional and global ischemia on cellular electrical activity and on arrhythmias induced by reperfusion were stud ied at different Mg^{2+} concentrations (Mg^{2+}_o, 0, 1.2, and 4.8 mM) in perfused rat hearts. Surface electrograms and transmembrane potentials were recorded during control, 10 min of ischemia (perfusion arrest or coronary ligation), and reperfusion. Increasing Mg^{2+}_o from 0–4.8 mM decreased heart rate, did not alter action potential morphology, and had a strong antiarrhythmic action on reperfusion following coronary ligation. At low and normal Mg^{2+}_o, the incidence of tachyarrhythmias was between 70 and 80%. Global ischemia led to progressive atrioventricular block and the final ventricular beating rate was similar at all Mg^{2+}_o despite unequal initial values. The severity of arrhythmias was similar to that found after regional ischemia in $Mg^{2+}_o = 0$, but much lower at normal and high Mg^{2+}_o. The resting depolarization induced by coronary ligation decreased as Mg^{2+}_o was raised, but such a relation was not seen during global ischemia where the depolarization was less marked. The action potential duration did not vary with the ventricular rate between 160 and 380 beats per min but increased considerably when sinus rate was markedly slowed (40 to 80 bpm) by raising Mg^{2+}_o to 9.6 mM. Our data show that a high Mg^{2+}_o exerts a strong protection against reperfusion arrhythmias regardless of the type of ischemia. Modulation of the sinus rhythm by Mg^{2+} may contribute to its protective effect by decreasing K^+_o accumulation and Na^+_i loading during ischemia. (Mol Cell Biochem **186**: 79–86, 1998)

Key words: reperfusion arrhythmias, global vs. regional ischemia, action potential, Mg^{2+}_o

Introduction

It is known that reperfusion of ischemic myocardium results in cell injury whose degree and reversibility are related to the severity of the preceding ischemic insult. Ventricular tachyarrhythmias appearing early on reflow are one of the manifestations of reperfusion injury. Although of multifactorial nature, most of the studies indicate that these arrhythmias belong to a non-reentrant, Ca^{2+} dependent type [1]. A transient increase in cellular Na^+ concentration (Na^+_i) leading to a cytosolic Ca^{2+} overload is postulated as a major mechanism of several manifestations of reperfusion injury [2, 3]. The Ca^{2+} overload triggers spontaneous release of Ca^{2+} from the sarcoplasmic reticulum [4] which in turn induces transient inward currents underlying delayed afterdepolarizations [5]. Reactivation of Na-H exchange by reperfusion has been signalled as a paramount factor in this sequence of events which is attenuated or abolished by amiloride and its analogues [2, 3]. However, we have shown that changes in action potential characteristics that may decrease Na^+ entry also exert a protective role independently of inhibition of Na-H exchange [6]. Moreover, increasing Mg^{2+} concentration (Mg^{2+}_o) in the perfusate also attenuates arrhythmias appearing on release of coronary ligation without altering cellular electrical activity, although the ischemic depolarization of the resting potential is reduced [7]. In

Address for offprints: A. Ponce Zumino, Cátedra de Fisiología, Facultad de C. Médicas, Universidad Nacional de Cuyo, CC 33, 5500 Mendoza, Argentina

addition, high Mg^{2+}_o decreases the heart rate [7] which would indirectly alleviate Na^+ loading because the fast inward current is a major source of Na^+ entry into the cell. This interpretation is consistent with the observation that the incidence and severity of reperfusion arrhythmias varies with the stimulation rate [8].

In order to further assess the importance of the beating rate in the protection exerted by Mg^{2+} we compared the effects of global to those of regional ischemia on the cellular electrical activity and on the arrhythmias appearing during reflow after both conditions. This approach was chosen because global ischemia induces atrioventricular conduction block of different degrees and therefore markedly slows the ventricular beating rate.

Materials and methods

Wistar rats (280–300 g) were stunned by a blow on the head. The hearts were rapidly removed, trimmed of pericardium and connective tissue in cold, oxygenated perfusion solution, and then perfused with the Langendorff technique (temperature between 35 and 36°C) at spontaneous rates of beating. To avoid temperature changes, the hearts were kept in a double walled tissue chamber superfused with the control solution at the same temperature as the coronary perfusate. The control solution contained (mM) 121 NaCl, 5 KCl, 2 $CaCl_2$, 1.2 $MgSO_4$, 1.2 NaH_2PO_4, 25 $NaHCO_3$, and 5 glucose. When equilibrated with 5% CO_2 in O_2, the pH was 7.4 ± 0.03. The Mg^{2+} concentrations tested were 0, 1.2 and 4.8 mM without correction for osmolarity and the same concentration was maintained throughout the experiment.

The surface electrogram was continuously recorded with two electrodes placed on the right atrium and near the apex of the heart (like in the conventional standard II lead) with a Sanborn Twin Viso electrocardiograph. The ECG was also synchronously displayed with the transmembrane potential recorded from subepicardial cells on the left ventricle with flexibly mounted glass microelectrodes filled with 3 M KCl. The electrode resistance was between 12 and 20 MΩ. The recording equipment consisted of a custom made micro-electrode amplifier, a Tektronix 565 oscilloscope, and a C4 Grass camera.

The experimental protocol consisted of a 30 min equilibration followed by a 15 min control period during which the stability of membrane potential recordings was assessed. In the experiments with regional ischemia, the left anterior descending coronary artery was ligated with a 6/0 silk thread, 1–2 mm below the point where it emerges from under the left atrial appendage. The thread was tied into a loop so that the ligation could later on be released without further manipulation of the artery. Global ischemia was performed by arresting the perfusion. Regional or global ischemia was maintained for 10 min and was followed by reperfusion. Under our experimental conditions, the incidence of arrhythmias on reperfusion is highest after 10 min of coronary occlusion [6], in agreement with other observations [9]. The ECG and the membrane potential were recorded throughout the different experimental periods: control, ischemia, and for 10 min during reperfusion. During coronary occlusion, the ischemic region was identified by addition of cardiogreen to the perfusate.

The heart rate was determined in the ECG from the number of QRS complexes occurring over one second. These values were corroborated by measuring the cycle length between two successive action potential upstrokes. The ECG was also used for continuous monitoring of cardiac electrical activity and the determination of incidence and duration of arrhythmic episodes. The type of arrhythmia (premature ventricular beats, PVB; ventricular tachycardia, VT, and ventricular fibrillation, VF) was established according to the Lambeth conventions [10] and based not only on the ECG but on intracellular recordings as well. None of the hearts included in this study presented overt ventricular tachyarrhythmias during the control period or ischemia. The following action potential characteristics were measured in the intracellular recordings: resting potential (RP), action potential amplitude (APA), and action potential duration determined from the onset of depolarization to 90% repolarization (APD_{90}).

Data are expressed as means ± S.E. from n hearts. The individual data entering these calculations were means from pooled values recorded during any given experimental period. The results were analyzed with ordinary one-way ANOVA followed by the Duncan post test. The influence of Mg^{2+}_o on the incidence of arrhythmias was tested with contingency tables using the chi-square test. The statistical significance level was $p < 0.05$.

Results

Two series of experiments were performed to compare the effects of regional (RI) versus global ischemia (GI) on action potential characteristics and incidence of arrhythmias on reperfusion after the ischemic episode. In each series, the effects of varying Mg^{2+}_o between zero and 4.8 mM were studied. This resulted in 6 groups of experiments consisting of 10 hearts each. In addition, 9.6 mM Mg^{2+} was used in a 7th group submitted to global ischemia to further explore the changes in action potential characteristics at very low rates of beating.

Electrical effects of RI and GI

Figure 1 shows representative recordings of membrane potential and ECG obtained in hearts exposed to normal (1.2 mM) or high (4.8 mM) Mg^{2}_o during control perfusion,

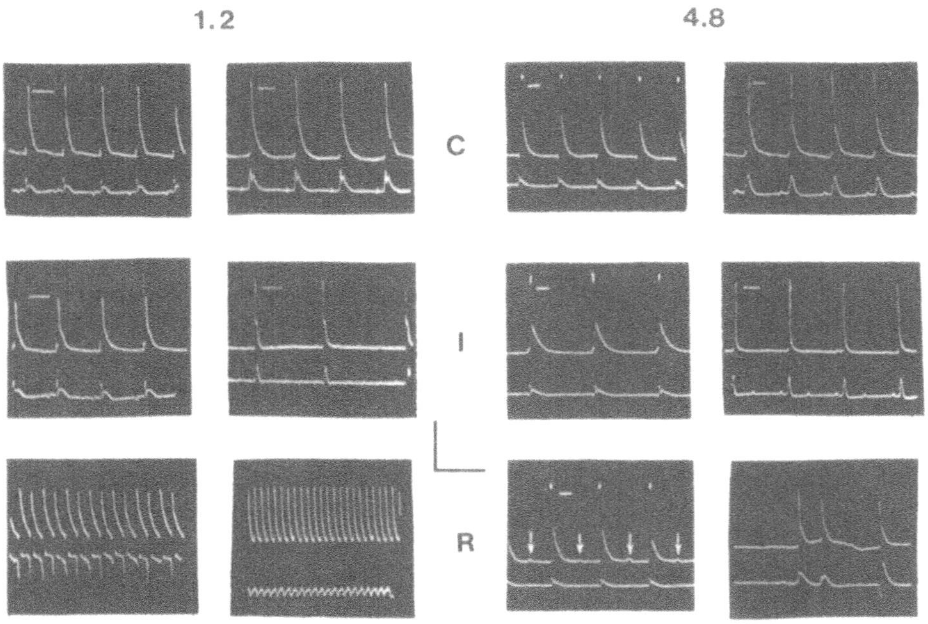

Fig. 1. Representative recordings of transmembrane potential (upper traces) and ECG (lower traces) during control perfusion (C), ischemia (I), and reperfusion (R) at two Mg^{2+}_o as indicated on top. For each concentration, the left column corresponds to coronary ligation and the right to global ischemia. Note the AV block during the latter (complete in 1.2 mM Mg^{2+}_o and 2:1 in 4.8 mM). Arrows indicate delayed afterdepolarizations. Calibration: 60 mV (for transmembrane potential only); horizontal, 600 ms during global ischemia (both concentrations) and 300 ms in all the other panels.

regional or global ischemia, and early reperfusion (within 1 min). Varying Mg^{2+}_o did not alter the action potential morphology. Regional ischemia induced a decrease in resting potential that was attenuated by increasing Mg^{2+}_o. As the cells depolarized the action potential amplitude was reduced and a slowly rising foot preceded the fast rising upstroke. The presence and amplitude of the foot was roughly related to the magnitude of the resting depolarization. Most probably, it reflected impaired conduction resulting from action potential depression. Overt ventricular dysrrhythmias were not observed during coronary occlusion at any Mg^{2+}_o but subliminal delayed afterdepolarizations (DADs) were found towards the end of ischemia in about half of the preparations. Reperfusion after coronary occlusion induced severe ventricular arrhythmias in all hearts perfused without Mg^{2+} or with 1.2 mM (a burst of ventricular tachycardia is illustrated in Fig. 1) whereas, when present, arrhythmias consisted mainly of single premature ventricular beats at 4.8 mM Mg^{2+} (not shown). However, subliminal DADs were frequently observed at this concentration (Fig. 1). Sinus rhythm was maintained throughout coronary occlusion.

The most conspicuous effect of global ischemia was the progressive development of atrioventricular conduction impairment leading to various degrees of AV block in all hearts studied. As a result, the ventricular rate of beating was markedly slowed throughout ischemia. Figure 1 illustrates a complete AV block at 1.2 mM Mg^{2+} and a second degree block (coupling ratio, 2:1) at 4.8 mM Mg^{2+}. Moreover, in half of the hearts, a ventricular standstill occurred by the 8th or 9th min of ischemia. Concomitantly, the ischemic resting depolarization was less marked and signs of intraventricular conduction impairment (as indicated by slowly rising feet at the onset of action potentials) were less conspicuous. On reperfusion, one to one atrioventricular conduction was almost immediately restored but ventricular arrhythmias consistently developed within 1 min at 0 and 1.2 mM Mg^{2+}. However, at the latter, the incidence of fibrillation was lower than that found after regional ischemia (see below). In addition, subliminal DADs that did not trigger overt arrhythmias were also observed following global ischemia at all Mg^{2+}_o. Tachyarrhythmias occurred very seldom at 4.8 mM, where premature beats arising from DADs were instead observed (Fig. 1).

Figure 2 further illustrates the differential effects of regional and global ischemia on the ventricular beating rate (VR) and on reperfusion arrhythmias. Increasing Mg^{2+}_o decreased the control beating rate in both series of experiments. Moreover, when Mg^{2+}_o was raised to 9.6 mM, the cardiac rate decreased to 134 ± 10 beats per min (n = 10), not shown. However, Mg^{2+}_o did not influence the bradycardia induced by regional ischemia that ranged between 35–45 beats per min. In contrast, the reduction in ventricular rate during

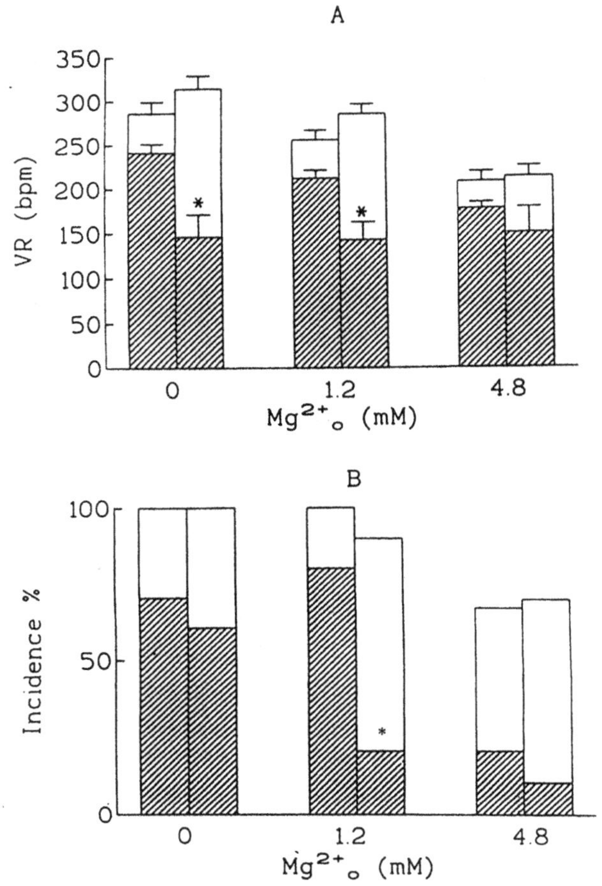

Fig. 2. Influence of Mg^{2+}_o on ventricular beating rate (VR) during ischemia and on the incidence of arrhythmias on reperfusion. Each pair of bars compares regional ischemia (left) to global ischemia (right) for the same Mg^{2+}_o. A, ventricular beating rate during control perfusion (total height of bars) and in ischemia (hatched portions). Mean ± S.E. B, global incidence of arrhythmias on reperfusion (including PVB, VT, and VF) depicted by total height of bars; hatched portion, incidence of ventricular fibrillation. * indicate significantly different from regional ischemia.

global ischemia (mainly due to the development of AV block) was attenuated as Mg^{2+}_o increased: VR decreased by 54% in Mg^{2+} free, by 40% at 1.2 mM, and by 30% at 4.8 mM. Regardless of the initial frequency during control perfusion, the ventricular rate during ischemia was roughly the same at 0, 1.2, or 4.8 mM Mg^{2+}, suggesting that the degree of AV block decreased as Mg^{2+}_o was raised, although all the hearts showed AV block at these concentrations and ventricular arrest was observed in half of them in the last two minutes of ischemia. At 9.6 mM Mg^{2+}, only 4 out of 10 hearts developed 2nd or 3rd degree AV block but the ventricular rate computed for all hearts showed a dramatic fall to 51 ± 5 beats per min (not illustrated in Fig. 2A).

Figure 2B shows that the global incidence of reperfusion arrhythmias did not vary with the type of the preceding ischemia. Only at 1.2 mM, was the severity of the ventricular arrhythmias (judged by the presence of ventricular fibrillation and the total duration of the arrhythmic episodes) substantially less after global ischemia than after coronary ligation. The average duration of the VT and VF episodes during the first 3 min of reperfusion was 85 ± 18 sec after RI and 35 ± 23 sec after GI ($p = 0.015$). Raising Mg^{2+}_o to 4.8 mM reduced the development of arrhythmias as well as the incidence of VF, although neither the latter nor the duration of the arrhythmic events (14 ± 6 sec after RI and 19 ± 16 sec after GI, $p = 0.26$) depended on the type of ischemia. Similarly, in the absence of Mg^{2+}, there was no difference in the high incidence of arrhythmias and VF between both ischemic conditions. These arrhythmias were often sustained and a return to sinus rhythm occurred only in 40% of these hearts. Recovery to regular heart rates at control levels was found in 70% of the experiments at 1.2 and 4.8 mM Mg^{2+}. The difference in the incidence and severity of arrhythmias following regional or global ischemia cannot be solely attributed to the slowing of the ventricular rate induced by the latter because: (a) the incidence of VF was not modified by the decrease in rate in the absence of Mg^{2+} and (b) the degree of protection against arrhythmias did not correlate with VR values that were similar during global ischemia at all Mg^{2+}_o. When global ischemia was performed in hearts exposed to 9.6 mM Mg^{2+}, tachyarrhythmias were not observed on reperfusion and only one heart out of ten presented ventricular extrabeats. Recovery of heart rate to control levels was observed in all the experiments.

Action potential characteristics

Table 1 compares the effects of regional and global ischemia on action potential characteristics, which were evaluated in all hearts throughout the experimental periods: Control perfusion, ischemia, and in the absence of arrhythmias on reperfusion.

In agreement with previous data [7], the resting potential showed a linear trend to hyperpolarize with the logarithm of Mg^{2+}_o during control perfusion in both series of experiments (post test for linear trends, INSTAT V 2.0, Graphical Software, CA, $p = 0.03$). As expected, the resting potential decreased in ischemic cells but this response varied with Mg^{2+}_o and the type of ischemia. Coronary ligation produced a marked depolarization whose magnitude decreased as Mg^{2+}_o was raised (16 mV at 0 mM and 10 mV at 4.8 mM). This resulted in a more significant trend between Mg^{2+}_o and RP values ($p = 0.0006$). In contrast, the depolarization caused by global ischemia was not significantly related to Mg^{2+}_o between 0 and 4.8 mM. At the latter, the RP values were similar to those found during regional ischemia, but they were significantly higher at lower Mg^{2+}_o. At 9.6 mM Mg^{2+}, ischemia produced only a 4 mV depolarization. Consistent with higher RP values

Table 1. Effects of regional and global ischemia on action potential characteristics at different Mg^{2+}_o.

	RI			GI		
Mg^{2+}_o (mM)	C	I	R	C	I	R
Resting potential (mV)						
0	−79.0 ± 1.7	−63.1 ± 1.8*	−81.7 ± 1.3	−79.5 ± 2.9	−72.2 ± 2.7*†	−74.7 ± 1.9
1.2	−80.5 ± 1.8	−66.9 ± 2.0*	−79.1 ± 3.0	−81.7 ± 2.0	−71.7 ± 1.9*†	−74.5 ± 2.7
4.8	−83.4 ± 1.7	−73.0 ± 2.2*	−81.7 ± 1.4	−84.2 ± 1.7	−74.2 ± 1.6*	−78.9 ± 1.2
9.6	–	–	–	−84.7 ± 1.7	−80.7 ± 3.2*	−83.9 ± 2.9
Action potential amplitude (mV)						
0	91.7 ± 2.4	66.9 ± 3.6*	90.3 ± 3.4	91.3 ± 3.3	81.2 ± 3.1*† (7)	85.3 ± 1.4 (7)
1.2	92.3 ± 2.5	72.5 ± 2.3*	89.0 ± 4.0	95.2 ± 2.5	84.2 ± 2.5*† (6)	89.6 ± 2.3 (8)
4.8	93.5 ± 1.6	77.4 ± 3.5*	92.0 ± 1.4	91.2 ± 1.9	83.1 ± 1.1* (7)	90.3 ± 2.4 (9)
9.6	–	–	–	99.5 ± 2.1	95.3 ± 4.4	99.8 ± 2.8
Action potential duration (ms)						
0	84.5 ± 2.9	89.0 ± 3.4	81.0 ± 6.0	73.6 ± 3.2	95.9 ± 11.2*	85.3 ± 3.2
1.2	78.5 ± 3.0	76.5 ± 2.5	78.6 ± 2.8	72.9 ± 2.5	73.0 ± 6.9	84.8 ± 3.4
4.8	84.9 ± 3.2	90.0 ± 5.0	80.0 ± 3.0	72.9 ± 2.3	80.3 ± 5.7	77.0 ± 3.1
9.6	–	–	–	101.5 ± 6.0	196.1 ± 13.0*	114.4 ± 5.9

Mean ± S.E.M. from 10 hearts for each protocol and experimental condition (unless otherwise indicated in parentheses under APA). Values were derived from pooled data recorded during control periods (C) at the steady state during ischemia (I) and in the absence of arrhythmias during reperfusion (R). RI and GI denote regional and global ischemia respectively. * indicate significantly different from control within the same group; †$p < 0.05$ with respect to corresponding value in regional ischemia.

a lesser depression of action potential amplitude was found during global ischemia.

It has been repeatedly reported that the action potential shortens during ischemia in multicellular as well as unicellular preparations [11]. However, we did not observe such effect in our experiments during regional ischemia either in previous work [6, 7] or in this study and the action potential duration did not vary with Mg^{2+}_o in any experimental condition in the hearts that were exposed to coronary ligation. However, the action potential lengthened during global ischemia at the two extreme Mg^{2+}_o. In addition, APD_{90} during control perfusion at 9.6 mM Mg^{2+} was 40% longer than at normal Mg^{2+}_o. During reperfusion, all the action potential characteristics returned toward control values, which were reached in some cases by the end of 10 min of observation.

The decrease in APD during ischemia of short duration relates to extracellular K^+ accumulation and cannot be explained by opening of ATP-sensitive K^+ channels [11]. We previously attributed the lack of change of APD_{90} during regional ischemia to the concomitant bradycardia whose effect on APD would offset the shortening induced by the increase in K^+_o [6]. The reduction in heart rate produced by increasing Mg^{2+}_o together with the AV block induced by global ischemia allowed us to examine the frequency dependence of APD_{90} over a wide range of ventricular rates occurring spontaneously in rat hearts. A preliminary plot of APD_{90} vs. VR values for all the experimental conditions (control, RI, GI, and reperfusion) and every Mg^{2+}_o showed that APD_{90} varied with frequency only at very low rates regardless of Mg^{2+}_o or the presence of ischemia. We then averaged similar APD_{90} values within frequency intervals whose size was selected for very small variations of APD_{90} within a given interval. The resulting data are shown in Fig. 3, where the ranges of ventricular rates corresponding to regional (RI) or global (GI) ischemia for each Mg^{2+}_o tested are also indicated. The action potential duration did not vary with the ventricular rate of beating over a very wide range (from 140–390 bpm), which covers the data recorded during regional ischemia at Mg^{2+}_o between 0–4.8 mM. A slight APD_{90} lengthening occurred for VR between 80–140 bpm. Below these ranges, APD_{90} rose steeply with decreasing frequencies to reach values comparable to those found in other mammalian ventricular cells.

The figure also illustrates the influence of Mg^{2+}_o on the ventricular rate, which during regional ischemia resulted from the concentration dependent slowing of sinus rhythm, without relation to the ischemic condition. In contrast, as also shown in Fig. 2A, the range of ventricular rate during global ischemia did not vary between 0–4.8 mM Mg^{2+} which indicates once more that the AV block existing under these conditions is the major determinant of the ventricular rate. Moreover, these observations demonstrate that unlike

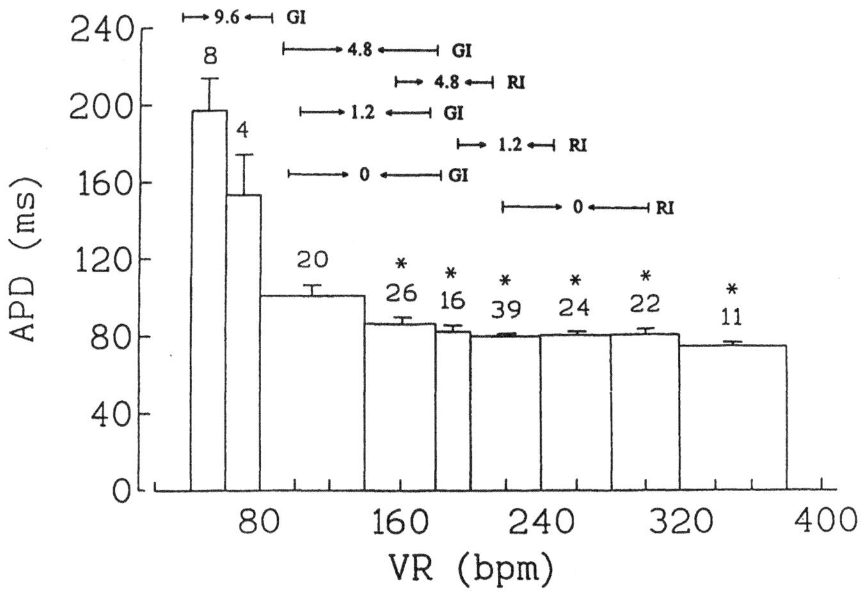

Fig. 3. Frequency dependence of action potential duration (APD_{90}) in spontaneously beating rat hearts. VR, ventricular rate. The data are mean ± S.E. (n on top of columns corresponds to number of experimental periods pooled within a given VR interval). The size of the intervals was chosen as to group similar APD_{90} values regardless of Mg^{2+}_o and the experimental condition (control, ischemia, or reperfusion) at which they were obtained. Each individual value was the average of 5–15 recordings. From a total of 210 experimental periods, data could not be obtained in 41 because of ventricular arrest or severe arrhythmias. The columns marked with * contain data from both experimental series, RI and GI. The frequency ranges corresponding to ischemia at different Mg^{2+}_o are indicated in the upper part of the figure (for individual effects of RI, GI, and Mg^{2+}_o on VR, see Fig. 2A).

ventricular muscle from other mammalian species, the action potential duration in rat myocardium is not influenced by the frequency of stimulation within the range likely to be found *in vivo*.

Discussion

The most striking difference between the effects of regional and global ischemia is the impairment of atrioventricular conduction during the latter that results in a drastic reduction of the ventricular rate. The appearance of AV block under these conditions is a common finding and is attributed to an early depression of the nodal action potential [12]. Atrioventricular conduction was not affected by ligation of the left anterior descending coronary artery whose territory of distribution is confined to the left ventricle. However, regional ischemia induced a slowing of the heart rate whose mechanism is presently unknown but has also been observed by us and others [6, 7, 13]. This effect was not modified by Mg^{2+}_o and did not depend on the initial rate of beating that decreased as Mg^{2+}_o was raised (Fig. 2A). The slowing of sinus rate by Mg^{2+} may reflect an inhibition of $I_{Ca,T}$. In cat atrial myocytes a Mg^{2+}_o of 4.2 mM reduces this current by 40% and markedly decreases the slope of the late diastolic depolarization in action potentials recorded from cells exhibiting pacemaker activity [14].

Even though Mg^{2+}_o similarly affected the control beating rate prior to global ischemia, it did not however influence the ventricular rate during ischemia. Inspection of Fig. 2A may suggest that increasing Mg^{2+}_o decreased the degree of AV block but this action would be opposite to the one expected from a reduction of $I_{Ca,L}$ by Mg^{2+} [14]. Therefore, the more profound block of AV conduction found in the absence of Mg^{2+} rather can be attributed to the higher initial rate of beating. It appears therefore that ischemia in itself limited the conduction capabilities of the AV node so that the coupling ratio between atrial and ventricular activations decreased more as the frequency of atrial stimulations was higher. This effect could be attributed to a depression of $I_{Ca,L}$ in AV nodal cells secondary to the early development of intra and extracellular acidosis during ischemia [15, 16].

Changes in membrane potential

Resting depolarization resulting from extracellular K^+ accumulation is the earliest electrical effect of ischemia [11, 15]. It leads to action potential depression and slowing of conduction. Several ionic transport processes are involved in the increased K^+ efflux induced by metabolic inhibition [17] and the participation of K^+ channels in this process has been repeatedly reported [11, 17–20]. Membrane depolarization produced by regional ischemia was reduced as Mg^{2+}_o was raised (Table 1), most probably because of a lesser K^+_o accumulation. This effect could result from modulation of K^+ channels by Mg^{2+} because a strong interaction between Mg^{2+}_o and K^+ efflux has been observed in rat myocardium [21]. On the other hand, it is known that cellular K^+ losses in hypoxia and ischemia are closely linked to the presence of electrical activity and the frequency of stimulation [22–24] and we found a concentration dependent decrease in beating rate as Mg^{2+}_o was raised in the experiments with regional ischemia. Moreover, resting depolarization was not related to Mg^{2+}_o during global ischemia, where the ventricular rate of beating was similar at all Mg^{2+}_o. Our data therefore suggest that the modulation of K^+_o accumulation by Mg^{2+}_o in our experiments was mediated by the variations in the beating rate rather than by a direct effect on K^+ channels.

Despite the resting depolarization (indicative of K^+_o accumulation in ischemia) and the wide range of ventricular rates found in our experiments we did not observe consistent changes in APD_{90} with ischemia or Mg^{2+}_o. Although all the current systems that underlie the repolarizing phase of the cardiac action potential (I_{to}, $I_{Ca,L}$, I_K, I_{K1}) have been found in rat ventricular cells [25–27], their relative contributions to APD are not yet fully elucidated. If time dependent currents were major determinants of repolarization, APD would vary with frequency, as in other species. Figure 3 shows that this was not the case, except at very low pacing rates where the increase in APD_{90} probably reflected variations in $I_{Ca,L}$, whose recovery from inactivation would be more complete at long diastolic intervals. Although indirect, our findings suggest that APD_{90} is mainly controlled by time independent current(s) over a wide range of frequencies and point to the main background current I_{K1} as a major determinant of APD_{90} in the rat. Moreover, we have shown that 100 μM Ba^{2+} lengthens APD_{90} both during control perfusion and regional ischemia [6]. The insensitivity of APD_{90} to Mg^{2+}_o in this study therefore suggests that Mg^{2+} does not modulate I_{K1} in the rat in agreement with data obtained in other species [28].

Reperfusion arrhythmias

Several factors are considered to play a role in post ischemic arrhythmogenesis such as K^+_o and H^+_o washout, α adrenergic stimulation, and formation of oxygen free radicals [1]. Except for K^+_o washout which could induce re-entry pathways, the other factors would be involved in a cascade of events leading to Ca^{2+} overload that triggers Ca^{2+} dependent dysrrhythmias. Reactivation of Na-H exchange on reflow has been signalled as a major mechanism mediating cytosolic Na^+ and Ca^{2+} accumulation but little attention has been paid to events underlying increases in Na^+ and Ca^{2+} levels prior to reperfusion. The frequent observation of subliminal DADs during the last 2 min of coronary occlusion [7, this work] indicates that Na^+ and Ca^{2+} loading are initiated before reflow. At this stage, Na^+ channels appear to be the main pathway for Na^+ entry into the cells because exposure to lidocaine before perfusion arrest substantially reduces ischemic Na^+ accumulation in rat hearts [29]. Consequently, the ventricular rate of beating as a determinant of Na^+ influx per unit time, is expected to relate to the incidence and/or severity of arrhythmias on reperfusion. This influence was found at normal Mg^{2+}_o where the ventricular rate during global ischemia as well as the occurrence and duration of ventricular fibrillation on reperfusion were much lower than observed with regional ischemia. On the other hand, high Mg^{2+}_o strongly reduced the incidence of ventricular fibrillation on release of coronary ligation without causing a major difference in the rate of beating. This antiarrhythmic action was most probably mediated by direct inhibition of spontaneous Ca^{2+} release by the sarcoplasmic reticulum [4] regardless of the type of ischemia. The data obtained with Mg^{2+} free medium support this view because the beating rate during global ischemia was as low as with the other Mg^{2+}_o but no protection against fibrillation was observed.

In summary, our study shows that in addition to its known effects on Ca^{2+} homeostasis, the protective effect of high Mg^{2+}_o on reperfusion arrhythmias in the isolated heart is partly mediated by the slowing of the sinus rhythm. Thus, by indirectly modulating the ionic fluxes through the channels involved in electrical activity, Mg^{2+} may decrease K^+_o accumulation and Na^+_i loading during ischemia.

Acknowledgements

A. Ponce Zumino is an established investigator from CONICET, Argentina. This work was supported by funds from CIUNC and CONICET to A. Ponce Zumino and from QHSF to O.F. Schanne and E. Ruiz Petrich. The authors thank Ms A Carrion for her technical assistance, Ms C. Ducharme for secretarial help, and Ms Y. Rioux for the artwork.

References

1. Whalley DW, Wendt DJ, Grant AO: Electrophysiologic effects of acute ischemia and reperfusion and their role in the genesis of cardiac arrhythmias. In: PJ Podrid, PR Kowey (eds). Cardiac Arrhythmias: Mechanisms, Diagnosis and Management. Williams and Wilkins, Philadelphia, 1995, Section III chapt 10 pp 109–130
2. Karmazyn M, Moffat MP: Role of Na^+/H^+ exchange in cardiac physiology and pathophysiology: mediation of myocardial reperfusion injury by the pH paradox. Cardiovasc Res 27: 915–924, 1993
3. Pierce GN, Czubryt MP: The contribution of ionic imbalance to ischemia/reperfusion-induced injury. J Mol Cell Cardiol 27: 53–63, 1995

4. Terada H, Hayashi H, Noda N, Satoh H, Katoh H, Yamazaki N: Effects of Mg^{2+} on Ca^{2+} waves and Ca^{2+} transients of rat ventricular myocytes. Am J Physiol 270: H907–H914, 1996
5. Kass RS, Lederer WJ, Tsien RW, Weingart R: Role of calcium ions in transient inward currents and aftercontractions induced by strophanthidin in cardiac Purkinje fibres. J Physiol (L) 281: 187–208, 1978
6. Ruiz Petrich E, Ponce Zumino A, Moffat MP, Rioux Y, Schanne OF: Modulation of the electrophysiological effects of ischemia reperfusion by methylisobutyl amiloride. J Mol Cell Cardiol 28: 1129–1141, 1996
7. Ponce Zumino A, Risler NR, Schanne OF, Ruiz Petrich E, Carrion A: Magnesium: Effects on reperfusion arrhythmias and membrane potential in isolated rat hearts. Mol Cel Biochem 171: 85–93, 1997
8. Manning AS, Hearse DJ: Reperfusion arrhythmias: Mechanisms and prevention. J Mol Cell Cardiol 16: 497–518, 1984
9. Kopacz M, Karwatowska-Prokopczuk E, Beresewicz A: Reperfusion arrhythmias and purine wash-out in isolated rat and rabbit heart. Effect of allopurinol, dimethylthiourea and calcium reduction. J Mol Cell Cardiol 25: 859–874, 1993
10. Walker MJA, Curtis MJ, Hearse DJ, Campbell RWF, Janse MJ, Yellow DM, Cobre SM, Coker SJ, Harness JB, Harron DWG, Higgins A, Julian DJ, Lab MJ, Manning AS, Northover BJ, Parrat JR, Riemersam RA, Riva E, Russel DC, Sheridan DJ, Winslow E, Woodward B: The Lambeth conventions: Guidelines for the study of arrhythmias in ischaemia, infarction, and reperfusion. Cardiovasc Res 22: 447–455, 1988
11. Yan G-X, Yamada KE, Kléber AG, McHowat J, Corr PB: Dissociation between cellular K^+ loss, reduction in repolarization time, and tissue ATP levels during myocardial hypoxia and ischemia. Circ Res 72: 560–570, 1993
12. Ponce Zumino A, Parisii IM, Ruiz P de Ceretti E: Effect of ischemia and low-sodium medium on atrioventricular conduction. Amer J Physiol 218: 1489–1494, 1970
13. Rees SA, Curtis MJ: Further investigations into the mechanism of antifibrillatory action of the specific I_{K1} blocker, RP58866, assessed using the rat dual coronary perfusion model. J Mol Cell Cardiol 27: 2595–2606, 1995
14. Wu J, Lipsius SL: Effects of extracellular Mg^{2+} on T- and L-type Ca^{2+} currents in single atrial myocytes. Am J Physiol 259: H1842–H1850, 1990
15. Fleet WF, Johnson TA, Graebner C, Engle CD, Gettes LS: Effect of serial brief ischemic episodes on extracellular K^+, pH and activation in the pig. Circulation 72: 922–932, 1985
16. Irisawa H, Sato R: Intra- and extra-cellular actions of proton on the calcium current of isolated guinea pig ventricular cells. Circ Res 59: 348–355, 1986
17. Mitani A, Shattock MJ: Role of Na-activated K channel, Na-K-Cl cotransport, and Na-K pump in $[K]_o$ changes during ischemia in rat heart. Am J Physiol 263: H333–H340, 1992
18. Dresdner KP, Wit AL: Barium ion blocks interstitial potassium ion accumulation during acute myocardial ischemia; role of inward rectifier (I_{K1}) and ATP-sensitive potassium channels (abstract). Biophys J 55: 45a, 1989
19. Jiang C, Crake T, Poole-Wilson PA: Inhibition by barium and glibenclamide of the net loss of $^{86}Rb^+$ from rabbit myocardium during hypoxia. Cardiovasc Res 25: 414–420, 1991
20. Ruiz Petrich E, de Lorenzi F, Chartier D: Role of the inward rectifier I_{K1} in the myocardial response to hypoxia. Cardiovasc Res 25: 17–26, 1991
21. Shine KI: Myocardial effects of Mg. Am J Physiol 237: H413–H423, 1979
22. Ruiz Petrich E, Schanne OF, Ponce Zumino A: Electrophysiological responses to ischaemia and reperfusion. In: M Karmazyn (ed). Myocardial Ischaemia: Mechanisms, Reperfusion, Protection. Birkhäuser, Basel, Switzerland. 1996.
23. Leblanc N, Ruiz Petrich E, Chartier D: Potassium loss from hypoxic myocardium: Influence of external K concentration. Can J Physiol Pharmacol 65: 861–866, 1987
24. Rau EE, Langer GA: Dissociation of energetic state and potassium loss from anoxic myocardium. Am J Physiol 235: H537–543, 1978
25. Payet MD, Schanne OF, Ruiz-Ceretti E, Demers JM: Slow outward and inward currents of rat ventricular fibers under anoxia. J Physiol (P) 74: 31–35, 1978
26. Fermini B, Schanne OF: Determinants of action potential duration in neonatal rat ventricle cells. Cardiovasc Res 25: 235–243, 1991
27. Apkon M, Nerbonne JM: Characterization of two distinct depolarization-activated K^+ currents in isolated adult rat ventricular myocytes. J Gen Physiol 97: 973–1011, 1991
28. Martin RL, Koumi S, Ten Eick RE: Comparison of the effects of internal $[Mg^{2+}]$ on I_{K1} in cat and guinea-pig cardiac myocytes. J Mol Cell Cardiol 27: 673–691, 1995
29. Van Emous JG, Nederhoff MGJ, Ruigrok TJC, Van Echbeld CJA: The role of the Na^+ channel in the accumulation of intracellular Na^+ during myocardial ischemia: Consequences for post-ischemic recovery. J Moll Cell Cardiol 29: 85–96, 1997

Molecular and Cellular Biochemistry **186**: 87–97, 1998.
© 1998 *Kluwer Academic Publishers.*

Ischemic preconditioning and superoxide dismutase protect against endothelial dysfunction and endothelium glycocalyx disruption in the postischemic guinea-pig hearts

Andrzej Beręsewicz,[1] Elżbieta Czarnowska[2] and Michał Mączewski[1]

[1]*Department of Clinical Physiology, Medical Centre of Postgraduate Education, Warsaw;* [2]*Department of Pathomorphology, The Children's Memorial Health Institute, Warsaw, Poland*

Abstract

The effect of ischemic preconditioning and superoxide dismutase (SOD) on endothelial glycocalyx and endothelium-dependent vasodilation in the postischemic isolated guinea-pig hearts was examined. Seven groups of hearts were used: group 1 underwent sham aerobic perfusion; group 2 was subjected to 40 min global ischemia without reperfusion; group 3, 40 min ischemia followed by 40 min reperfusion; group 4 was preconditioned with three cycles of 5 min global ischemia followed by 5 min of reperfusion (IPC), prior to 40 min ischemia; group 5 was subjected to IPC prior to standard ischemia/reperfusion; group 6 underwent standard ischemia/reperfusion and SOD infusion (150 U/ml) was begun 5 min before 40 min ischemia and continued during the initial 5 min of the reperfusion period; group 7 was subjected to 80 min aerobic perfusion with NO-synthase inhibitor, L-NAME, to produce a model of endothelial dysfunction independent from the ischemia/reperfusion. Coronary flow responses to acetylcholine (ACh) and sodium nitroprusside (SNP) were used as measures of endothelium-dependent and endothelium-independent vascular function, respectively. Reduction in coronary flow caused by NO-synthase inhibitor, L-NAME, served as a measure of a basal endothelium-dependent vasodilator tone. After completion of each experimental protocol, the hearts were stained with ruthenium red or lanthanum chloride for electron microscopy evaluation of the endothelial glycocalyx. While ischemia led only to a slightly flocculent appearance of the glycocalyx, in ischemia/reperfused hearts the glycocalyx was disrupted, suggesting that it is the reperfusion injury which leads to the glycocalyx injury. Moreover, the coronary flow responses to ACh and L-NAME were impaired, while the responses to SNP were unchanged in the ischemia/reperfused hearts. The disruption of the glycocalyx and the deterioration of ACh and L-NAME responses was prevented by IPC. In addition, the alterations in the glycocalyx and the impairment of ACh responses were prevented by SOD. The glycocalyx appeared to be not changed in the hearts subjected to 80 min aerobic perfusion with L-NAME. In conclusion: (1) the impairment of the endothelium-dependent coronary vasodilation is paralleled by the endothelial glycocalyx disruption in the postischemic guinea-pig hearts; (2) both these changes are prevented by SOD, suggesting the role of free radicals in the mechanism of their development; (3) both changes are prevented by IPC. We hypothesize, therefore, that alterations in the glycocalyx contribute to the mechanism of the endothelial dysfunction in the postischemic hearts. (Mol Cell Biochem **186**: 87–97, 1998)

Key words: isolated guinea-pig heart, endothelial dysfunction, glycocalyx, ischemic preconditioning, free radicals, superoxide dismutase, endothelium

Address for offprints: A. Beręsewicz: Department of Clinical Physiology, Medical Centre of Postgraduate Education, Marymoncka 99, 01-813 Warszawa, Poland

Introduction

The injury of the hearts caused by ischemia/reperfusion is not limited to myocardial cells but extends also to coronary vasculature, including endothelium. Indeed, in various experimental models, ischemia/reperfusion has been shown to impair endothelium-dependent but not endothelium independent coronary vasodilation, indicating a selective endothelial dysfunction whilst vascular smooth muscle function remains intact [1–9]. Possible adverse consequences of this endothelial dysfunction include a reduction in coronary flow and an increased accumulation of polymorphonuclear leukocytes within the reperfused myocardium [10, 11]. Oxygen-derived free radicals have been implicated in the mechanism of the postischemic endothelial dysfunction because the impairment of the endothelium-dependent vasorelaxation can be prevented by treatment with superoxide dismutase, catalase and number of free radical scavengers [4, 12–14].

Many recent studies on ischemia/reperfusion have focused on the protective effects of ischemic preconditioning (IPC) where it has been shown that brief periods of ischemia and reperfusion reduce myocardial infarct size caused by a subsequent period of ischemia [15, 16]. Other beneficial effects of IPC, including the enhanced recovery of myocardial contractile function, reduced severity of reperfusion-induced arrhythmias and preserved autonomic reflexes, have been also described [16, 17]. Recent studies on coronary vasculature in dog [5] and rat heart [8, 18, 19], and on vasculature in rat hindquarters [20] suggest that IPC protection extends also to endothelial cells.

A very early ultrastructural manifestation of endothelial damage in ischemic [21, 22] and hypoxic heart [23] is the disruption of the endothelial glycocalyx, an injury which seems to precede further injury to the endothelial cells [23, 24], and which has been implicated in the mechanism of an increased vascular permeability in the postischemic hearts [23]. Recently we have demonstrated that the onset of the reperfusion is the critical time for the injury to the endothelial glycocalyx in the isolated rat hearts. Moreover we have provided evidence for the involvement of oxygen free radicals in this injury [24].

Accordingly, one hypothesis addressed in this study is that it is the disruption of the endothelial glycocalyx caused by free radicals which accounts for the impairment of the endothelium-dependent vasodilation in the postischemic heart, and consequently, that IPC protects against the endothelial dysfunction because it protects the endothelial glycocalyx. In particular, the present study was designed to assess: (1) whether ischemia/reperfusion produces endothelial dysfunction and glycocalyx disruption in isolated guinea-pig heart; (2) whether IPC would reduce the extent of these changes (3) whether a superoxide anion scavenging enzyme, superoxide dismutase, would mimic the protective effects of IPC and (4) whether the impairment of the endothelium-dependent vasorelaxation not related to ischemia/reperfusion, like that produced by NO-synthase inhibitor, L-NAME, would cause a similar injury to the endothelial glycocalyx.

Materials and methods

Isolated heart preparation

The investigation conforms with the Guide for the care and use of laboratory animals (US National Institutes of Health publication No 85-23, revised 1985)

Guinea pigs (300–380 g) of either sex were injected with 500 units of heparin sulphate, i.p., 20 min before being killed by a blow on the head. Hearts were removed and placed in an ice-cold perfusion solution. The aorta was cannulated and the hearts were perfused aerobically in the Langendorff mode with a prefiltered (5.0 μM Millipore filter) perfusion fluid containing, in mmol/l: 118 NaCl; 23.8 $NaHCO_3$; 4.7 KCl; 1.2 KH_2PO_4; 2.5 $CaCl_2$; 1.2 $MgSO_4$ and 11 glucose and bubbled with 95% O_2 + 5% CO_2 gas mixture giving pH 7.4 and pO_2 580-620 mmHg at 37°C. Perfusion pressure was set at 65 mmHg by adjusting the height of the reservoir. A fluid-filled latex balloon, connected by a stainless steel catheter to a pressure transducer (P23 Pressure Transducer, Gould Statham Instruments Inc.), was inserted into the left ventricle via the left atrium and inflated to set an end-diastolic pressure of 4–7 mmHg. Left ventricular developed pressure (LVDP) was continuously recorded throughout each experiment. The hearts were enclosed in a small, water-jacketed chamber. The temperature of the perfusate, as well as that of the atmosphere surrounding the heart, was thermostatically controlled and checked at regular intervals to ensure 37°C irrespective of coronary flow. The hearts were not stimulated. Global ischemia and reperfusion were induced by clamping and unclamping the aortic inflow line just above the aortic cannula. LVDP and its first-time derivative (dP/dt) were continuously recorded throughout each experiment with an Elema Schönander Mingograph-81 polygraph (Stockholm, Sweden). Coronary flow was quantified by a timed collection and weighing of perfusate exiting the right heart.

Agents used

Acetylcholine chloride (ACh), sodium nitroprusside (SNP), N^{ω}-nitro-L-arginine methyl ester (L-NAME), glutaraldehyde, lanthanum chloride and ruthenium red were purchased from Sigma, and superoxide dismutase (SOD, from bovine erythrocytes) from Boehringer Mannheim. ACh, SNP, L-NAME and SOD were dissolved in the perfusing fluid immediately before use. The glassware and tubing containing SNP were protected from light. Concentrations of ACh, SNP and L-NAME used in the study were selected in preliminary dose-response studies in the isolated guinea pig hearts. The concentrations of ACh (5 nM) and SNP (20 nM) were chosen because they produced approximately 75% of the maximum increase in coronary flow (EC 75). L-NAME (10 μM) was used because it produced the maximum reduction in coronary flow.

Experimental protocols

The guinea-pigs were assigned to seven experimental groups (Fig. 1). All the hearts had an initial 20 min equilibration perfusion followed by:

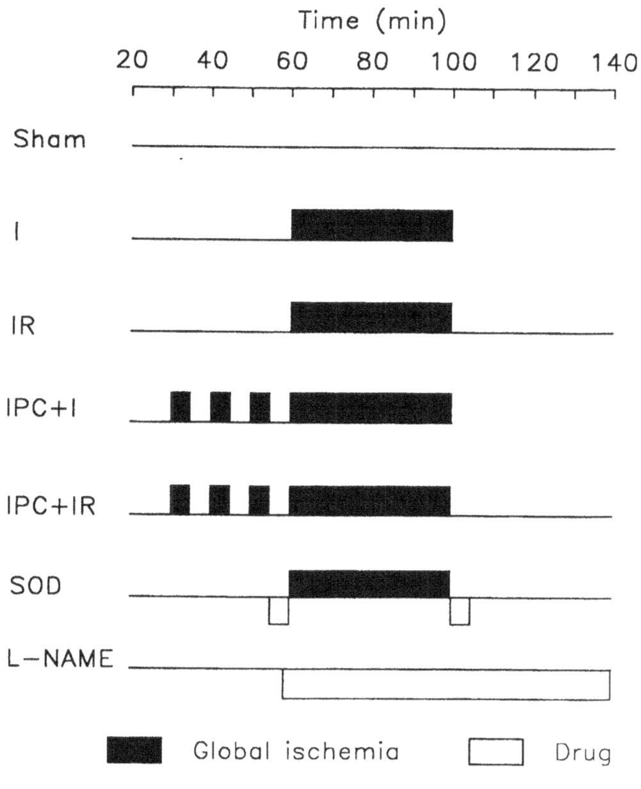

Fig. 1. Experimental protocols. Each experiment started with a 20-min stabilization period. Sham hearts were used as time-matched controls; I, standard 40 min global ischemia without reperfusion; IR, 40 min period of ischemia + 40 min of reperfusion; IPC + I, ischemic preconditioning stimulus consisting of three cycles of 5 min of global ischemia/5 min of reperfusion applied prior to a standard 40 min period of ischemia; IPC +IR, ischemic preconditioning applied prior to a standard IR; SOD, superoxide dismutase infusion beginning 5 min prior to a standard ischemic period and continuing through the initial 5 min of the reperfusion period; L-NAME, 80 min aerobic perfusion in the presence of the drug.

Protocol 1 *(Sham)* - a further 120 min aerobic perfusion.
Protocol 2 *(I, Ischemia)* - a further 40 min aerobic perfusion followed by 40 min global ischemia without reperfusion.
Protocol 3 *(IR)* - the 40 min aerobic perfusion + 40 min global ischemia followed by 40 min reperfusion.
Protocol 4 *(IPC + I)* - three cycles of a preconditioning ischemia (3 ´ 5 min global ischemia, each followed by 5 min of reperfusion), immediately prior to the standard 40-min ischemia, as in group 2.
Protocol 5 *(IPC + IR)* - three cycles of the preconditioning ischemia, prior to the standard 40-min ischemia + 40-min reperfusion, as in group 3.
Protocol 6 *(SOD)* - standard ischemia/reperfusion as in group 3 with addition of SOD infusion (150 U/ml) beginning 5 min before the 40-min ischemia and continuing through the initial 5 min of reperfusion.
Protocol 7 *(L-NAME)* - 40 min aerobic perfusion and then L-NAME, 10 μM, was administered for 80 min with no ischemia.

After the completion of each perfusion protocol, representative hearts were fixed for the morphological studies. Within each experimental protocol the hearts were separated into groups either designated to study coronary flow responses to ACh, or to SNP, or to L-NAME (only in groups 1, 3 and 5), or designated for ultrastructural examinations. The reasons for performing only 1 test in one heart were as follows. ACh and SNP responses were evaluated in the separate hearts in order to minimize their potential preconditioning action on the endothelium, an effect demonstrated by others in rat hearts [25, 26]. Preliminary experiments revealed that ACh and SNP, injected to the reperfused hearts, strongly exacerbated the postischemic ultrastructural alterations in the endothelium. To avoid this effect, the tests with ACh and SNP were avoided in the hearts designated for the morphological examination.

Evaluation of coronary vascular function

Coronary flow response to a bolus of ACh (5 nM ACh in 50 μL, injected into the aortic cannula) was used as a measure of an agonist-inducible endothelium-dependent vascular function. Reduction in coronary flow produced by NO-synthase inhibitor, L-NAME (10 μM) served as a measure of a basal endothelium- and NO-dependent vasodilator tone. A bolus of the NO donor SNP (20 nM in 50 μL) was injected to assess endothelium-independent vascular function. In experimental protocols, the test was performed at 20 min (baseline) and, except of groups 2 and 4, it was repeated at 130 min (Fig. 1). To evaluate ACh or SNP coronary flow response, steady-state coronary flow was assessed first by collecting and weighing 2-min sample of the effluent and then dividing by 2. Then ACh or SNP bolus injection was performed while 5 sec samples of the effluent were being collected and weighed over the next 60 sec. Data from these measurements were used to calculate the 1-min increase in flow produced by the drug and a normalized drug-induced vasodilator response (drug-induced coronary flow increase at 130 min of the perfusion/the increase in flow at 20 min × 100%). To evaluate the vascular effects of L-NAME, the drug was included at 130 min of the experimental protocol (only in protocols 1, 3, and 5) and coronary flow and other hemodynamic variables were assessed 10 min later.

Evaluation of ventricular rhythm disturbances

Continuous recordings of LVDP were used for the evaluation of ventricular arrhythmias. In the hearts subjected to the standard ischemia/reperfusion, the arrhythmias developed only upon the reperfusion and were absent prior to or during the 40 min period of ischemia. To evaluate the reperfusion arrhythmias, the overall incidence of ventricular tachycardia (VT) and ventricular fibrillation (VF) occurring within the whole 40 min reperfusion period was determined as a percent of all hearts included into a given study protocol. The mean duration of reperfusion-induced VT and VF was calculated from the formula: total duration of VT + total duration of VF divided by a number of those hearts in the study protocol that developed the reperfusion-induced VT and/or VT.

Morphological examination

The method described by Ward and Donnely was used to visualize the glycocalyx [23]. The hearts were fixed by perfusion with 2.5% glutaraldehyde and 1% ruthenium red or 1% lanthanum chloride in 0.1 M sodium cacodylate HCl buffer (pH 7.4) at room temperature, as described before [24]. Endomyocardial specimens were divided into 0.5–1 mm^3 pieces and processed for electron microscopy. Three of five blocks randomly selected from each heart were sectioned for electron microscopy. For the ultrastructural observation, the area presenting undamaged myocardium was selected from the semithin sections stained with toluidine blue. During electron microscopical examination, whole capillary profiles were photographed systematically so that all capillaries in a section were photographed with the exception of those which were collapsed. This was because only the capillaries with the open lumen were uniformly stained with the glycocalyx tracers and presented fairly uniform changes. Approximately 40 capillaries were photographed in the sections obtained from the three blocks selected from each heart. Ruthenium red and lanthanum treated tissue was assessed qualitatively only. The evaluation was performed without knowledge of the group to which each photograph originated. Initially, each photograph was evaluated on two separate occasions by the same person. Later on, the repeated evaluations were discontinued because both evaluations produced consistent results.

To evaluate ultrastructure of the cardiomyocytes, the left ventricular specimens were minced in 2.5% buffered glutaraldehyde and processed for electron microscopy as described before [27]. Sections from randomly selected blocks from each hearts were examined under a light microscope for quantification of cell damage. Representative areas from blocks from each group were thin sectioned and stained for electron microscopy. A semiquantitative estimate of cell damage was carried out on 1 μm thick sections from five randomly chosen blocks from each heart. Approximately 2000 cells per heart were analyzed, and the degree of cell injury was assigned to each cell according to criteria proposed by Miki *et al.* [28]: Normal, compact myofibers with uniform staining of nucleoplasm, well defined rows of stained mitochondria between the myofibrils, without any separation of opposing intercalated discs; Mild injury, as above, except for some vacuoles adjacent to mitochondria; Severe injury, reduced staining of cytoplasmic organelles, clumped chromatin material, contraction bands, and granularity of cytoplasm. The morphological evaluations were performed without knowledge of the group to which each sample originated.

Statistics

The parametric data are presented as mean ± S.E.M. Significant differences among groups were calculated by one-way analysis of variance followed by Dunnet's procedure. Paired or non-paired Student's *t*-test was also used when appropriate. To compare frequency distribution of normal, mildly injured and severely injured myocytes and the incidence of VT and VF between groups an overall C test for 2× n table was constructed followed by a sequence of 2 × 2 of Fisher's exact test to compare the individual groups. The values were considered to differ significantly if $p < 0.05$.

Results

Effect of IPC and SOD on the postischemic endothelial dysfunction and hemodynamic recovery

It is evident from Fig. 2 (upper panel), that ACh responses were significantly reduced (by 41%, as compared to sham group) in the isolated guinea-pig hearts subjected to 40 min ischemia + 40 min reperfusion, that IPC completely and SOD partially prevented this deterioration of ACh response, and

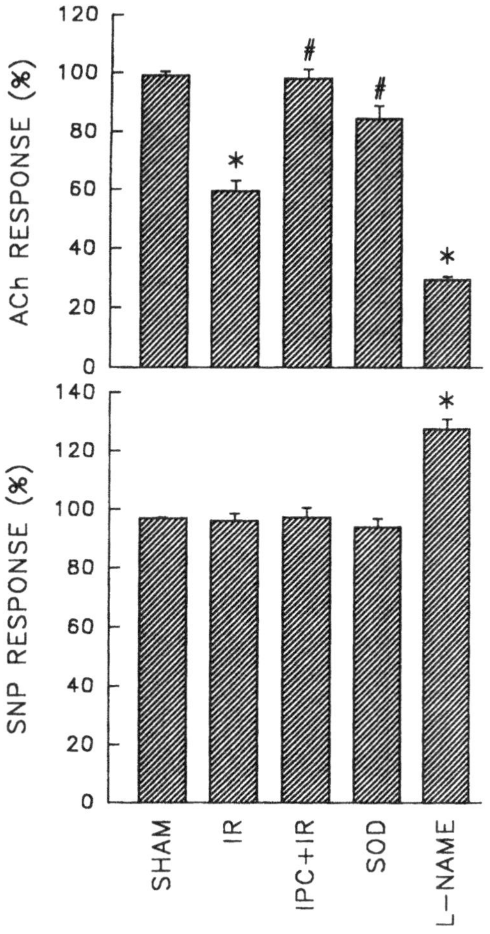

Fig. 2. Normalized ACh (upper panel) and SNP responses (bottom panel) in the hearts subjected to sham perfusion (sham), to a standard 40 min period of ischemia + 40 min reperfusion (IR), to ischemic preconditioning, prior to the standard IR (IPC + IR), to SOD infusion just before and just after the standard IR (SOD), and to 80 min aerobic perfusion with L-NAME (L-NAME). *p < 0.05, vs. sham; #p < 0.05, vs. IR; n = 5-6 hearts per column.

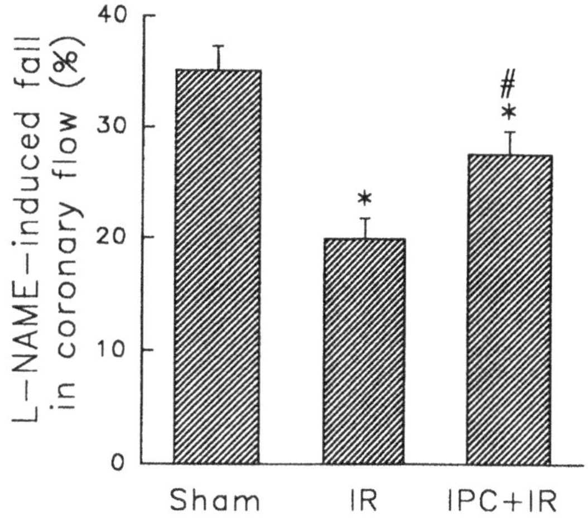

Fig. 3. Vasoconstrictor response to L-NAME in isolated guinea-pig hearts. Before the administration of L-NAME, 10 μM, the hearts were subjected either to a prolonged aerobic perfusion (sham) or to a standard 40 min period of ischemia + 40 min reperfusion (IR) or ischemic preconditioning, prior IR (IPC+IR). *p < 0.05, vs. sham; #p < 0.05, vs. IR; n = 5 hearts per column.

Table 1. Heart rate and reperfusion arrhythmias

		Heart Rate (beats/min)		Incidence (%)		Mean duration of VF+VT (sec)
	N	30 min	140 min	VT	VF	
Sham	23	258 ± 10	261 ± 11			
IR	28	263 ± 15	255 ± 18	77.8	94.4	19 ± 12
IPC+IR	28	249 ± 14	253 ± 12	72.2	88.9	29 ± 20
SOD	18	261 ± 13	255 ± 16	73.3	100.0	11 ± 8
L-NAME	18	253 ± 15	235 ± 12*			

Values for heart rate and mean duration of ventricular tachycardia (VT) + ventricular fibrillation (VF) are mean ± S.E.M.; N = number of hearts; Terms: Sham, IR, IPC+IR, SOD, and L-NAME refer to the experimental protocols delineated in Fig. 1. *p < 0.05, 140 min vs. 30 min.

that the endothelium-independent vasodilation to SNP was comparable in sham, IR, IPC + IR and SOD groups (Fig. 2, bottom panel). Moreover, L-NAME-induced drop in coronary flow was significantly reduced in IR group as compared to the sham group and this deterioration of L-NAME response was partially prevented by IPC (Fig. 3). The ACh responses were reduced by 71% and SNP responses were increased by 21% in L-NAME perfused hearts (L-NAME group), indicating that it is an increased production of NO which mainly accounts for the coronary vasodilation to ACh in guinea pig hearts.

There were no significant differences in baseline values for coronary flow, left ventricular developed pressure (LVDP) (not shown) or heart rate (Table 1) between any of the experimental groups. The postischemic hemodynamic recoveries of coronary flow, LVDP (Fig. 4) and heart rate (Table 1) were similar in all the experimental groups and amounted approximately to 90, 65 and 100%, respectively, as compared to the baseline values. The majority of hearts developed short-lasting reperfusion-induced VT and VF. However, there were no significant differences in the incidence and the duration of these arrhythmias between any of the experimental groups (Table 1).

Endothelial glycocalyx

In the hearts from sham group, both tracers were bound to the glycocalyx, giving a more or less smooth continuous electron dense layer on the luminal surface of the endothelium (Fig. 5, A, A′). Neither marker had any effect on the structure

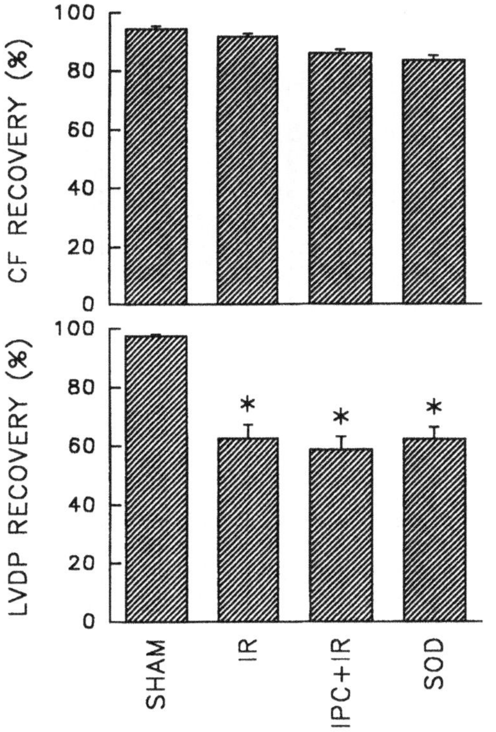

Fig. 4. Postischemic recoveries of coronary flow (upper panel) and LVDP (bottom panel) in the hearts subjected to sham perfusion (sham), to a standard 40 min period of ischemia + 40 min reperfusion (IR), to ischemic preconditioning, prior to the standard IR (IPC + IR), and to SOD infusion just before and just after the standard IR (SOD). $^*p < 0.05$, vs. sham; $^\#p < 0.05$, vs. IR; n = 18-28 hearts per column.

of the endothelial cells. In the hearts subjected to the ischemia only, the glycocalyx, which still remained continuous, had a slightly flocculent appearance, with occasional large protrusions and thinning (Fig. 5, B, B′). The ultrastructure of the endothelial cells was still normal. The changes in the glycocalyx were even more advanced in the hearts subjected to the ischemia/reperfusion, suggesting that they are, in fact, the manifestation of the reperfusion injury. Thus, both markers demonstrated lack of continuity and redistribution of the glycocalyx with the appearance of large gaps between the clumps (Fig. 5, D, D′). The ultrastructure of the endothelial cells remained still normal, although some capillaries presented swollen endothelial cells.

Both ischemia-induced and reperfusion-induced injury to the glycocalyx was largely prevented by IPC. This is evidenced by the fact that in the IPC + I group (Fig. 5, C, C′) as well as in the IPC + IR group (Fig. 5, E, E′), the glycocalyx was continuous and its protrusions, if present, were relatively small. Similarly as IPC, also SOD, administrated just before the ischemia and at the beginning of the reperfusion, prevented the ischemia/reperfusion changes in the glycocalyx (Fig. 6, A, A′). The glycocalyx appeared unchanged in hearts subjected to 80 min aerobic perfusion with L-NAME (Fig. 6, B, B′).

Cardiomyocytes

The ultrastructure of the cardiomyocytes subjected to ischemia/reperfusion was relatively well preserved compared to the sham perfused hearts, with only 16 ± 4% and 4.5 ± 2% of the myocytes presenting mild and severe injury, respectively ($p < 0.05$). Similar numbers of mildly and severely injured cells were found in the ischemia/reperfusion and IPC + IR groups (Fig. 7), suggesting that the ultrastructural changes induced by ischemia/reperfusion were not affected by IPC.

Discussion

The major finding of this study is that IPC protects not only against the endothelial dysfunction but also against the ultrastructural alterations in the endothelial glycocalyx and that these protective effects of IPC are mimicked by SOD. From this we hypothesize a cause and effect relation between the free radical-induced damage to the endothelial glycocalyx and the impairment of the endothelium-dependent vasorelaxation in the postischemic hearts.

Postischemic injury to the endothelium and the effects of IPC

Ischemia/reperfusion has been frequently reported to impair the coronary endothelium-dependent vasorelaxation [1–9] as well as to disrupt the coronary endothelial glycocalyx [21, 22, 24] and the role of free radicals in the mechanism of these alterations has been suggested [4, 12–14, 24]. Here, the morphological and functional manifestations of the endothelial ischemia/reperfusion injury were correlated in one study.

The vasodilation to ACh and the vasoconstriction to L-NAME were used as indices of receptor-stimulated and basal endothelial release of NO, respectively. Under our experimental conditions both these indices were significantly impaired in the postischemic hearts, whereas the vasodilation response to endothelium-independent vasodilator, SNP, was retained. This indicates that, in our model, ischemia/reperfusion resulted in an endothelial dysfunction characterized by a selective impairment of the basal as well as the receptor-stimulated production of biologically active NO, whilst coronary smooth muscle function remained intact. Together with these functional alterations, the postischemic endothelium demonstrated severe disruption of the endothelium glycocalyx, while the ultrastructure of the endothelial cells remained

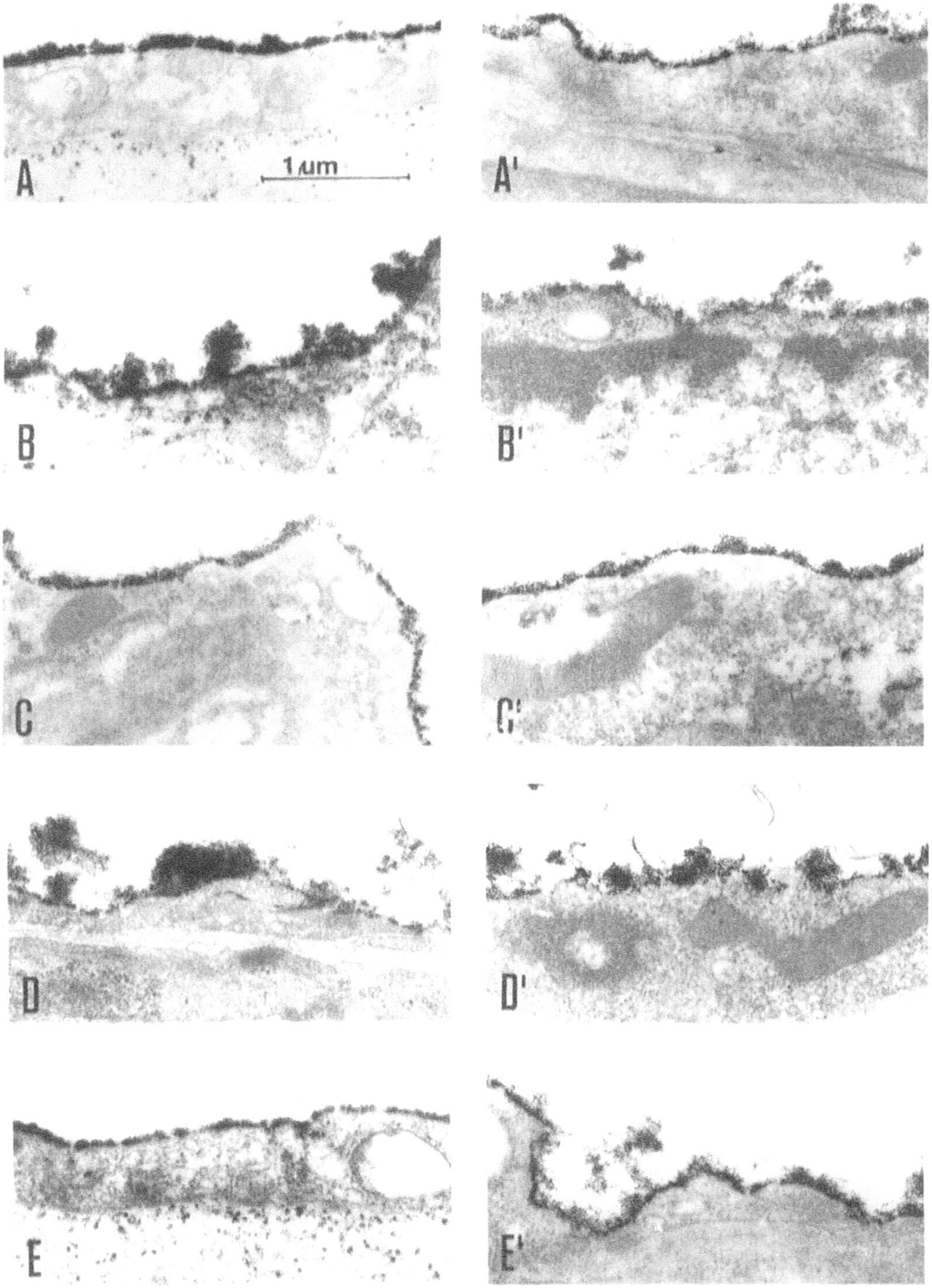

Fig. 5. Electron micrographs of isolated left ventricular capillaries from isolated guinea-pig hearts. Endothelial glycocalyx delineated with ruthenium red (A–E) and with lanthanum chloride (A′-E′) in the hearts subjected to: sham perfusion (A, A′), the standard 40 min ischemia without reperfusion (B, B′), ischemic preconditioning, prior to the standard 40 min ischemia (C, C′), standard 40 min ischemia + 40 min reperfusion (D, D′), ischemic preconditioning, prior to the standard ischemia/reperfusion (E, E′). Each experimental group consisted of 8 hearts (4 hearts for each glycocalyx tracer) and the representative pictures from each group are presented. Marker bar = 1 μm. Magnification × 26 000.

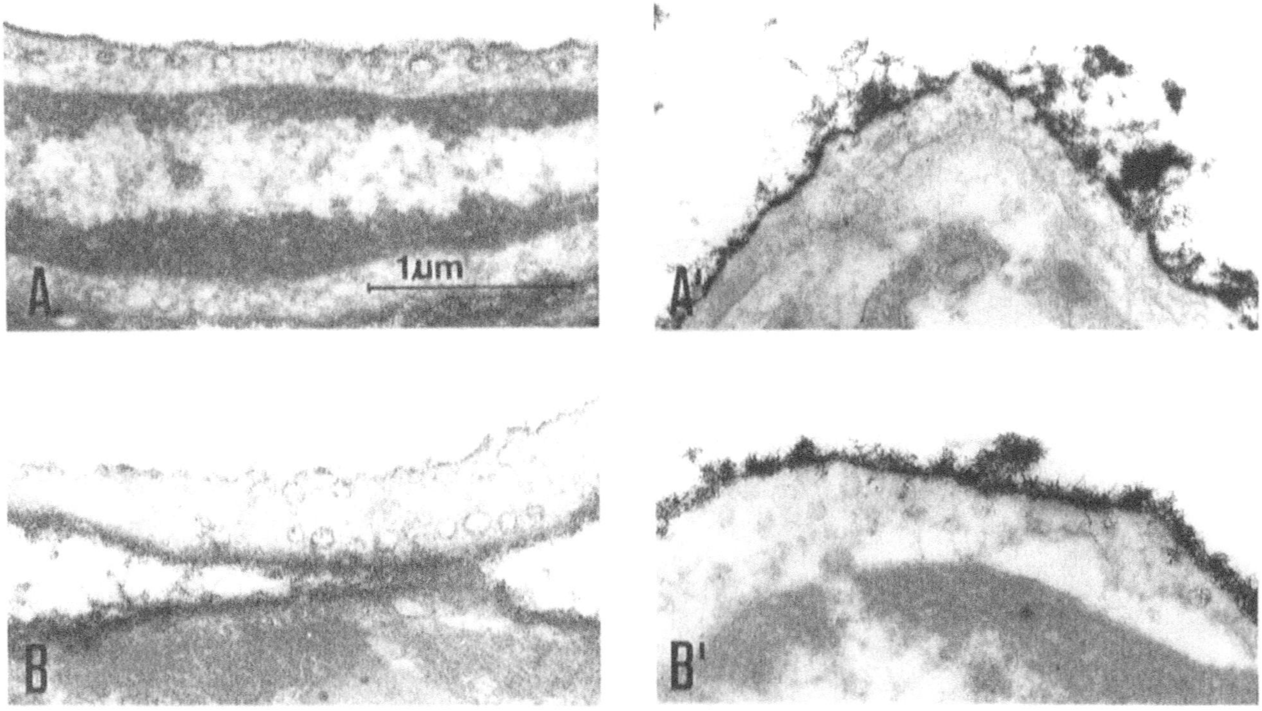

Fig. 6. Endothelial glycocalyx delineated with ruthenium red (A–B) and with lanthanum chloride (A′–B′) in isolated guinea-pig hearts subjected to: SOD infusion just before and just after the standard 40 min ischemia + 40 min reperfusion (A, A′) and 80-min aerobic L-NAME infusion (B, B′). Each experimental group consisted of 8 hearts (4 hearts for each glycocalyx tracer) and representative pictures from each group are presented. Marker bar = 1 μm. Magnification × 26 000.

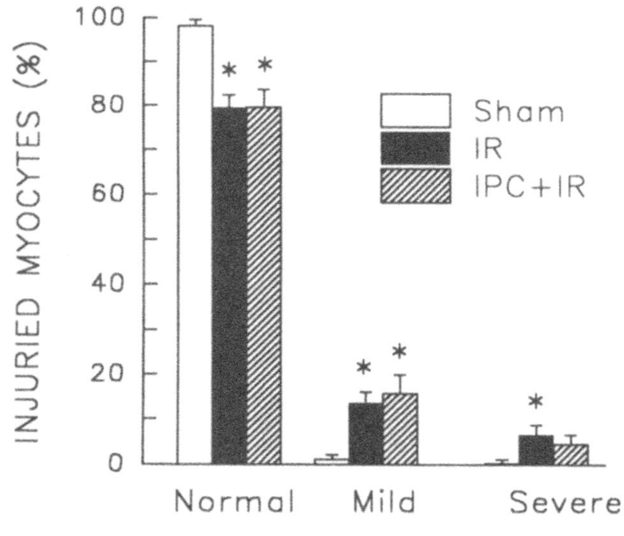

Fig. 7. Frequency distribution of normal, mildly injured, and severely injured cardiomyocytes in the hearts subjected to sham perfusion (sham, open boxes), to standard 40 min ischemia + 40 min reperfusion (IR, closed boxes), and to ischemic preconditioning, prior to the standard IR (IPC + IR, hatched boxes). Each group consisted of 4 hearts and approximately 2000 myocytes per each heart were analyzed. $^{*}p < 0.05$, vs. sham.

unchanged. Thus, similarly as in isolated rat hearts [24], in guinea pig-hearts, the disruption of the glycocalyx seems to be the earliest structural manifestation of the postischemic endothelial injury.

In our experimental model, the impaired vasodilator response to ACh was completely restored and that to L-NAME restored partially by IPC. Moreover, we demonstrate for the first time, that in addition to the endothelial dysfunction, the disruption of the glycocalyx is also prevented by IPC. The IPC protection against the endothelial dysfunction has been already demonstrated in other species [5, 8, 18, 19]. In most of these studies, IPC has been found also to protect the cardiomyocytes [8, 18, 19]. However in our guinea-pig experiments, the IPC protection seemed to be limited to the endothelium. This is evidenced by the fact that the postischemic recoveries of LVDP (Fig. 4), the reperfusion arrhythmias (Table 1) and the degree of the ultrastructural damage to the cardiomyocytes (Fig. 7) did not differ between IR and IPC+IR groups. The effect of IPC on guinea-pig heart has not been studied thoroughly. Only recently, Miyamae *et al.* [29] have reported IPC to attenuate the postischemic deterioration of LVDP and the postischemic creatine kinase release in isolated guinea-pig hearts, which suggests that the injury to the cardiomyocytes is protectable by IPC also in this species. However, while the postischemic recovery of LVDP amounted to approximately 65% in our study, only 30% recovery of LVDP occurred in the study of Miyamae *et al.* [29]. Thus, we speculate that the discussed discrepancy in the effectiveness of IPC may be related to the fact that in our study the postischemic injury was much milder (stunning vs. irreversible injury?). Of interest, the effectiveness of IPC to protect against myocardial stunning seems to be species dependent. Thus, IPC has been demonstrated to be protective in rabbit and rat hearts [30, 31] and not protective in dog and pig hearts [32, 33].

An advantage offered by our experimental model is that it allowed dissociation of the injury to the endothelium from that to the cardiomyocytes. The observation that IPC and SOD afforded protection against the ultrastructural and functional alterations in the endothelium, but did not protect the cadiomyocytes indicates that, at least in our experimental model of crystalloid perfused guinea-pig heart: (i) the injury to the endothelium does not contribute to the mechanism of the postischemic cardiomyocytes' injury, and (ii) that it is some mechanism intrinsic to the endothelium itself (i.e., independent from the cardiomyocytes' injury) which is responsible for the endothelial protection by IPC and SOD.

Are structural and functional endothelial alterations causally related?

The disruption of the endothelial glycocalyx has been implicated in the mechanism of the increased vascular permeability and tissue edema which occurs in the ischemic myocardium [23]. A question arises as to a possible role of the glycocalyx in the mechanism of the impaired endothelium-dependent vasodilation in the postischemic heart.

Several aspects of this study indicate that the functional and structural postischemic endothelial alterations have several features in common. First, both seem to be the manifestation of the reperfusion injury. The evidence regarding the glycocalyx comes from this and our earlier study in isolated rat hearts [24]. The reperfusion-induced mechanism of the postischemic endothelial dysfunction has been demonstrated by other authors [4, 8]. Second, both changes are prevented by the pretreatment with SOD. The beneficial effects of SOD on the endothelial dysfunction [4, 12–14], and of hydroxyl radical scavenger, MPG, on the glycocalyx disruption [24], have been previously reported. Altogether, these data strongly support the role of oxygen-derived free radicals in the mechanism of the structural as well as the functional injury in the postischemic coronary endothelium. Third, as it is demonstrated in this study, the disruption of the glycocalyx as well as the endothelial dysfunction are prevented by IPC. Finally, taken together, the data from our SOD and IPC experiments strongly suggest that IPC affords the structural and functional protection of the endothelium because it somehow limits the net vascular production of free radicals during the reperfusion. In fact, IPC has been reported to attenuate the postischemic free radical production in isolated rat hearts [34], and superoxide anion generation in the mitochondria of the postischemic rat hearts [35] and also to increase myocardial antioxidant activity in dog [36], but not in rabbit [37] and pig hearts [38].

Although the discussed 'features in common' can not be regarded as a proof for the causal relationship between the structural and the functional alterations in the postischemic endothelium, they are very suggestive of such a relationship. Consequently, this raises a question as to whether the injury to the glycocalyx causes the attenuated endothelial NO production or *vice versa*? This was addressed in the experiment in which we attempted to assess whether the inhibition of the endothelial NO-production by means of the NO-synthase inhibitor (a model of endothelial dysfunction independent from the ischemia/reperfusion) would cause any injury to the endothelial glycocalyx. The answer was negative. Extrapolating this result to the ischemia/reperfusion, we speculate that the postischemic injury to the glycocalyx is not secondary to the endothelial dysfunction, at least in our types of the 'acute' endothelial dysfunction. We believe rather, but not have proof for it, that the opposite may be true. It has been suggested that the endothelial glycocalyx is instrumental in transducing increased coronary flow and shear stress signals to increased glycolytic flux in isolated guinea pig hearts [39]. In this context, very intriguing are reports suggesting that the impaired vasodilatory responses to agonists like ADP, thrombin [3] and ACh [40] in canine coronary arteries exposed to brief ischemia/reperfusion may not necessary reflect a true defect in NO synthase activity but rather a specific impairment of the transduction mechanism linking these humoral signals to NO synthase. Similarly, evidence indicates that the impairment of the transduction mechanism plays a role in the mechanism underlying the endothelial dysfunction in hypercholesterolemia and atherosclerosis [41]. We hypothesize that one possible mechanism for such impairment of the transduction mechanism would be the injury to the glycocalyx.

In conclusion, this study provides the evidence that: (1) the impairment of endothelium-dependent coronary vasodilation in the postischemic guinea-pig heart is accompanied by endothelial glycocalyx disruption; (2) both these changes are prevented by SOD, suggesting a role of oxygen free radicals in the mechanism of their development; (3) both changes are prevented by IPC.

These results and data from the literature are compatible

with the hypothesis that the alterations in the glycocalyx contribute to the mechanism of the endothelial dysfunction in the postischemic hearts. The hypothesis remains to be critically tested.

Acknowledgements

We thank Ms. Alicja Protasowicka and Mr. Marek Woźniak for technical assistance. The study was supported by the CMKP 501-1-1-05-16/97 grant.

References

1. VanBenthuysen KM, Mcmurtry IF, Horwitz LD: Reperfusion after coronary occlusion in dogs impairs endothelium-dependent relaxation to acetylcholine and augments contractile reactivity *in vitro*. J Clin Invest 79: 265–274, 1987
2. Mehta JL, Nichols WW, Donnelly WH, Lawson DL, Saldeen TGP: Impaired canine coronary vasodilator response to acetylcholine and bradykinin after occlusion-reperfusion. Circ Res 64: 43–54, 1989
3. Pearson PJ, Schaff HV, Vanhoutte PM: Long-term impairment of endothelium dependent relaxations to aggregating platelets after reperfusion injury in canine coronary arteries. Circulation 81: 1921–1927, 1990
4. Tsao PS, Aoki N, Lefer DJ, Johnson G III, Lefer AM: Time-course of endothelial dysfunction and myocardial injury during ischemia and reperfusion in the cat. Circulation 82: 1402–1412, 1990
5. Defily DV, Chilian WM: Preconditioning protects coronary arteriolar endothelium from ischemia-reperfusion injury. Am J Physiol 265: H700–H706, 1993
6. Ma X-L, Weyrich AS, Lefer DJ, Lefer AM: Diminished basal nitric oxide release after myocardial ischemia and reperfusion promotes neutrophil adherence to coronary endothelium. Circ Res 72: 403–412, 1993
7. Horwitz LD, Kaufman D, Keller MW, Kong YO: Time course of coronary endothelial healing after injury due to ischemia and reperfusion. Circulation 90: 2439–2447, 1994
8. Richard V, Kaeffer N, Tron C, Thuillez C: Ischemic preconditioning protects against coronary endothelial dysfunction induced by ischemia and reperfusion. Circulation 89: 1254–1261, 1994
9. Ehring T, Krajcar M, Baumgart D, Kompa S, Hummelgen M, Heusch G: Cholinergic and alpha-adrenergic coronary constriction with increasing ischemia-reperfusion injury. Am J Physiol 37: H886–H894, 1995
10. Sobey CG, Woodman OL: Myocardial ischaemia - what happens to the coronary arteries? Trends Pharmacol Sci 14: 448–453, 1993
11. Lefer AM, Lefer DJ: The role of nitric oxide and cell adhesion molecules on the microcirculation in ischaemia-reperfusion. Cardiovasc Res 32: 743–751, 1996
12. Mehta JL, Nichols WW, Donnelly WH, Lawson DL, Thompson L, ter Riet M, Saldeen TGP: Protection by superoxide dismutase from myocardial dysfunction and attenuation of vasodilator reserve after coronary occlusion and reperfusion in dog. Circ Res 65: 1283–1295, 1989
13. Dauber IM, Lesnefsky EJ, VanBenthuysen KM, Weil JV, Horwitz LD: Reactive oxygen metabolite scavengers decrease functional coronary microvascular injury due to ischemia-reperfusion. Am J Physiol 260: H42–H49, 1991
14. Gross GJ, Orourke ST, Pelc LR, Warltier DC: Myocardial and endothelial dysfunction after multiple, brief coronary occlusions - role of oxygen radicals. Am J Physiol 263: H1703–H1709, 1992
15. Murry CE, Jennings RB, Reimer KA: Preconditioning with ischemia: A delay of lethal cell injury in ischemic myocardium. Circulation 74: 1124–1136, 1986
16. Lawson CS, Downey JM: Preconditioning - state of the art myocardial protection. Cardiovasc Res 27: 542–550, 1993
17. Parratt JR: Protection of the heart by ischaemic preconditioning - mechanisms and possibilities for pharmacological exploitation. Trends Pharmacol Sci 15: 19–25, 1994
18. Kaeffer N, Richard V, Francois A, Lallemand F, Henry JP, Thuillez C: Preconditioning prevents chronic reperfusion-induced coronary endothelial dysfunction in rats. Am J Physiol 40: H842–H849, 1996
19. Kolocassides KG, Galinanes M, Hearse DJ: Ischemic preconditioning, cardioplegia or both? differing approaches to myocardial and vascular protection. J Mol Cell Cardiol 28: 623–634, 1996
20. Loke KE, Woodman OL: Effect of ischaemic preconditioning on vascular dysfunction induced by ischaemia and reperfusion in rat hindquarters. Cardiovasc Res 32: 1081–1087, 1996
21. Ishiharajima S, Aida T, Nakagawa R, Kameyama K, Sugano K, Oguro T, Asano G: Early membrane damage during ischemia in rat heart. Exp Mol Pathol 44: 1–6, 1986
22. Haack DW, Bush LR, Schlafer M, Lucchesi BR: Lanthanum staining of coronary microvascular endothelium. Effects of ischemia reperfusion, propranolol and atenolol. Microvasc Res 21: 362–376, 1981
23. Ward BJ, Donnelly JL: Hypoxia induced disruption of the cardiac endothelial glycocalyx: Implications for capillary permeability. Cardiovasc Res 27: 384–389, 1993
24. Czarnowska E, Karwatowska-Prokopczuk E: Ultrastructural demonstration of endothelial glycocalyx disruption in the reperfused rat heart. Involvement of oxygen free radicals. Basic Res Cardiol 90: 357–364, 1995
25. Richard V, Blanc T, Kaeffer N, Tron C, Thuillez C: Myocardial and coronary endothelial protective effects of acetylcholine after myocardial ischaemia and reperfusion in rats: role of nitric oxide. Br J Pharmacol 115: 1532–1538, 1995
26. Bilińska M, Mączewski M, Beręsewicz A: Donors of nitric oxide mimic effects of ischaemic preconditioning of reperfusion induced arrhythmias in isolated rat heart. Mol Cell Biochem 160/161: 265–271, 1996
27. Karwatowska-Prokopczuk E, Czarnowska E, Beręsewicz A: Iron availability and free radical induced injury in isolated ischemic/reperfused rat heart. Cardiovasc Res 26: 1–9, 1992
28. Miki S, Ashraf M, Salka S, Sperelakis N: Myocardial dysfunction and ultrastructural alterations mediated by oxygen metabolites. J Mol Cell Cardiol 20: 1009–1024, 1988
29. Miyamae M, Diamond I, Weiner MW, Camacho SA, Figueredo VM: Regular alcohol consumption mimics cardiac preconditioning by protecting against ischemia-reperfusion injury. Proc Natl Acad Sci USA 94: 3235–3239, 1997
30. Urabe K, Miura T, Iwamoto T, Ogawa T, Goto M, Sakamoto J, Iimura O: Preconditioning enhances myocardial resistance to postischaemic myocardial stunning via adenosine receptor activation. Cardiovasc Res 27: 657–662, 1993
31. Steenbergen C, Perlman ME, London RE, Murphy E: Mechanism of preconditioning. Ionic alterations. Circ Res 72: 112–125, 1993
32. Miyamae M, Fujiwara H, Kida M, Yokota R, Tanaka M, Katsuragawa M, Hasegawa K, Ohura M, Koga K, Yabuuchi Y, Sasayama S: Preconditioning improves energy metabolism during reperfusion but does not attenuate myocardial stunning in porcine hearts. Circulation 88: 223–234, 1993

33. Ovize M, Przyklenk K, Hale SL, Kloner RA: Preconditioning does not attenuate myocardial stunning. Circulation 85: 2247–2254, 1992
34. Tosaki A, Cordis GA, Szerdahelyi P, Engelman RM, Das DK: Effects of preconditioning on reperfusion arrhythmias, myocardial functions, formation of free radicals, and ion shifts in isolated ischemic/reperfused rat hearts. J Cardiovasc Pharmacol 23: 365–373, 1994
35. Park JW, Chun YS, Kim YH, Kim CH, Kim MS: Ischemic preconditioning reduces O_2^- generation and prevents respiratory impairment in the mitochondria of post-ischemic reperfused heart of rat. Life Sci 60: 2207–2219, 1997
36. Hoshida S, Kuzuya T, Fuji H, Yamashita N, Oe H, Hori M, Suzuki K, Taniguchi N, Tada M: Sublethal ischemia alters myocardial antioxidant activity in canine heart. Am J Physiol 264: H33–H39, 1993
37. Turrens JF, Thornton J, Barnard ML, Snyder S, Liu G, Downey JM: Protection from reperfusion injury by preconditioning hearts does not involve increased antioxidant defenses. Am J Physiol 262: H585–H589, 1992
38. Fu LX, Kirkeboen KA, Liang QM, Sjogren KG, Hjalmarson A, Ilebekk A: Free radical scavenging enzymes and g-protein mediated receptor signalling systems in ischaemically preconditioned porcine myocardium. Cardiovasc Res 27: 612–616, 1993
39. Suarez J, Rubio R: Regulation of glycolytic flux by coronary flow in guinea pig heart. Role of vascular cell glycocalyx. Am J Physiol 261: H1994–H2000, 1991
40. Kim YD, Fomsgaard JS, Heim KF, Ramwell PW, Thomas G, Kagan E, Moore SP, Coughlin SS, Kuwahara M, Analouei A, Meyers AK: Brief ischemia-reperfusion induces stunning of endothelium in canine coronary artery. Circulation 85: 1473–1482, 1992
41. Flavahan NA: Atherosclerosis or lipoprotein-induced endothelial dysfunction. Potential mechanisms underlying reduction in EDRF/nitric oxide activity. Circulation 85: 1927–1938, 1992

Molecular and Cellular Biochemistry **186**: 99–106, 1998.
© 1998 *Kluwer Academic Publishers.*

Role of H_2O_2 in changing β-adrenoceptor and adenylyl cyclase in ischemia-reperfused hearts

Sujata Persad, Vincenzo Panagia and Naranjan S. Dhalla

Institute of Cardiovascular Sciences, St. Boniface General Hospital Research Centre and Department of Physiology, Faculty of Medicine, University of Manitoba, Winnipeg, Canada

Abstract

In view of the accumulation of H_2O_2 in the myocardium due to ischemia-reperfusion and changes in β-adrenoceptor mechanisms in the ischemic-reperfused heart, we investigated the effects of H_2O_2 on the β-adrenoceptor, G-protein and adenylyl cyclase complex. Rat hearts were perfused with 1 mM H_2O_2 for 10 min before isolating membranes for measuring the biochemical activities. The stimulation of adenylyl cyclase by different concentrations of isoproterenol was depressed upon perfusing hearts with H_2O_2. Both the affinity and density of β_1-adrenoceptors as well as the density of the β_2-adrenoceptors were decreased whereas the affinity of β_2-adrenoceptors was increased by H_2O_2 perfusion. Competition curves did not reveal any effect of H_2O_2 on the proportion of coupled receptors in the high affinity state. The basal as well as forskolin-, NaF- and Gpp(NH)p-stimulated adenylyl cyclase activities were depressed by perfusing the heart with H_2O_2. Catalase alone or in combination with mannitol was able to significantly decrease the magnitude of alterations due to H_2O_2. The positive inotropic effect of 1 μM isoproterenol was markedly attenuated upon perfusing hearts with 200–500 μM H_2O_2 for 10 min. These results suggest that H_2O_2 may depress the β_1-adrenoceptor, G_s-proteins and catalytic subunit of the adenylyl cyclase enzyme and thus may play an important role in attenuating the β-adrenoceptor linked signal transduction due to ischemia-reperfusion injury. (Mol Cell Biochem **186**: 99–106, 1998)

Key words: cardiac β-adrenoceptors, cardiac adenylyl cyclase, oxidative stress, H_2O_2 and the heart, inotropic effect of isoproterenol

Introduction

Several studies have reported an increase in the formation of H_2O_2 under different pathological conditions including ischemia-reperfusion in the heart [1–3]. The generation of this potent oxidant of cellular lipids and proteins is known to occur in the mitochondria as well as due to dismutation of the superoxide radicals in the myocardium. In fact, H_2O_2 has been well documented to induce cardiac dysfunction and injury by promoting the occurrence of intracellular Ca^{2+} overload. Modification of Ca^{2+}-handling mechanism due to this oxidant has been observed at the level of the sarcolemma and sarcoplasmic reticulum [4–10]. Since the β-adrenoceptor, G-proteins, and adenylyl cyclase system is involved in the regulation of intracellular Ca^{2+} and cardiac function [11–13], it is likely that this sarcolemmal signal transduction pathway may also be affected by H_2O_2. In this regard, while only a decrease in the affinity of β-adrenoceptors has been shown to be due to H_2O_2, the density of these receptors has been reported to be increased, decreased and unchanged upon treating cardiac membranes with H_2O_2 [14, 15]. Likewise, H_2O_2 was observed to decrease and increase the adenylyl cyclase activity in cardiac membranes [16] and vascular smooth muscle cells [17], respectively. Although H_2O_2 has been found to exert no effect on G-proteins in cardiac membranes and vascular smooth muscle cells [16, 17], the results are of preliminary nature and thus no meaningful conclusions can be made. In view of the conflicting and inconclusive results regarding the effects of H_2O_2 on the β-adrenoceptor, G-proteins and adenylyl cyclase system as

Address for offprints: N.S. Dhalla, Institute of Cardiovascular Sciences, St. Boniface General Hospital Research Centre, 351 Tache Avenue, Winnipeg, Manitoba, Canada, R2H 2A6

well as the lack of information concerning the action of H_2O_2 on cardiac β_1- and β_2-adrenoceptors, this study was undertaken to provide detailed information on changes in the β-adrenoceptor linked signal transduction mechanism due to H_2O_2. For this purpose, alterations in β_1- and β_2-adrenoceptors and adenylyl cyclase activities were monitored upon perfusing isolated rat hearts with H_2O_2. In addition, changes in the inotropic effects of isoproterenol, a β-adrenoceptor agonist, were investigated in hearts perfused with H_2O_2.

Materials and methods

Isolated heart preparations

Adult male Sprague-Dawley rats (250–300 g) were sacrificed by decapitation and their hearts were rapidly removed and perfused according to the Langendorff procedure at a constant flow of 10 ml/min using Krebs-Henseleit buffer (K-H buffer) oxygenated with 95% O_2 and 5% O_2, pH 7.4 [18]. The composition of K-H buffer was in mM: NaCl 120.0, $NaHCO_3$ 25.0, KCl 4.7, KH_2PO_4 1.2, $CaCl_2$ 1.25, glucose 11.0. The hearts were electrically stimulated (Phipps and Bird stimulator) at 300 beats/min via a square wave of 1.5 ms duration at twice the threshold voltage. The left ventricular developed pressure (LVDP), the rate of change in developed pressure (+dP/dt) and the rate of change in relaxation (–dP/dt) were measured by using a water-filled latex balloon inserted into the left ventricle. The balloon was connected to a pressure transducer (Model 1050BP, Biopac System, Inc., Goleta, CA) and the volume of the balloon was adjusted to maintain the left ventricular end-diastolic pressure (LVEDP) of 10 mm Hg at the beginning of the experiment. Data were recorded online through an analogue-digital interface (MP100, Biopac System, Inc., Goleta, CA) stored and processed with Acknowledge 3.0.1 for windows (Biopac System, Inc., Goleta, CA). All hearts were allowed to equilibrate for 20–30 min with normal oxygenated K-H buffer before perfusing with H_2O_2 at the desired concentration for 10 min. In a separate series of experiments, hearts were perfused with H_2O_2 for 10 min in the presence or absence of catalase (CAT) (1.0×10^5 U/l) plus D-mannitol (MAN) (20 mM); control hearts were perfused for comparable periods with K-H buffer. H_2O_2 was purchased from Mallinckrodt (Chesterfield, MO) whereas CAT and MAN were obtained from Sigma (St. Louis, MO). At the end of each perfusion period, the hearts were removed and the ventricular tissue was used to prepare crude membranes according to the method described by Dixon *et al.* [19] and used for biochemical analysis. In some experiments, hearts perfused without (control) or with H_2O_2 for 10 min were used for studying the effect of isoproterenol (1 μM) infusion on contractile parameters.

β-Adrenergic receptor binding

To determine β_1- and β_2-adrenergic receptor binding, aliquots of control or experimental membrane preparations (0.1 mg/ml) were incubated for 60 min at 37°C with various concentrations (5–400 pM) of [^{125}I]-cyanopindolol (^{125}I-CYP) (2200 Ci/mmol) in the absence or presence of either 100 μM CGP 20712A (a highly selective β_1-antagonist) or 100 μM ICI 118,551 (a highly selective β_2-antagonist). Incubations were stopped by rapid vacuum filtration through Whatman GF/C filters. Specific binding to the β_1-adrenoceptors was calculated as the difference between [^{125}I]-CYP binding values in the absence (total) and presence (non-specific) of ICI-118,551, whereas β_2-adrenoceptors specific binding was the difference between [^{125}I]-CYP binding values in the absence (total) or presence (non-specific) of CGP-20712A. The values for maximal binding (B_{max}) and dissociation constant (K_d) were calculated from the Scatchard plot analysis of the binding data according to the interactive LIGAND program of Munson and Rodbard [20]. In agonist competition studies, a concentration of ^{125}I-CYP (0.07 nM) was used in the presence of increasing concentrations of β-agonist, isoproterenol (10^{-10} to 10^{-4} M). In this set of experiments, assays were performed by incubating membranes for 40 min at 37°C and terminating the reaction by rapid filtration on Whatman GF/C filters. The agonist binding curves for each preparation were plotted by subtracting the non-specific binding (displaced by 0.1 mM isoproterenol) from the total ^{125}I-CYP bound at each concentration of isoproterenol.

Determination of adenylyl cyclase activity

Adenylyl cyclase activity was determined by measuring [^{32}P] cAMP formation from [α-^{32}P] ATP as described previously [21]. Unless otherwise indicated the incubation assay medium contained 50 mM glycylglycine (pH 7.5), 0.5 mM MgATP, [^{32}P] ATP (1–1.5 × 10^6 cpm), 5 mM $MgCl_2$ (in excess of the ATP concentration), 100 mM NaCl, 0.5 mM cAMP, 0.1 mM EGTA, 0.5 mM 3-isobutyl-1-methylxanthine and 10 U/ml adenosine deaminase and an ATP-regenerating system comprising of 2 mM creatine phosphate, 0.1 mg creatine kinase/ml in a final volume of 200 μl. Incubations were initiated by the addition of membranes (30–70 μg) to the reaction mixture, which had been equilibrated for 3 min at 37°C. The incubation time was 10 min at 37°C and the reaction was terminated by the addition of 0.6 ml of 120 mM zinc acetate. It should be mentioned that 0.5 mM unlabelled cAMP was also added at this time in order to monitor the recovery of [^{32}P]-cAMP by measuring absorbancy at 259 nm. The determination of cAMP formation was carried out by co-precipitation of other nucleotides with $ZnCO_3$ by the addition of 0.5 ml of 144 mM Na_2CO_3 and subsequent chromatography by a double column system as described by Salomen *et al.* [22]. Under the assay conditions used, the adenylyl cyclase activity was linear with respect to protein concentration and time of incubation.

Statistical analysis of the data

All results are expressed as mean ± S.E. Difference between control and experimental groups was analysed statistically by using the Students *t*-test. For multi-group comparison, the data was evaluated by the analysis of variance (ANOVA) and Duncan's multiple-range test. A p value < 0.05 was taken to reflect a significant difference between the control and experimental preparations.

Results

Alteration in contractile response to isoproterenol

A time course effect of isoproterenol (1 μM) infusion on LVDP, +dP/dt, and –dP/dt was studied in isolated rat hearts perfused with 200 and 500 μM H_2O_2. The results in Fig. 1 show that perfusion with 200 μM H_2O_2 produced a significant decrease ($p < 0.05$) in the isoproterenol-induced increase of LVDP at all time points between 10–30 sec whereas significant ($p < 0.05$) depressions in +dP/dt and –dP/dt were seen at 25 and 30 sec only. On the other hand, perfusion with 500 μM H_2O_2 resulted in an almost complete loss of the positive inotropic action of isoproterenol. It should be pointed out that perfusion of rat hearts with H_2O_2 for 10 min produced a marked depressant action on the contractile activity. In this regard, the values for LVDP (mm Hg), +dP/dt (mm Hg/sec) and –dP/dt (mm Hg/sec) before starting isoproterenol infusion were 78.2 ± 4.0, 1324 ± 64 and 1306 ± 80 for control hearts; and 33.7 ± 6.5, 572 ± 132 and 448 ± 136 for hearts perfused with 200 μM H_2O_2 and 3.6 ± 0.1, 48 ± 8 and 36 ± 6 for hearts perfused with 500 μM H_2O_2, respectively. This depression in the cardiac contractile parameters including the response of isoproterenol upon H_2O_2 perfusion was 75–85% prevented when CAT was present in the perfusion medium. Addition of MAN in medium containing CAT did not afford further protection (data from 3 experiments not shown).

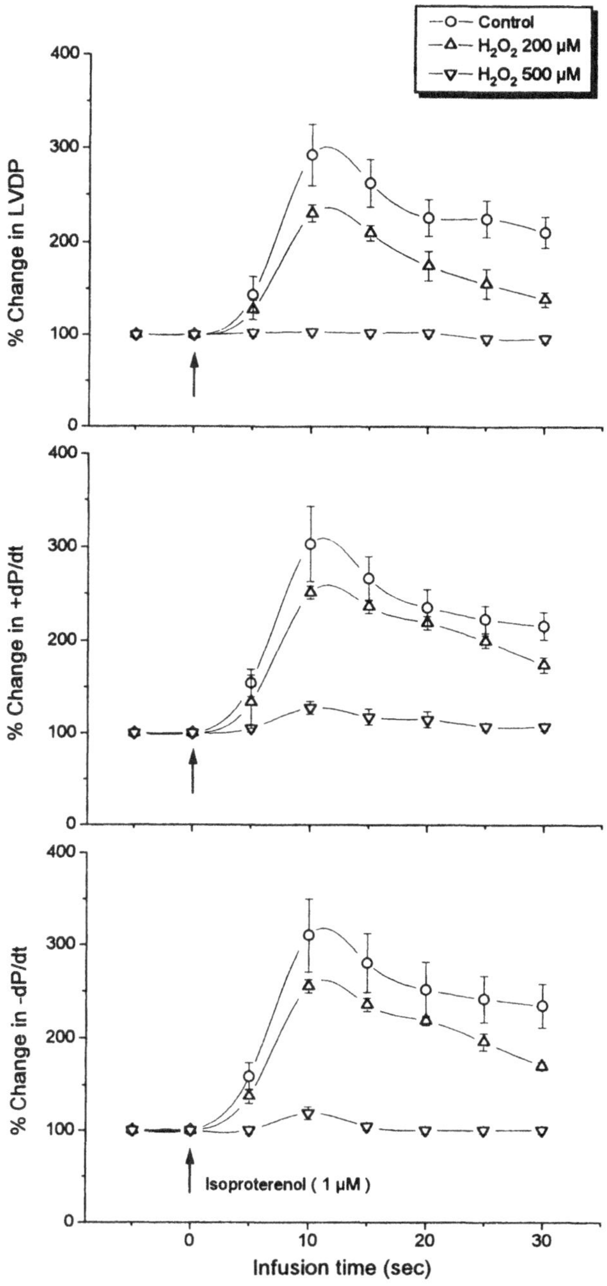

Fig. 1. Effect of infusion of isoproterenol (1 μM) on contractile parameters in control hearts and hearts perfused with H_2O_2 (200 or 500 μM) for 10 min. Each value is a mean ± S.E. of 5–6 separate experiments. LVDP: left ventricular developed pressure; +dP/dt: rate of change of developed pressure; –dP/dt: rate of change of relaxation.

Alteration in adenylyl cyclase response to isoproterenol

Preliminary experiments determined that 50% of the maximal activity of adenylyl cyclase in the presence of isoproterenol was decreased when hearts were perfused with 1 mM H_2O_2 for 10 min. Accordingly, the effect of H_2O_2 upon adenylyl cyclase activity in the presence of different concentrations of isoproterenol was studied by measuring the enzyme activity in membranes from control hearts and hearts perfused for 10

Table 1. Adenylyl cyclase activity in the presence of different concentrations of isoproterenol in membrane preparations from control hearts and hearts perfused with H_2O_2.

[Isoproterenol]	Adenylyl cyclase activity (pmol/mg protein/min) Control	H_2O_2	H_2O_2 + CAT + MAN
None	105 ± 7.0	65 ± 5.0*	97 ± 7.2#
0.1 μM	131 ± 9.1	89 ± 7.1*	119 ± 8.0#
1 μM	154 ± 11.1	104 ± 6.1*	140 ± 9.3#
10 μM	162 ± 11.0	109 ± 7.0*	145 ± 10.1#
100 μM	182 ± 12.2	113 ± 8.2*	167 ± 11.2#

Values are the mean ± S.E. of 6 separate experiments. The assay medium contained 10 μM Gpp(NH)p and 0.3% ascorbic acid. Adenylyl cyclase assays were performed on membranes isolated from rat ventricles perfused with 1 mM H_2O_2 in the absence or presence of catalase (CAT) and D-mannitol (MAN) for 10 min. Control hearts were perfused for 10 min with normal perfusion medium. Concentrations of CAT and MAN used were 1 × 10^5 U/L and 20 mM, respectively. *Significantly different from control at the respective concentration ($p < 0.05$); #Significantly different from its respective H_2O_2 group ($p < 0.05$).

min with 1 mM H_2O_2. The results in Table 1 indicate that there was a marked attenuation of isoproterenol stimulated adenylyl cyclase activity in membranes from hearts perfused with H_2O_2. The presence of CAT plus MAN was able to significantly prevent these alterations due to H_2O_2 (Table 1); removal of MAN from the medium did not decrease this preventive action.

Alterations in β-adrenoceptors

In order to demonstrate if β-adrenergic receptors were altered in membranes from hearts perfused for 10 min with 1 mM H_2O_2, the specific binding of ^{125}I-CYP to both β_1- and β_2-adrenoceptors was studied in cardiac membranes. Fig. 2 shows the specific binding data for β_1-adrenoceptors as well as Scatchard plot analysis of ^{125}I-CYP binding to β_1-receptors in control membranes and membranes from H_2O_2-perfused hearts. Both the density and affinity ($^1/K_d$) of the β_1-receptors were significantly reduced in membranes from H_2O_2-perfused

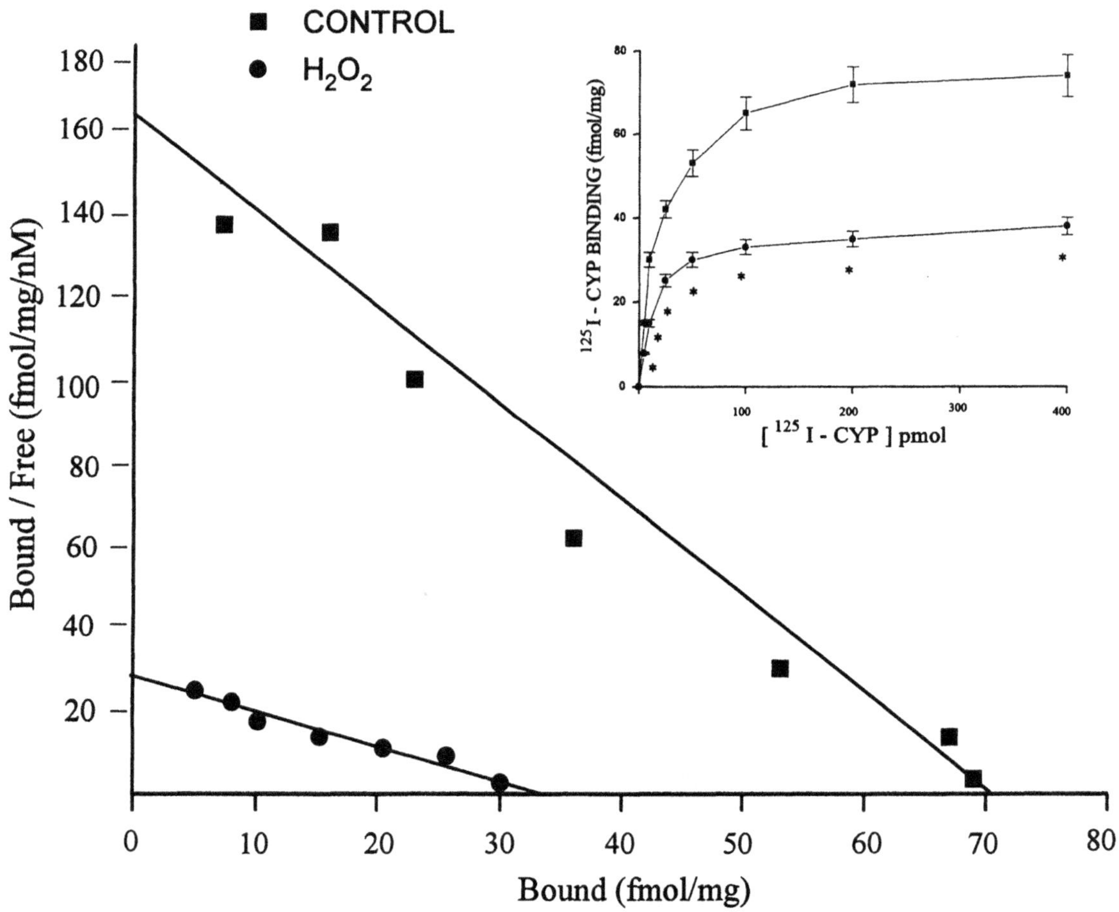

Fig. 2. Scatchard plot analysis of [^{125}I]-CYP binding in membranes from control hearts (■) and hearts perfused with 1 mM H_2O_2 for 10 min (●). Data represents a typical experiment performed in triplicate. Inset: Equilibrium specific binding of ^{125}I-CYP with membranes using ICI-118,551 (100 μM), from 5–6 separate preparations. *Significantly different from control ($p < 0.05$). B/F: Bound/Free [^{125}I]-CYP (iodocyanopindolol).

Table 2. Binding characteristics of ^{125}I-cyanopindolol to crude membrane preparations from rat hearts perfused with H_2O_2.

	Control	H_2O_2	H_2O_2 + CAT + MAN
β_1-adrenergic receptors			
K_d (pmol)	33 ± 2.3	66 ± 5.0*	43 ± 3.0#
B_{max} (fmol/mg)	71 ± 5.4	35 ± 2.8*	56 ± 3.2#
β_2-adrenergic receptors			
K_d (pmol)	13 ± 1.0	8.5 ± 0.6*	13 ± 1.0#
B_{max} (fmol/mg)	23 ± 2.0	14 ± 1.1*	22 ± 2.0#

Each value is a mean ± S.E. of 6 separate experiments. Membranes were isolated from rat hearts perfused with H_2O_2 (1 mM) for 10 min. Control hearts were perfused with normal medium for 10 min. Specific binding of ^{125}I-CYP at different concentrations with β_1- and β_2-adrenergic receptors were determined in the presence of ICI-118,551 and CGP-20712A compounds, respectively. K_d and B_{max} values were determined from Scatchard plot analysis of the data. The concentrations of catalase (CAT) and mannitol (MAN) were 1 × 10^5 U/l and 20 mM, respectively. *Significantly different from control ($p < 0.05$); #Significantly different from H_2O_2 group ($p < 0.05$).

hearts compared to controls (Fig. 2 and Table 2). Although Scatchard plot analysis for data on ^{125}I-CYP binding to β_2-adrenoceptors also revealed a depression in the density of this receptor subtype, an increase in the affinity ($^1/K_d$) of β_2-adrenoceptors was evident in the membranes from H_2O_2-perfused hearts (Table 2). The presence of CAT plus MAN in the perfusion medium prevented the H_2O_2-induced alteration in the β_1- and β_2-adrenoceptors (Table 2). The presence of MAN with CAT did not afford any greater protection than that seen with CAT alone. Agonist competition curves, using isoproterenol (Fig. 3), revealed no shift in the curve upon 10 min exposure to H_2O_2 compared to control suggesting the absence of any alterations to the status of coupled β-adrenoceptors with G_s-proteins in the H_2O_2-perfused hearts compared to control.

Adenylyl cyclase activity

The basal adenylyl cyclase activity in membranes from hearts perfused with 1 mM H_2O_2 was decreased significantly compared to control (Table 3). This was the case when the activity was measured in the presence of various stimulants such as forskolin, NaF or Gpp(NH)p (Table 3). These alterations due to H_2O_2 were markedly attenuated in the presence of CAT plus MAN (Table 3). The degree of protection against H_2O_2-induced changes in adenylyl cyclase activities was similar when CAT was used in the absence of MAN (data not shown).

Discussion

Perfusion of the heart with 1 mM H_2O_2 was observed to produce a depression of the isoproterenol-stimulated adenylyl cyclase activity in cardiac membranes. A decrease of isoproterenol-stimulated adenylyl cyclase activity in H_2O_2-treated cardiac membranes has also been reported by

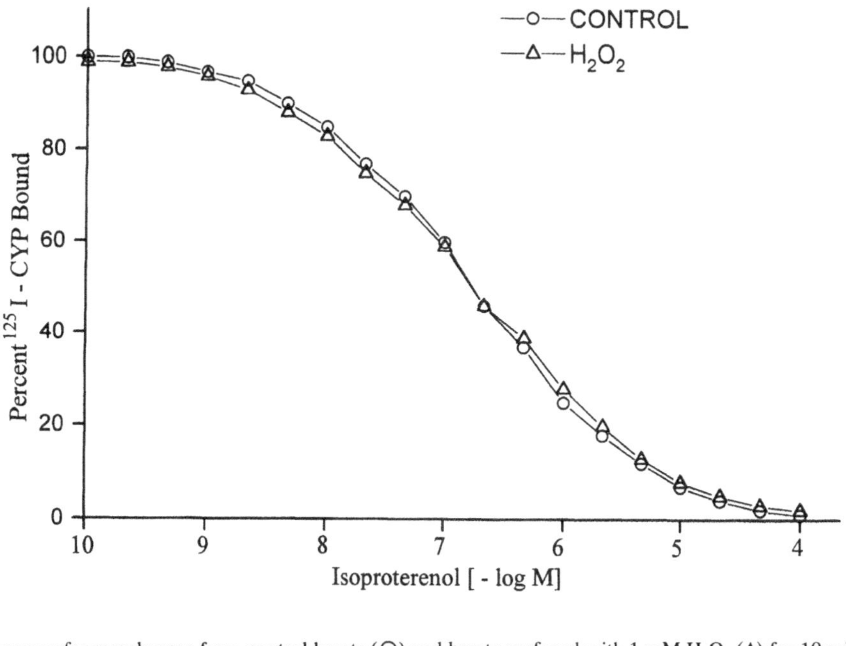

Fig. 3. Agonist competition curves for membranes from control hearts (O) and hearts perfused with 1 mM H_2O_2 (Δ) for 10 min. Data represents a typical of 3 experiments using 0.07 nM [^{125}I]-CYP as a ligand and different concentrations of isoproterenol as a competing agonist. [^{125}I]-CYP (iodocyanopindolol).

Table 3. Effect of forskolin, NaF, and Gpp(NH)p on the adenylyl cyclase activity in membrane preparation from control hearts and hearts perfused with H_2O_2 for 10 min.

	Adenylyl cyclase activity (pmol/mg protein/min)		
	Control	H_2O_2	H_2O_2 + CAT + MAN
Basal	42 ± 3.0	30 ± 2.0*	39 ± 2.1#
+ Forskolin (100 μM)	350 ± 30.0	250 ± 20.0*	308 ± 23.0#
+ NaF (5 mM)	161 ± 12.1	110 ± 7.1*	141 ± 11.0#
+ Gpp(NH)p (30 μM)	180 ± 12.2	138 ± 11.0*	163 ± 13.0#

Values are the mean ± S.E. of 5 separate experiments. Adenylyl cyclase assays were carried out with membranes isolated from rat hearts perfused with 1 mM H_2O_2 in the presence or absence of catalase (CAT) and D-mannitol (MAN) for 10 min. Control hearts were perfused for 10 min with normal perfusion medium. Concentrations of CAT and MAN used were 1.0×10^5 U/L and 20 mM, respectively. *Significantly different from its respective control ($p < 0.05$); #Significantly different from its respective H_2O_2 group ($p < 0.05$).

other investigators [16]. The depressant effect of H_2O_2 on the isoproterenol-stimulated adenylyl cyclase activity is unlikely to be an artefact because it was evident at various concentrations of isoproterenol in membranes obtained from hearts perfused with 1 mM H_2O_2. Such depression in the isoproterenol-stimulated adenylyl cyclase activity may be related to alterations of β-adrenoceptors in cardiac membranes by H_2O_2. This view is supported by our observations showing a decrease in both the density and affinity of β_1-adrenoceptors in cardiac membranes from hearts perfused with H_2O_2. Although the density of β_2-adrenoceptors in these cardiac membranes was also decreased, the affinity ($1/K_d$) of this subtype of β-adrenoceptors was increased. Such effects of H_2O_2 on the density and affinity of β_2-adrenoceptors would be of doubtful significance since opposite changes due to decreased density and increased affinity may cancel each other in terms of the isoproterenol-stimulated adenylyl cyclase activity. Although decreased density without any changes in the affinity or vice versa with respect to the β-adrenoceptors in cardiac membranes treated with H_2O_2 in earlier studies [14, 15] are at variance with the data reported in the present study, these differences appear to be due to the nature of ligand used for receptor binding studies as well as the type of membranes and experimental conditions employed. The observed changes in the β_1- and β_2-adrenoceptors are not due to any non-specific action of H_2O_2 on the membrane because Ca^{2+}-channel antagonist binding in cardiac membranes was not affected by H_2O_2 treatment under similar conditions [23].

The observed changes in the isoproterenol-stimulated adenylyl cyclase due to H_2O_2 cannot be considered to be due to any alterations in the proportion of coupled β-adrenoceptors in the high affinity state because no shift in the competition curves was evident upon treatment of cardiac membranes with H_2O_2. Nonetheless, the evidence presented in this study indicate that alterations in both the catalytic activity of adenylyl cyclase enzyme and its activation via G_s-proteins may explain the observed depression in the isoproterenol-stimulated adenylyl cyclase upon perfusing the heart with H_2O_2. In this regard, adenylyl cyclase activities in the absence (basal) and presence of forskolin, which is known to stimulate the catalytic subunit of the enzyme directly [24], were found to decrease upon perfusing the hearts with H_2O_2. Furthermore, adenylyl cyclase activities in the presence of Gpp(NH)p and NaF, which are known to stimulate the enzyme by interacting with G_s-proteins directly or indirectly [21], were depressed in the H_2O_2-perfused hearts. Thus it appears that the depression in the isoproterenol-stimulated adenylyl cyclase activity in hearts perfused with H_2O_2 may be due to changes in the characteristics of β_1-adrenoceptors, G_s-proteins and catalytic subunits of adenylyl cyclase.

A depression in the β_1-adrenoceptor-G_s-protein-adenylyl cyclase activity due to H_2O_2 can be seen to decrease the formation of cAMP [11–13] and this may attenuate the positive inotropic effect of catecholamines in the heart. In fact perfusing the hearts with H_2O_2 was found to depress the increase in LVDP, +dP/dt and –dP/dt due to isoproterenol infusion. However, it should be pointed out that a complete loss of the positive inotropic responses to isoproterenol was evident upon perfusing the heart with 500 μM H_2O_2 whereas perfusion of the heart with 1 mM H_2O_2 produced only a 50% depression of the isoproterenol-induced increase in the adenylyl cyclase activity in the membrane preparations. An analysis of the results in this study reveals that the depression in the isoproterenol-induced changes in contractile parameters upon perfusing the hearts with 500 μM H_2O_2 was of greater magnitude in comparison to that observed in the β_1-adrenoceptor-adenylyl cyclase complex in cardiac membranes obtained from hearts perfused with 1mM H_2O_2. This greater sensitivity of H_2O_2 in attenuating the inotropic responses of hearts to isoproterenol may be due to the fact that other cellular mechanisms which may be involved in contractile responses, are also affected by H_2O_2 [2, 5, 9, 23]. Alternatively, this mismatch in concentrations of H_2O_2 for isoproterenol-induced changes in cardiac contractile force development and activation of adenylyl cyclase raises questions regarding the contribution of β-receptor modification to the depression in the inotropic responses upon perfusing the heart with H_2O_2. Nonetheless, the observed depression in the β_1-adrenoceptor-G_s-protein-adenylyl cyclase complex cannot be considered to participate in raising the intracellular concentrations of Ca^{2+} in cardiomyocytes exposed to H_2O_2 as it will decrease both the influx of Ca^{2+} as well as release of Ca^{2+} from the intracellular Ca^{2+}-stores [12]. Furthermore, the results presented in this study indicate that the effects of H_2O_2 on the β-adrenoceptor-adenylyl cyclase complex are probably of a direct nature and not mediated through the highly reactive hydroxyl radicals generated by the interaction of H_2O_2 with redox-active transitional metals [2, 3]. This is evident from our observations that CAT was able to prevent the alterations effected by H_2O_2 and these beneficial effects of CAT were not further enhanced by the presence of MAN.

Although the exact mechanisms by which H_2O_2 may alter the β-adrenoceptor signal transduction system are unclear, it seems appropriate to indicate that the reduced β_1-adrenoceptor-G_s protein-adenylyl cyclase activity may be caused by the effect of H_2O_2-induced lipid peroxidation [2, 3] on the physical state of the membrane. This view is based on the fact that the trans-membrane β-adrenoceptor signal transduction has been observed to be depressed by a reduction in membrane fluidity [25], and lipid peroxidation has been reported to decrease the membrane fluidity [26]. The nature of the modifications of β_1-adrenoceptor-G_s-protein-adenylyl cyclase may also be suggestive of changes occurring at the level of the protein components of the β-adrenoceptor complex. This view is supported by the fact that aldehydes (malondealdehyde and 4-hydroxyneonatal) formed during lipid peroxidation have been shown to influence the function of β-adrenoceptor-adenylyl cyclase system in sarcolemma by reacting with NH_2- and/or SH-groups of these components [26, 27]. In fact both the adenylyl cyclase enzyme as well as β-adrenergic receptors are known to possess sulfhydryl groups in their active sites [28, 29], the modification of which may alter the characteristic of the proteins. Irrespective to the exact mechanisms, the results regarding the inhibitory effect of H_2O_2 on the β-adrenoceptor-G_s-protein-adenylyl cyclase pathway may be of some pathophysiological significance because the concentrations of H_2O_2 used in this study are comparable to those that have been reported to occur in the myocardium during ischemia-reperfusion injury [1, 30]. The involvement of H_2O_2 in inducing β-adrenoceptor signal transduction abnormalities in the ischemic-reperfused hearts was evident from the fact that these changes were markedly prevented by the presence of CAT in the perfusion medium [21]. Although the responses to catecholamines and the β-adrenoceptor mechanisms are attenuated in the ischemic-reperfused heart [21, 31, 32], the pattern of abnormalities in the β-adrenoceptors, G-proteins and adenylyl cyclase system may not be exactly similar to that seen upon perfusing the heart with H_2O_2. This view is based on the fact that different free radicals, oxidants and other metabolites, generated in the ischemic-reperfused hearts, may also affect the β-adrenoceptor-linked signal transduction mechanism. Thus, accumulation of H_2O_2 in the ischemic-reperfused hearts may be regarded as one of the several pathophysiologic factors which could produce defects in the β-adrenoceptor signal transduction pathway under this condition.

Acknowledgements

The research reported in this paper was supported by a grant from the Medical Research Council of Canada (MRC Group in Experimental Cardiology). Sujata Persad was a predoctoral fellow of the Heart and Stroke Foundation of Canada during the tenure of this study.

References

1. Cochrane CG: Cellular injury by oxidants. Am J Med 91: 235–305, 1991
2. Kaul N, Siveski-Iliskovic N, Hill M, Slezak J, Singal PK: Free radicals and the heart. J Pharmacol Toxicol Meth 30: 55–67, 1993
3. Kukreja RC, Hess ML: The oxygen free radical system: From equation through membrane-protein interactions to cardiovascular injury and protection. Cardiovasc Res 26: 641–655, 1992
4. Fliss H, Masika M, Eley DW, Korecky B: Oxygen radical mediated protein oxidation in heart. In: PK Singal (ed). Oxygen Radicals in the Pathophysiology of Heart Disease. Kluwer Academic Publishers, Boston, 1988, pp 71–90
5. Hata T, Kaneko M, Beamish RE, Dhalla NS: Influence of oxygen free radicals on heart sarcolemmal Na^+-Ca^{2+} exchange. Coronary Artery Disease 2: 397–407, 1991
6. Hayashi H, Miyata H, Watanabe H, Kobayashi A, Yamazaki N: Effects of hydrogen peroxide on action potentials and intracellular concentrations of guinea pig heart. Cardiovasc Res 23: 767–773, 1989
7. Kaminishi T, Matsuoka T, Yanagishita T, Kako KJ: Increase vs. decrease of calcium uptake by isolated heart cells induced by H_2O_2 and HOCl. Am J Physiol 25: C598–C607, 1989
8. Kaminishi K, Yanagishita T, Kako KJ: Oxidant injury to isolated heart cells. Can J Cardiol 5: 168–174, 1989
9. Kaneko M, Beamish RE, Dhalla NS: Depression of heart sarcolemmal Ca^{2+}-pump activity by oxygen free radicals. Am J Physiol 256: H368–H374, 1989
10. Rowe GT, Manson NH, Caplan M, Hess ML: Hydrogen peroxide and hydroxyl radical mediation of activated leukocyte depression of cardiac sarcoplasmic reticulum. Circ Res 53: 584–591, 1983
11. Bristow MR, Kantrowitz NE, Ginsburg R, Fowler MR: Beta adrenergic function in heart muscle disease and heart failure. J Mol Cell Cardiol 17, Suppl 2: 41–52, 1985
12. Dhalla NS, Pierce GN, Panagia V, Singal PK, Beamish RE: Calcium movements in relation to heart function. Basic Res Cardiol 77: 117–139, 1982
13. Stiles GL: Adrenergic receptor responsiveness and congestive heart failure. Am J Cardiol 67: 13C–17C, 1991
14. Haenen GRMM, Van Dansik P, Vermeulen NPE, Timmerman H, Bast A: The effect of hydrogen peroxide on β-adrenoceptor function in the heart. Free Rad Res Commun 4: 243–249, 1988
15. Kaneko M, Chapman DC, Ganguly PK, Beamish RE, Dhalla NS: Modification of cardiac adrenergic receptors by oxygen free radicals. Am J Physiol 260: H821–H826, 1991
16. Masuda H, Kaneko M, Hong RB, Ikegaya T, Hayashi H, Kobayashi A, Yamazaki N: Effects of hydrogen peroxide on stimulatory guanine nucleotide-binding protein in rat heart. Jpn Circ J 57: 1007–1015, 1993
17. Tan LM, Xenoyannis S, Feldman RD: Oxidant stress enhances adenylyl cyclase activation. Circ Res 77: 710–717, 1995

18. Persad S, Gupta KK, Dhalla NS: Status of Ca^{2+}-channels in hearts perfused with Ca^{2+}-free medium as well as upon reperfusion (Ca^{2+}-paradox). J Mol Cell Cardiol 27: 513–522, 1995
19. Dixon IMC, Lee S-L, Dhalla NS: Nitrendipine binding in congestive heart failure due to myocardial infarction. Circ Res 66: 782–788, 1990
20. Munson PJ, Rodbard D: Ligand: A versatile computerized approach for characterization of ligand binding systems. Anal Biochem 107: 220–239, 1980
21. Persad S, Takeda S, Panagia V, Dhalla NS: β-adrenoceptor-linked signal transduction in ischemic-reperfused heart and scavenging of oxyradicals. J Mol Cell Cardiol 29: 545–558, 1997
22. Salomon Y, Londos C, Rodbell M: A highly sensitive adenylate cyclase assay. Anal Biochem 58: 541–548, 1974
23. Kaneko M, Lee S-L, Wolf CM, Dhalla NS: Reduction of calcium channel antagonist binding sites by oxygen free radicals in rat heart. J Mol Cell Cardiol 21: 235–243, 1989
24. Bristow MR, Ginsberg R, Strasberg A, Montgomery W, Minobe W: Pharmacology and inotropic potential of forskolin in the human heart. J Clin Invest 74: 212–223, 1984
25. Peters R: Lateral mobility of proteins and lipids in the radical membrane and the activation of adenylyl cyclase by adrenergic receptors. FEBS Lett 243: 1–7, 1988
26. Paradisi L, Parangini C, Parola M, Barreva G, Dianzani MU: Effects of 4-hydroxyneonatal on adenylate cyclase and 5'-nuelcotidase activities in rat liver plasma membranes. Chem Biol Interac 53: 209–217, 1985
27. Haenen GRMM, Plug HJM, Vermeulen NPE, Timmerman H, Bast A: Contribution of 4-hydroxy-2,3-trans-neonatal to the reduction of β-adrenoceptor function by oxidative stress. Life Sci 45: 71–76, 1989
28. Padersen SE, Ross EM: Functional activation of β-adrenergic receptors by thiol in the presence or absence of agonist. J Biol Chem 260: 14150–14157, 1985
29. Skurat AV, Yurkova MS, Varanova LA, Gulyaev NN, Bulargina TV, Severin ES: Evidence for the existence of a sulfhydryl group in the adenylate cyclase active site. Biochem Int 10: 451–461, 1985
30. Turrens BF, Beconi M, Barilla J, Chavez VB, McCord JM: Mitochondrial generation of oxygen radicals during reoxygenation of ischemic tissue. Free Rad Res Commun 12: 681–689, 1991
31. Homcy CJ, Vatner SF, Vatner DE: β-adrenergic receptor regulation in the heart in pathophysiological states: abnormal adrenergic responsiveness in cardiac disease. Annu Rev Physiol 53: 137–159, 1991
32. Megg LG, Huang H, Li P, Capasso JM, Anversa P: Chronic nonocclusive coronary artery constriction in rats. β-adrenoceptor signal transduction and ventricular failure. J Clin Invest 88: 1940–1946, 1991

Molecular and Cellular Biochemistry **186**: 107–115, 1998.
© 1998 *Kluwer Academic Publishers.*

PBN spin trapping of free radicals in the reperfusion-injured heart. Limitations for pharmacological investigations

Norbert Vrbjar,[1] Stefan Zöllner,[2] Reiner F. Haseloff,[3] Margit Pissarek[4] and Ingolf E. Blasig[3]

[1]*Institute for Heart Research, Slovak Acadamy of Sciences, Bratislava, Slovak Republic;* [2]*Berlex Biosciences, Richmond, CA, USA;* [3]*Forschungsinstitut für Molekulare Pharmakologie, Berlin;* [4]*Institut für Pharmakologie und Toxikologie, Universität Leipzig, Leipzig, Germany*

Abstract

Post-ischemic reperfusion causes cardiac dysfunction and radical-induced lipid peroxidation (LPO) detectable by ESR spin trapping. This study deals with the applicability of the spin trapping technique to pharmacological investigations during myocardial reperfusion injury. The use of the spin trap phenylbutylnitrone (PBN, 3 mM) in isolated rat hearts demonstrated the release of alkoxyl radicals ($a_N = 1.39$ mT, $a_H^\beta = 0.19$ mT) formed particularly within the first 15 min of reperfusion following 30 min of ischemia. The decline of radicals, after 10 min of reperfusion, was accompanied by recovery of function in 80% of the hearts. The radical concentration in the coronary effluent (maximum after 7.5 min) was reduced by the infusion of 1 mM mercaptopropionylglycine (MPG, 2.7 ± 0.5 U/ml, $p < 0.001$) or 5 μM vitamin E (11.7 ± 0.8 U/ml, $p < 0.001$), compared to the (PBN-containing) control (29.7 ± 4.3 U/ml). Moreover, functional recovery (left ventricular developed pressure, LVDP 91.6 ± 20% of pre-ischemic level, $p < 0.05$) was improved by the hydrophilic radical scavenger MPG, compared to the (PBN-containing) control (LVDP 50.5 ± 15.7% of baseline). PBN alone led to higher functional recovery ($p < 0.05$) and reduced VF (duration of ventricular fibrillation; 7.10 ± 0.36 min/30 min, $p < 0.05$), compared to the untreated (PBN-free) control (LVDP 26.6 ± 11.8%; VF 19.42 ± 3.64 min/30 min). The Ca antagonist verapamil (0.1 μM), MPG, and the lipophilic vitamin E showed cardioprotection in the absence of PBN: post-ischemic recovery of LVDP was 25.4 ± 6.8% ($p < 0.05$), 39.6 ± 12.7% ($p < 0.05$) and 52.4 ± 2.6% ($p < 0.01$), respectively, compared to the corresponding untreated control (13.3 ± 6.6%). Whereas verapamil and vitamin E were able to protect the heart when present alone, they offered no additive effect in the presence of PBN. Therefore, PBN can be used to estimate the radical scavenger properties of an agent in the heart. However, because of the protective properties of PBN itself, the results of simultaneous investigations of the effects of other compounds, such as Ca antagonists or lipophilic radical scavengers, on heart function may be limited. (Mol Cell Biochem **186**: 107–115, 1998)

Key words: antioxidants, ESR spin trapping, ischemia/reperfusion, isolated rat heart, pharmacological investigations, radical-induced lipid peroxidation

Introduction

It is generally accepted that calcium overload and reactive species, such as radical intermediates of lipid peroxidation (LPO), contribute to myocardial injury induced during ischemia and reperfusion [1–5]. Use of calcium antagonists [6] is known to reduce the reperfusion-induced injury. The destructive effect of free radicals can be prevented by a variety of compounds which act as antioxidants either by scavenging the radicals or by inhibiting radical generation. Antioxidants, such as vitamin E, have been shown to reduce infarct size *in vivo* in the rat [7] and may also improve contractility upon reperfusion following global ischemia in isolated rat hearts [8]. In addition to this indirect evidence for

Address for offprints: I.E. Blasig, Forschungsinstitut für Molekular Pharmakologie, Alfred-Kowalke-Str. 4, D-10315 Berlin, Germany

an involvement of free radicals in reperfusion-induced damage the combination of electron spin resonance (ESR) spectroscopy with the spin trapping procedure has demonstrated radical production during the first few minutes of reperfusion [9–13]. Applying this technique using phenylbutylnitrone (PBN) as the spin trap a strong correlation between the time course of radical production and incidence of arrhythmias has been observed [12]. Thus, the spin trapping method gives information on radical formation which is related to the degree of the myocardial reperfusion injury.

The present study deals with the influence of cardioprotective agents, such as vitamin E, mercaptopropionylglycine (MPG) and verapamil, on the time-course of free radical production monitored by the formation of PBN radical adducts and the relation of this production to post-ischemic functional deterioration in the isolated rat heart. The aim of this investigation was to prove the applicability of the spin trap PBN for pharmacological studies of cardioprotective agents.

Material and methods

Heart perfusion technique

Male Sprague-Dawley rats (280–350 g) fed a standard diet (Alltromin, Germany) and water *ad libitum* were used. Animals were anesthetized with diethyl ether. Then, 60 sec after the injection of heparin (500 IU, intravenously) hearts were excised and placed immediately in ice-cold (4°C) perfusion medium. Hearts were cannulated via the aorta and perfused at 37°C in a non-recirculating Langendorff mode at a constant pressure of 100 cm H_2O (10 kPa). An incision was made in the pulmonary artery to ensure the ejection of the coronary effluent. Perfusion was performed with a Krebs-Henseleit medium gassed with 95% O_2 and 5% CO_2 (pH 7.4), containing (in mM): NaCl 118; KCl 5.8; $MgCl_2$ 1.2; $NaHCO_3$ 25; $CaCl_2$ 1.8; NaH_2PO_4 1.2 and glucose 11. To remove particulate impurities perfusion fluid was filtered through a cellulose nitrate filter (5 μM pore size) [12].

Heart function and electrophysiological disturbances

Heart function was evaluated throughout the experiment by measuring left ventricular pressure (LVP), electrical activity of the myocardium and coronary flow (CF). CF was measured by timed collection of the effluent from the heart. LVP was measured by means of a fluid-filled balloon catheter inserted into the left ventricle via the left atrium. The balloon was initially inflated until the left vetricular enddiastolic pressure (LVEDP) reached 0.4 kPa. The catheter was connected to a pressure transducer (W 112, Biomonitor BMT 3111, Zwönitz, Germany). LVDP (left ventricular developed pressure = left ventricular systolic peak pressure - LVEDP), LVEDP, rate of maximum and minimum pressure development (dP/dt_{max}, dP/dt_{min}, for contractility and relaxation, respectively) were recorded with a precalibrated multichannel physiograph (3 NEK-1, Zwönitz, Germany). Pressure-rate product (PRP) was calculated from the product of LVDP and heart rate (HR) [14]. An epicardial electrocardiogram (ECG) was recorded by using two silver electrodes attached to the apex of the heart and the cannula near to the root of the aorta. The ECG was used for the analysis of cardiac rhythm and HR. Rhythm disturbances were analyzed according to the Lambeth conventions [15] and classified as follows: normal sinus rhythm (SR); ventricular premature beats (VPB) of duration less than 20 sec; bigeminy (BG) or salvos (S) or VPB lasting more than 20 sec; ventricular tachycardia (VT) or bradycardia (BR); non-sustained (< 2 min) ventricular fibrillation (VF); sustained (> 2 min) VT or VF. Hearts with sustained VF or VT persisting beyond 20 min of reperfusion were classified as non-recovering. The proportion of recovering hearts in an experimental group is expressed as the incidence of functional recovery and is given in percent.

After a stabilization phase of 15–20 min, hearts were perfused aerobically for 30 min (pre-ischemic control period). Three hearts did not achieve a steady or regular level of function and were excluded from this study (injury during preparation resulted in BR in one, cardiac arrest in a second, and aortic laceration in the third). Global normothermic ischemia was then induced by clamping aortic inflow for 30 min. After ischemia aortic inflow was resumed for 30 min (reperfusion period). During the preischemic control period (before any treatment), 3 baseline values of heart function were measured, averaged and used for normalizing the data. Absolute baseline values in untreated controls were (mean ± S.E.M.): HR, 287 ± 10 beats/min; CF, 11.2 ± 0.4 ml/min; LVDP, 20.7 ± 0.6 kPa; LVEDP, 0.6 ± 0.1 kPa; $LVdp/dt_{max}$, 763 ± 29 kPa/s; dp/dt_{min}, 432 ± 15 kPa/s; PRP, 5862 ± 127 kPa/min (n = 13) (the factor to convert kPa to mmHg is 7.5).

Sample collection and radical detection

Free radicals were detected by the spin trapping procedure and subsequent ESR measurement of the spin adduct formed [13, 16]. *N-tert*-Butyl-α-phenylnitrone (PBN, Sigma) was used as spin trap. All procedures using PBN were done in the dark to prevent light effects on the spin trap or its adducts. An aqueous solution of 120 mM PBN containing 40 mM NaCl was infused into the perfusion medium (infusion pump, Lineomat, MLW, Germany) just above the heart. PBN was infused during the last 10 min prior to ischemia and during reperfusion. The infusion speed was adjusted to 1/40 of CF, ensuring a final PBN concentration of 3 mM. Before the onset of ischemia and during reperfusion 5 ml aliquots of coronary effluent were collected for ESR measurements. For positive control experiments (identification of the radical adduct) 1 mM cumene hydroperoxide (Sigma) was added to 0.2 mM $FeSO_4$ in the perfusion buffer containing 10 mM PBN. The spin adducts were extracted into 0.75 ml of toluene by mixing for 30 sec (Vortex VF2, Janke and Kunkel, Germany) and centrifugation (1500 × g, Janetzki T32, Germany) for 5 min at room temperature. After the toluene extracts were bubbled with N_2 for 2 min, spectra were recorded in 4 mm quartz tubes at room temperature on a Bruker ECS 106 spectrometer with modulation frequency 100 kHz, microwave power 10 mW, sweep width 50.0 G, sweep time 670 sec, modulation amplitude 1.0 G, and receiver gain $1 \cdot 10^6$. The signal intensity (in U) representing the concentration of the spin adduct (PBN/RO·) was adjusted for the sample volume of the coronary effluent and expressed as concentration of PBN/RO· (in U/ml). Myocardial release of PBN/RO· at a given time of reperfusion was calculated as the product of radical concentration and CF.

Pharmacological investigations

The cardioprotective concentration of the compounds investigated was determined in preceding dose-response experiments. (±)-Verapamil hydrochloride (Sigma) was infused for the last 10 min before ischemia and during the first 15 min of reperfusion. Concentrations ≥ 0.1 mM improved funtional recovery (LVDP, PRP, $LVdp/dt_{max}$, $LVdp/dt_{min}$); 0.01–0.1 mM reduced the duration of VF and VT. Therefore, 0.1 mM verapamil was selected for further studies. N-(2-mercaptopropionyl)-glycine (MPG, Fluka) infused for 20 min before ischemia and during reperfusion) showed protective effects between 0.1 and 1 mM (improved recovery of CF, $LVdp/dt_{max}$; reduction in duration of VT). Therefore, 1 mM MPG was selected. Vitamin E (DL-α-tocopherol, Serva) was not soluble in sufficient amounts in the medium. Therefore, vitamin E (5 μM) was infused during the last 20 min before and during the first 20 min after ischemia as a solution containing 3.8 μM bovine serum albumin (BSA, Sigma). In separate experiments, BSA was studied alone. Stock solutions of all agents investigated were prepared in concentrations exceeding 80 times the required final concentration, and they were added by a second infusion pump.

Statistics

Measures of myocardial function (HR, CF, LVDP, LVdp/dt, LVEDP, PRP) and radical concentration are expressed as mean ± S.E.M. The mean values of a group were compared by the parameter-free Wilcoxon test. For incidence of functional recovery, a one-way analysis of variance was first carried out to test for any difference between the mean values of all groups. If differences were established, an overall χ^2 test for 2 × n table was constructed followed by a sequence of 2 × 2 χ^2 tests to compare individual groups. A p value of $p < 0.05$ was considered significant.

Results

Effect of cardioprotective agents on post-ischemic heart function

Ischemia/reperfusion induced profound disturbances to untreated control hearts. Only 4 of 13 hearts showed recovery of heart work during reperfusion (Fig. 1). Functional parameters (LVDP, LV dP/dt_{max} LV dP/dt_{min}, PRP) recovered to values of approximately 10% after 30 min of reperfusion (Tables 1 and 2) when compared to the pre-ischemic period. Verapamil, MPG, BSA and vitamin E+BSA increased the incidence of functional recovery to at least 67% (Fig. 1A) and the recovery of functional parameters up to 52% (most

Table 1. Effect of verapamil (0.1 μM, n = 12), mercaptopropionyl glycine (MPG, 1 mM, n = 12), bovine serum albumin (BSA, 3.8 μM, n = 9), and BSA plus vitamin E (BSA+VitE; 3.8 μM BSA + 5μM vitamin E, n = 9) on functional recovery of isolated rat hearts after 30 min of reperfusion following 30 min of global ischemia.

% recovery of:	Control	Verapamil	MPG	BSA	BSA+VitE
HR	33.6 ± 12.8	71.6 ± 10.4*	64.5 ± 14.0*	64.4 ± 16.2	93.5 ± 3.3*
LVDP	13.3 ± 6.6	25.4 ± 6.8*	39.6 ± 12.7*	30.0 ± 9.3*	52.4 ± 2.6**#
$LVdp/dt_{max}$	9.3 ± 5.4	18.7 ± 6.4*	38.3 ± 8.8**	24.5 ± 8.8*	40.7 ± 3.4**#
$LVdp/dt_{min}$	12.3 ± 6.3	23.5 ± 7.4*	41.9 ± 9.5**	31.0 ± 10.3*	51.2 ± 3.7**
LVEDP	2331 ± 196	1723 ± 321*	1231 ± 275**	2267 ± 221	1731 ± 83*#
CF	63.4 ± 5.2	64.8 ± 5.6	70.9 ± 2.7	60.6 ± 2.7	60.7 ± 2.1
PRP	10.4 ± 6.4	22.4 ± 6.4*	38.3 ± 8.6*	27.6 + 9.2*	47.0 ± 2.7**#

Data are given in % (related to the pre-ischemic values) and represent means ± S.E.M. *, **p < 0.05 and p < 0.01, respectively, compared to the control (no treatment, n = 13). #p < 0.05 compared to the BSA group.

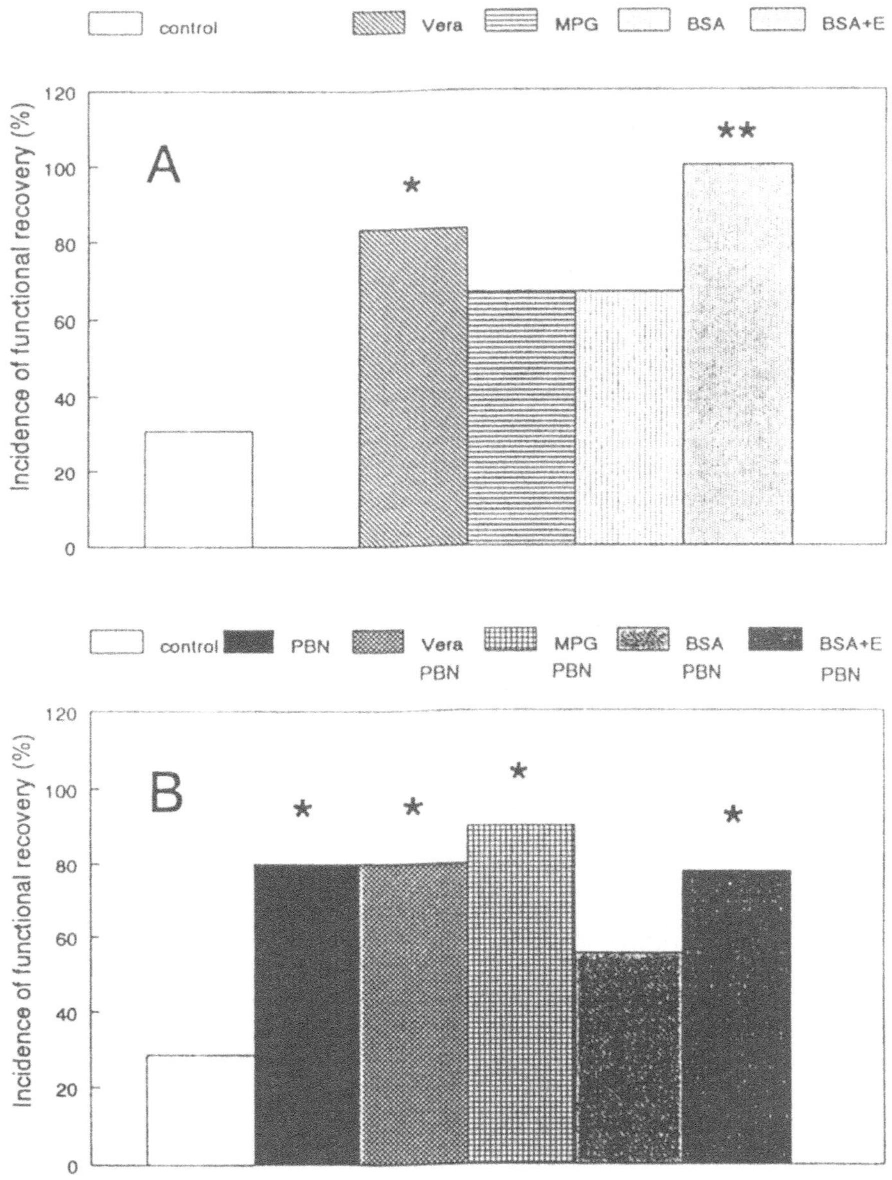

Fig. 1. Effect of verapamil (Vera, 0.1 μM), mercaptopropionyl glycine (MPG, 1 mM), bovine serum albumin (BSA, 3.8 μM) and BSA + vitamin E (BSA+E; 3.8 μM BSA + 5μM vitamin E) on the incidence of functional recovery of isolated rat hearts after 30 min of reperfusion following 30 min of global ischemia (A) in the absence and (B) in the presence of 3 mM phenylbutylnitrone (PBN). Data are given in % (incidence of functional recovery represents the proportion of hearts showing contractile function during reperfusion, related to the total number of hearts in the same group). A: control, no treatment, n = 13; Vera, n = 12; MPG, n = 12; BSA, n = 9; BSA+E, n = 9. B: control, no treatment, n = 14; PBN, n = 10; Vera and PBN, n = 10; MPG and PBN, n = 10; BSA and PBN, n = 9; BSA+VitE and PBN, n = 9. *$p < 0.05$ and **$p < 0.01$, respectively, compared to the respective control.

values between 25–40%) (Table 1). The effect of BSA was moderate and significantly less than that of BSA+vitamin E (Fig. 1, Table 1). CF of all PBN-free groups recovered during reperfusion to 60–70% of that of the respective pre-ischemic values (Table 1). Without treatment, functional recovery occurred 15–20 min after the onset of reperfusion. This interval was shortened to about 8 min by MPG and BSA, to 3–4 min by vitamin E+BSA and to the first 2 min by verapamil. In all these groups the LVEDP showed the highest values at the beginning of reperfusion (data not shown).

PBN significantly increased the proportion of recovering hearts to 80% compared to the respective control (29%, p <

Table 2. Effect of verapamil (0.1 μM, n = 10), mercaptopropionyl glycine (MPG, 1 mM, n = 10), bovine serum albumin (BSA, 3.8 μM, n = 9), and BSA + vitamin E (BSA+VitE; 3.8 μM BSA + 5 μM VitE, n = 9) on functional recovery of isolated rat hearts after 30 min of reperfusion following 30 min of global ischemia in the presence of 3 mM phenylbutylnitrone (PBN).

% recovery of:	Control	PBN	Verapamil +PBN	MPG +PBN	BSA +PBN	BSA+VitE +PBN
HR	27.9 ± 12.4	65.3 ± 12.9*	52.3 ± 11.0	69.6 ± 9.7**	40.4 ± 15.8	57.7 ± 16.5*
LVDP	26.6 ± 11.8	50.5 ± 15.7*	60.3 ± 12.0*	91.6 ± 20.0**§	42.7 ± 17.2	70.3 ± 20.7**
$LVdp/dt_{max}$	24.9 ± 11.5	45.8 ± 9.6*	57.1 ± 17.7**	76.7 ± 14.1**§	39.6 ± 14.5	60.0 ± 16.4**
$LVdp/dt_{min}$	21.7 ± 9.9	47.4 ± 12.6**	46.6 ± 13.1**	69.9 ± 13.4**§	45.1 ± 16.8	57.4 ± 14.4**
LVEDP	1869 ± 596	1214 ± 395	784 ± 218	1087 ± 344	1691 ± 914	1363 ± 741
CF	72.2 ± 5.7	44.3 ± 5.5**	52.5 ± 10.0**	45.5 ± 7.2**	40.2 ± 5.1**	39.9 ± 7.3**
PRP	25.8 ± 11.4	35.3 ± 8.5	39.8 ± 9.1*	61.3 ± 13.9**	34.0 ± 14.1	40.3 ± 12.5*

Data are given in % (related to the pre-ischemic values) and represent means ± S.E.M. *, **p < 0.05 and p < 0.01, respectively, compared to the control (no treatment, n = 14). §p < 0.05 compared to the PBN group (n = 10).

0.05, Fig. 1B) and increased the recovery of functional parameters to about 50% (Table 2). Among the cardioprotective agents tested in the presence of PBN, only MPG induced a notable additional increase (to 90%) in the proportion of recovering hearts (Fig. 1B) as well as the extent of recovery of functional parameters (60–90%, Table 2). CF recovered to values of about 40–50% (Table 2).

Effect of cardioprotective agents on post-ischemic rhythm disturbances

All groups developed severe rhythm disturbances (VF, VT) within the first 5 min of reperfusion, and they revealed a high incidence of BG. Between 5 and 10 min of reperfusion, the recovering hearts stabilized to sinus rhythm with occasional occurrence of VPB or BG. By contrast, in non-recovering hearts frequent and sustained VT or VF, respectively, persisted till the end of the experiment. The compounds tested showed antiarrhythmic efficacy similar to that observed when analyzing functional parameters. The duration of VF was reduced by all agents studied, but this effect was statistically significant only in the presence of BSA when compared to the untreated (PBN-free) control hearts (Table 3). This effect was most dramatic when BSA was combined with vitamin E. PBN also significantly reduced the duration of VF. An additive effect was observed when verapamil, MPG, BSA or vitamin E+BSA was given simultaneously with PBN (Table 4). Hearts treated with PBN and verapamil at the same time showed bradycardia in combination with VT or VF during the infusion (data not shown).

Table 3. Effect of verapamil (0.1 μM, n = 12), mercaptopropionyl glycine (MPG, 1 mM, n = 12), bovine serum albumin (BSA, 3.8 μM, n = 9), and BSA + vitamin E (BSA+VitE; 5 μM VitE + 3.8 μM BSA, n = 9) on duration of ventricular fibrillation (VF) and ventricular tachycardia (VT) in isolated rat hearts during 30 min of reperfusion following 30 min of global ischemia.

Duration of:	Control	Verapamil	MPG	BSA	BSA+VitE
VF	19.24 ± 3.40	11.64 ± 2.68	16.58 ± 3.61	8.39 ± 2.65*	3.49 ± 1.47**
VT	1.73 ± 0.82	1.24 ± 0.48	1.11 ± 0.92	3.54 ± 1.22	0.86 ± 0.33

Data are given in min and represent mean ± S.E.M. *, **p < 0.05 and p < 0.01, respectively, compared to control (no treatment, n = 13).

Table 4. Effect of verapamil (0.1 μM, n = 10), mercaptopropionyl glycine (MPG, 1 mM, n = 10), bovine serum albumin (BSA, 3.8 μM, n = 9), and BSA + vitamin E (BSA+VitE; 3.8 μM BSA + 5 μM VitE, n = 9) on duration of ventricular fibrillation (VF) and ventricular tachycardia (VT) in isolated rat hearts during 30 min of reperfusion following 30 min of global ischemia in the presence of 3 mM phenylbutylnitrone (PBN).

Duration of:	Control	PBN	Verapamil +PBN	MPG +PBN	BSA +PBN	BSA+VitE +PBN
VF	19.42 ± 3.64	7.10 ± 0.36*	2.86 ± 1.51*§	0.92 ± 0.60**§	2.54 ± 0.82*§	2.35 ± 1.32**§
VT	0.98 ± 0.33	3.44 ± 2.25	3.86 ± 2.62	3.82 ± 2.87	0.84 ± 0.42§	1.90 ± 1.09

Data are given in min and represent mean ± S.E.M. *, **p < 0.05 and p < 0.01, respectively, compared to control (no treatment, n = 14). §p < 0.05 compared to the PBN group (n = 10).

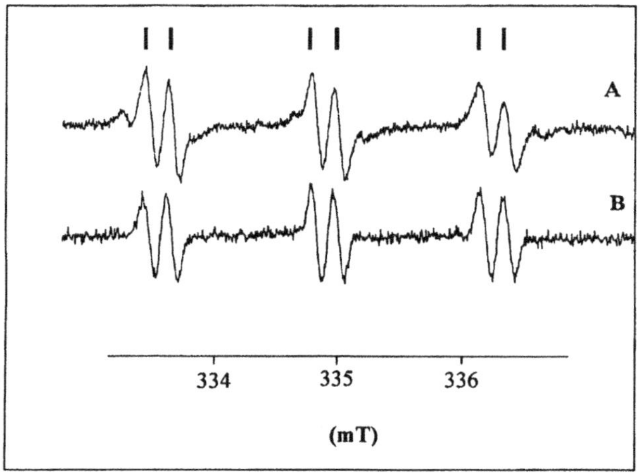

Fig. 2. (A) ESR spectrum of an alkoxyl radical adduct extracted into toluene after addition of 1 mM cumene hydroperoxide to a buffer solution in the presence of 3 mM PBN containing 0.2 mM $FeSO_4$. (B) Representative ESR spectrum of an alkoxyl radical adduct of PBN observed in a toluene extract of the coronary effluent of a heart 1 min after the onset of reperfusion (no additional treatment). Hyperfine splitting constants of the PBN radical adducts (indicated by I) were $a_N = 1.39$ mT and $a_H{}^\beta = 0.19$ mT; g = 2.0063.

Effect of cardioprotective agents on post-ischemic radical formation

The presence of PBN adducts in coronary effluent is documented by a representative ESR spectrum of a sample taken during the first min of reperfusion (Fig. 2B, no additional treatment). Positive control experiments were performed to identify the spin trap adduct observed in the coronary effluent. The formation of an alkoxyl radical, known to be produced after addition of ferrous ion to cumene hydroperoxide in the presence of PBN [9, 17], resulted in a spectrum (Fig. 2A) similar to that registered in the perfusate, showing comparable hyperfine splitting constants of $a_N = 1.39$ mT, $a_H{}^\beta = 0.19$ mT and a g-value of 2.0063. These signals were not observed under the same conditions when cumene hydroperoxide was omitted or in the perfusate before it entered the heart.

The concentration of PBN adducts in the effluent showed 2 maxima (after 1 and 7.5 min of reperfusion) similar to those found in earlier studies [17]. The concentration of PBN/RO· was significantly lowered by MPG and by BSA plus vitamin E, especially between 7.5 and 15 min of reperfusion (Fig. 3), whereas verapamil had little effect. The values of radical concentration were generally higher in non-recovering hearts than in recovering hearts; moreover, the latter showed no or only a small second peak of the adduct concentration (U/ml), both in the presence of PBN alone and in the presence of PBN plus a radical scavenger (data not shown). Changes in the myocardial release fraction of PBN/RO· generally paralleled alterations in effluent concentration (data not shown).

Discussion

The aim of this study was to investigate the applicability of the spin trap compound PBN for pharmacological investigations

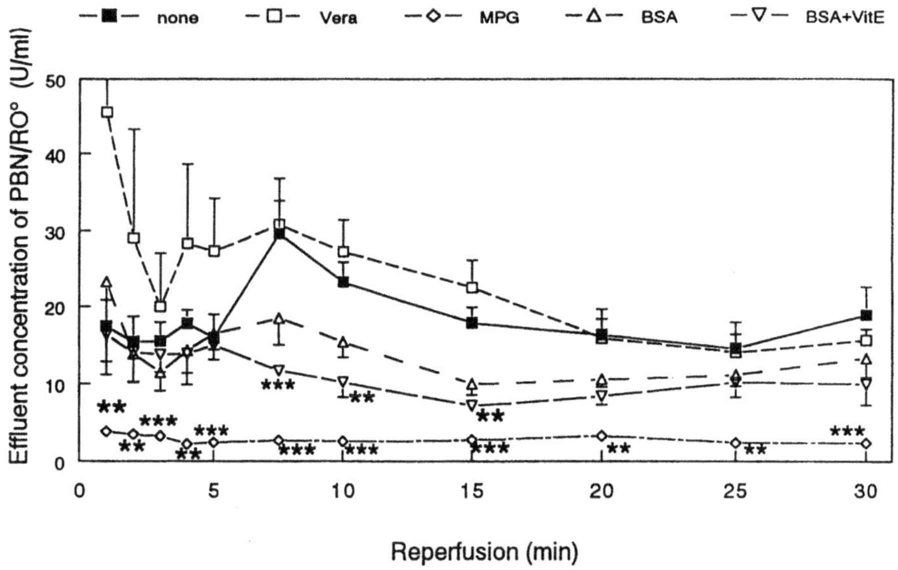

Fig. 3. Effect of verapamil (Vera, 0.1 μM, n = 10), mercaptopropionyl glycine (MPG, 1 mM, n = 10), bovine serum albumin (BSA, 3.8 μM, n = 9) and BSA + vitamin E (BSA+E; 3.8 μM BSA + 5μM vitamin E, n = 9) on the concentration of PBN/RO· in the coronary effluent of isolated perfused rat hearts during reperfusion following 30 min of global ischemia. *p < 0.05, **p < 0.01, and ***p < 0.001, respectively, compared to control treatment (none).

of additional agents. In the post-ischemic myocardium PBN has successfully been used to detect radicals formed via LPO [13, 17]. In the present experiments alkoxyl radicals have been identified. This is in agreement with our positive control experiments and those of others [18] showing lipid soluble alkoxyl radical adducts of PBN with ESR characteristics nearly identical to those of the adduct formed during reperfusion injury. Alkoxyl radicals are very reactive intermediates known to occur during LPO [19]. Recently it has been demonstrated that the detection of LPO-derived free radicals provides a reliable index of oxidative injury to the post-ischemic myocardium [9, 17]. These lipid radicals represent the destruction of membrane-derived lipids attacked by other more directly formed radicals [19]. Formation of these radicals in the heart implies that the endogenous radical defenses have been overcome. Therefore, these radicals are able to induce myocardial injury.

The results of this investigation agree with previous *in vitro* reports [9, 12, 20, 21] which demonstrate a transient burst of LPO-derived free radicals during the early minutes of reperfusion following ischemia. Similar patterns of post-ischemic free radical production have also been reported following coronary artery ligation in animal models [10, 18] and in patients after post-operative restoration of blood flow following coronary bypass surgery [22] and transluminal coronary angioplasty [23].

Moreover, the study shows that PBN is suitable to monitor the radical scavenger activity of cardioprotective agents. This can be demonstrated for water-soluble radical scavengers, such as MPG, and lipid-soluble radical scavengers, such as vitamin E. Both compounds are able to diminish the concentration of lipid radicals in the coronary effluent indicating radical scavenger activity or inhibition of radical generation in the myocardium. The reduction of radical concentration in the reperfused heart by MPG has been also reported by other authors using intact animals [10]. As expected the calcium antagonist verapamil has not shown antioxidative effects when administered in cardioprotective, submicromolar concentrations. Only concentrations several orders of magnitude higher effectively scavenge free radicals [24]. However, such high dosages have been found to be cardiotoxic because of the specific calcium channel blocking action of verapamil.

It has been shown recently that spin trapping compounds may affect heart function in a concentration-dependent manner. *In vivo* studies in dogs showed that PBN can be used as a spin trap at concentrations of 2.5 mM or less, but at 5 mM it can induce toxic effects [25]. In Langendorff-perfused rat hearts, PBN in concentrations up to 3 mM induces slight vasodilation. Increasing the PBN concentration to 10 mM significantly decreases left ventricular developed pressure and coronary flow [26]. Therefore, in our experiments the concentration of PBN entering the rat hearts was kept at 3 mM to avoid PBN-dependent toxicity. Nevertheless, this concentration has been shown to be sufficient for identification and detection of PBN adducts in the coronary effluent.

On the other hand, the investigation of cardioprotective properties of a test compound is limited in the presence of PBN. This is due to the powerful protective effect of PBN itself which is in the same range as that of the other agents tested. Only MPG shows an additive protection of heart function in the presence of PBN. This is possibly related to its high water solubility, thus making it particularly protective of hydrophilic structures, while PBN cannot easily act in an aqueous environment. Compared to MPG, PBN is a rather lipophilic agent with a partition coefficient of 15 [27] which would favor an action on hydrophobic structures. Therefore, it is concluded that pharmacological studies of hydrophilic agents are possible in the presence of PBN; functional investigations of lipophilic compounds, such as vitamin E which is insoluble in water, with properties similar to those of the spin trap are assumed to be of limited value. Moreover, it has been demonstrated that PBN may possess calcium antagonist activity [28] which may additionally explain the failure of the calcium antagonist verapamil to demonstrate an additive effect. The combined calcium channel blockade may also explain the bradycardia induced during reperfusion when both verapamil and PBN were infused.

PBN significantly increases the probability for functional recovery of the hearts during the post-ischemic period, due to the reduction of incidence and duration of severe ventricular arrhythmias. Significant improvement has also been observed for contractile parameters when comparing the untreated control group with those groups treated with PBN. These observations suggest a potent protective action of PBN related to the radical scavenger-like activity of the spin trap itself. This effect implies the importance of lipid free radicals in the reperfusion-induced injury of the heart since lipid-derived radicals have been detected by PBN. In all groups, the LVEDP was enhanced during the first minutes of reperfusion when the respective arrhythmia score was high (due to sustained VT or VF). Simultaneously the concentration of lipid radical adducts was elevated suggesting a relationship between the radical concentration and arrhythmias. Our finding is in agreement with that of others [29] showing that VF may be responsible for the formation of hydroxyl radicals in aerobically perfused rat hearts, which, in turn, may induce LPO [24] as detected in this study.

The calcium antagonist verapamil, known as an anti-arrhythmic drug, does not reduce the radical concentration in the post-ischemic coronary effluent of the heart. It tends to increase the PBN/RO· concentration, but it still protects the hearts. It is conceivable that cellular calcium accumulation, in addition to reactive species, plays a very important role in the induction of disturbances caused by reperfusion. Another reason for this paradoxical effect on radical concentration might be that verapamil-treated hearts have started to work already in the early period of reperfusion. It is known that contractile function may favor radical formation during this interval. Thus, hearts of other experimental groups without contractile function due to VT or VF during the same period show less radical formation.

Finally, it can be concluded that the application of the spin trap PBN and probably of other nitrone spin traps as well is a suitable tool to determine the radical scavenger potential of cardioactive agents in the myocardium. However, in the presence of PBN, the simultaneous demonstration of cardioprotective properties of a compound is limited to water soluble substances or compounds with a mode of action different from that of PBN. This limitation is due to the powerful anti-arrhythmic effect and protective functional activity of PBN during ischemia/reperfusion, especially in the lipophilic compartment of heart tissue.

Acknowledgments

The help of Dr. Katrin Schönheit and Dr. Volodja Lakomkin during preliminary experiments and the technical assistance of Gislinde Hartmann are gratefully acknowledged.

References

1. Buja LM, Hagler HK, Willerson JT: Altered calcium homeostasis in the pathogenesis of myocardial ischemic and hypoxic injury. Cell Calcium 9: 205–217, 1988
2. Nayler WG, Elz JS, Buckley DJ: Dissociation of hypoxia-induced calcium gain and rise in resting tension in isolated rat hearts. Am J Physiol 254: H678–H685, 1988
3. Ferrari R, Ceconi C, Curello S, Guarnieri C, Caldarera CM, Albertini A, Visioli O: Oxygen-mediated myocardial damage during ischemia and reperfusion: Role of the cellular defences against oxygen toxicity. J Mol Cell Cardiol 17: 937–945, 1985
4. Kehrer JP: Concepts related to the study of reactive oxygen and cardiac reperfusion injury. Free Rad Res Comms 5: 305–314, 1988
5. Marklund SL: Role of toxic effects of oxygen in reperfusion damage. J Mol Cell Cardiol 20: 23–30, 1988
6. Nayler WG, Ferrari R, Williams A: Protective effect of pretreatment with Verapamil, Nifedipine and Propranolol on mitochondrial function in the ischemic and reperfused myocardium. Am J Cardiol 46: 242–248, 1980
7. Kramer JH, Misik V, Weglicki WB: Magnesium-deficiency potentiates free radical production associated with postischemic injury to rat hearts: Vitamin E affords protection. Free Rad Biol Med 16: 713–723, 1994
8. Janero DR: Therapeutic potential of vitamin E against myocardial ischemic-reperfusion injury. Free Rad Biol Med 10: 315–324, 1991
9. Kramer JH, Misik V, Weglicki WB: Lipid peroxidation-derived free radical production and postischemic myocardial reperfusion injury. Ann NY Acad Sci 723: 180–196, 1994
10. Bolli R, Jeroudi MO, Patel BS, Aruoma OI, Halliwell B, Lai EK, McCay PB: Marked reduction of free radical generation and contractile dysfunction by antioxidant therapy begun at the time of reperfusion: Evidence that myocardial 'stunning' is manifestation of reperfusion injury. Circ Res 65: 607–622, 1989
11. Zweier JL, Kuppusami P, Williams R, Rayburn BK, Smith D, Weisfeldt ML, Flaherty JT: Measurement and characterization of postischemic free radical generation in the isolated perfused heart. J Biol Chem 264: 18890–18895, 1989
12. Blasig IE, Shuter S, Garlick P, Slater T: Relative time-profiles for free radical trapping, coronary flow, enzyme leakage, arrhythmias and function during myocardial reperfusion. Free Rad Biol Med 16: 35–41, 1994.
13. Blasig IE, Dickens BF, Weglicki WB, Kramer JH: Uncoupling of mitochondrial oxidative phosphorylation alters lipid peroxidation-derived free radical production but not recovery of postischemic rat hearts and posthypoxic endothelial cells. Mol Cell Biochem 160/161: 167–177, 1996
14. Blasig IE, Grune T, Schönheit K, Rohde E, Jakstadt M, Haseloff RF, Siems WG: 4-Hydroxynonenal, a novel indicator of lipid peroxidation for reperfusion injury of the myocardium. Am J Phys 269: H14–H22, 1995
15. Walker MJA, Curtis MJ, Hearse DJ, Campbell RWF, Janse MJ, Yellon DM, Cobbe SM, Coker SJ, Harness JB, Harron DWG, Higgins AJ, Julian DG, Lab MJ, Manning AS, Northover BJ, Parrat JR, Riemersma RA, Riva E, Russel DC, Sheridan DJ, Winslow E, Woodward B: The Lambeth Conventions: guidelines for the study of arrhythmias in ischaemia, infarction, and reperfusion. Cardiovasc Res 22: 447–455, 1988
16. Pissarek M, Jänichen F, Blasig IE, Haseloff R, Keller T, Tapp E, Krause E-G: Cardioprotective potency of the radical scavenger S-2-(3 aminopropylamino) ethylphosphorothioic acid in the post-ischaemic rat heart. Mol Cell Biochem 145: 121–129, 1995
17. Kramer JH, Dickens BF, Misik V, Weglicki WB: Phospholipid hydroperoxides are precursors of lipid alkoxyl radicals produced from anoxia/reoxygenated endothelial cells. J Mol Cell Cardiol 27: 371–381, 1995
18. Mergner GW, Weglicki WB, Kramer JH: Postischemic free radical production in the venous blood of the regionally ischemic swine heart: Effect of deferoxamine. Circulation 84: 2079–2090, 1991
19. Esterbauer H, Schaur R and Zollner H: Chemistry and biochemistry of 4-hydroxynonenal, malondialdehyde and related aldehydes. Free Radic Biol Med 11: 81–128, 1991
20. Mak IT, Kramer JH, Weglicki WB: Antioxidant properties of active and inactive isomers of nicardipin in cardiac membranes, endothelial cells, and perfused rat hearts. Coronary Artery Disease 3: 1095–1103, 1992
21. Baker JE, Konorev EA, Tse SYH, Joseph J, Kalyanaraman B: Lack of protection of PBN in isolated heart during ischemia and reperfusion: Implications for radical scavenging mechanism. Free Rad Res 20: 145–163, 1994
22. Tortolani AJ, Powell SR, Misik V, Weglicki WB, Pogo GJ, Kramer JH: Detection of alkoxyl and carbon-centered free radicals in coronary sinus blood from patients undergoing elective cardioplegia. Free Rad Biol Med 14: 421–426, 1993
23. Coghlan JG, Flitter WD, Holley AE, Norell M, Mitchell AG, Ilsley CD, Slater TF: Detection of free radicals and cholesterol hydroperoxides in blood taken from the coronary sinus of man during transluminal coronary angioplasty. Free Rad Res Commun 14: 409–417, 1991
24. Blasig IE, Ebert B, Hanke T, Löwe H: Hydroxyl radical scavenging action of cardioactive drugs compared to standard antioxidants: An ESR-spin trap study. Pharmazie 43: 139–140, 1988

25. Li X-Y, Sun J-Z, Bradamante S, Piccinini F, Bolli R: Effects of the spin trap alpha-phenyl N-tert-butyl nitrone on myocardial function and flow: A dose-response study in the open-chest dog and in the isolated rat heart. Free Rad Biol Med 14: 277–285, 1993
26. Konorev EA, Baker JE, Joseph J, Kalyanaraman B: Vasodilatory and toxic effects of spin traps on aerobic cardiac function. Free Rad Biol Med 14: 127–137, 1993
27. Hinton RD, Janzen, EG: Synthesis and characterization of phenyl-sustituted C-phenyl-N-tert-butyl nitrones and some of their radical adducts. EPR parameters for hydroxyl and hydroperoxyl spin adducts. J Org Chem 54: 2915–2920, 1992
28. Anderson DE, Yuan X-J, Tseng C-M, Rubin LJ, Rosen GM, Tod ML: Nitrone spin-traps block calcium channels and induce pulmonary relaxation independent of free radicals. Biochem Biophys Res Commun 193: 878–885, 1993
29. Ferdinandy P, Das DK, Tosaki A: Pacing-induced ventricular fibrillation leading to to oxygen free radical formation in aerobically perfused rat hearts. J Mol Cell Cardiol 25: 683–692, 1993

Molecular and Cellular Biochemistry **186:** 117–121, 1998.
© 1998 *Kluwer Academic Publishers.*

Comparison of the effects of ACE inhibition with those of Angiotensin II receptor antagonism on systolic and diastolic myocardial stunning in isolated rabbit heart

Celina Morales, Manuel Rodríguez, Omar Scapín, and Ricardo J. Gelpi*
Laboratorio de Fisiopatología Cardiovascular, Departamento de Patología, Facultad de Medicina, Universidad de Buenos Aires, Argentina and Grupo de Estudios Multicéntricos de Argentina (GEMA)

Abstract

The aim was to determine whether enalaprilat (0.08 mg/kg/min) or losartan (0.01 mg/kg/min) administration before ischemia can improve postischemic systolic and diastolic dysfunction ('stunned myocardium') and attenuate the 'hyperfunction' phase at the beginning of reperfusion. An isolated isovolumic rabbit heart preparation was subjected to 15 min of ischemia followed by 30 min of reperfusion without (group 1) or with pretreatment with enalaprilat (group 2) or losartan (group 3). Left ventricular developed pressure and end-diastolic pressure (diastolic stiffness) were measured and the time constant of isovolumic relaxation (τ, Tau) and the ratio between +dP/dt and –dP/dt were calculated. In comparison to the stunned group (group 1) both enalaprilat (group 2) and losartan (group 3) exerted a significant protective effect on postischemic recovery of contractile state and diastolic stiffness. Only enalaprilat attenuated the 'hypercontractile' phase. However, both enalaprilat and losartan failed to improve myocardial relaxation. In summary, these data strongly suggest a direct deleterious action of the local renin- angiotensin system on ischemic myocardium and diminution of myocardial stunning with its successful blockade. Although, we can not exclude the possibility that bradykinin has some cardioprotective effect, these data suggest that angiotensin exacerbates myocardial injury. (Mol Cell Biochem **186**: 117–121, 1998)

Key words: myocardial stunning, losartan, enalaprilat, diastole, isovolumic relaxation

Introduction

A significant role of the renin-angiotensin system (RAS) in the pathogenesis of myocardial ischemia-reperfusion injury has been postulated on the basis of studies of angiotensin-converting enzyme (ACE) inhibition during and after myocardial ischemia [1–3]. Angiotensin II (Ang) has important physiologic and metabolic effects in the RAS. The cardiac actions of Ang are initiated when it interacts with two pharmacologically distinct subtypes of cell-surface receptors. The roles of type 2 receptors in the heart are still unknown. However, type 1 receptors mediate the major cardiovascular effects of Ang [4, 5]. It has been shown that ACE inhibitors may attenuate ischemic myocardial damage and improve systolic postischemic recovery of stunned myocardium when ACE inhibitors are administered before ischemia [6–11]. Studies showing the protective effects of ACE inhibitor therapy during ischemia have yielded a diverse array of possible explanations of the underlying mechanisms [7, 11, 12]. Whereas some studies have suggested that an increase in bradykinin levels may be at least partially responsible for the cardioprotective activity of ACE inhibitors, other work has indicated that Ang suppresion by ACE inhibition and a concomitant increase in prostacyclin release in myocardium are chiefly responsible [6, 10]. These studies examined protective effects only on systolic alterations, and failed to

*Member of Research Career, CONICET, Argentina
Address for offprints: R.J. Gelpi, Laboratorio de Fisiopatología Cardiovascular, Departmento de Patología, Facultad de Medicina, Uriburu 950 - Piso 2, 1114 Buenos Aires, Argentina

evaluate possible salutary actions in early and late diastole. The effects of these agents on the diastolic alterations of myocardial stunning have not been well described [13, 14]. The role of the RAS on diastolic alterations is particularly important since it has been demonstrated that ACE inhibitors may attenuate the relaxation abnormalities in hypertensive patients with left ventricular hypertrophy [15]. For these reasons our first objective was to study the effects of enalaprilat on systolic and diastolic alterations of stunned myocardium. A second goal was to determine the separate effect of Ang inhibition. To accomplish this aim we administered losartan, an Ang type 1 receptor antagonist. An additional goal was to compare the effects of these compounds on the 'hypercontractile state' seen at the beginning of reperfusion.

Materials and methods

Perfused heart model

Adult rabbits (2–3 kg) were killed with thiopental and ketamine. The thorax was rapidly opened and the heart excised. After removal of fat and connective tissue the heart was perfused with Ringer's solution equilibrated with 95% O_2–5% CO_2 (pH = 7.42 ± 0.04, 37°C) in a non-recirculating Langendorff preparation. A steady state was reached in 10 min and the preparation was stable for 2 h. Hearts were paced at 184 ± 4 beats/min by atrial pacing. A latex balloon tied to a rigid tube was passed into the left ventricle through the mitral valve; the opposite end of the tube was then connected to a Deltran II pressure transducer. The balloon was filled with aqueous solution to an end-diastolic pressure of 8–12 mmHg, and the volume was not changed for the remainder of the experiment. Coronary flow rate (23 ± 3 ml/min) was controlled with a peristaltic pump, and coronary perfusion pressure (CPP) was monitored at the point of cannulation of the aorta.

Data Analysis

Left ventricular pressure and CPP were recorded with a PC computer. By subtracting end-diastolic pressure from the peak systolic pressure left ventricular developed pressure (LVDP) was obtained. Because left ventricular volume was kept constant in these rabbits hearts left ventricular end-diastolic pressure (LVEDP) was used as an index of diastolic stiffness [14]. The time constant of pressure decay was calculated by using a monoexponential model with asymptotic offset to evaluate myocardial relaxation [16]:

$$P = P_0 \times e^{-t/\tau} + P_b$$

Where P is the left ventricular pressure measured during the isovolumic phase of relaxation from the peak of $-dP/dt_{max}$ to a pressure equal to end-diastolic pressure plus 5 mmHg. P_b is the asymptotic value, P_0 is the left ventricular pressure at time = 0 during relaxation, t is the elapsed time during the isovolumic relaxation period, and τ is the relaxation time constant. The ratio between $+dP/dt_{max}$ and $-dP/dt_{max}$ (+P/–P) was used as a second index to evaluate relaxation [17].

Results are given as mean ± standard error. Data were analyzed by repeated measures ANOVA followed by Bonferroni *t*-test. (Primer of Biostatistics : the program, version 3.01, 1992, McGraw-Hill). A P value smaller than 0.05 was considered to be of significant difference.

Experimental protocol

Following 10 min of drug-free perfusion under control conditions to allow the hearts to stabilize, drugs were added to the perfusate for 15 min before ischemia. Global ischemia was then induced by stopping perfusion for 15 min followed by 30 min of reperfusion. The rabbits were randomly assigned to three groups: (a) group 1 (n = 9): untreated control group; (b) group 2 (n = 12): enalaprilat (0.08 mg/kg/min) was given before ischemia and throughout the experiment. This dose was chosen on the basis of previous work by Li and Chen [9] and Abe *et al.* [18]; (c) group 3 (n = 10) losartan (0.01 mg/kg/min) was given before ischemia and throughout the experiment. The 0.01 mg/kg/min of losartan used in the experiments was chosen to ensure saturation of the type 1 receptor [19, 20].

Results

Temporal changes of LVDP and $+dP/dt_{max}$ of group 1 (stunned group); group 2 (enalaprilat group) and group 3 (losartan group) are shown in Fig. 1. LVDP and $+dP/dt_{max}$ in group 1 decreased during the ischemic period but recovered promptly after reperfusion. After approximately 1 min of reperfusion LVDP and $+dP/dt_{max}$ reached 71 ± 6% and 80 ± 9% of the preischemic values, respectively. Both parameters quickly decreased and fell to 62 ± 5% and 70 ± 4% of the preischemic value after 30 min of reperfusion. Thus, contractility showed biphasic behavior during reperfusion with an early phase showing substantial recovery followed by a later decrease in function to about 65% of control values. When enalaprilat was administered, the hyperfunction phase early in reperfusion was abolished and developed pressure and $+dP/dt_{max}$ reached 82 ± 7% and 90 ± 8% of preischemic values after 30 min of reperfusion. Thus, enalaprilat abolished the biphasic behavior of myocardial stunning, and diminished the contractile dysfunction observed late in reperfusion. Although losartan also decreased the contractile dysfunction of myocardial

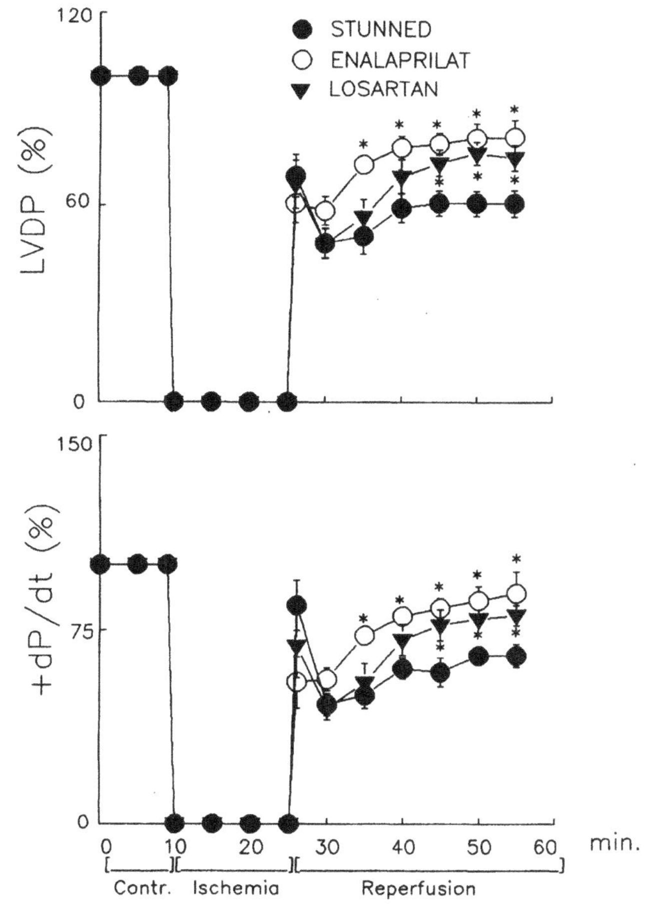

Fig. 1. Time course for the recovery of LVDP and +dP/dt in 3 groups of isolated rabbit hearts following 15 min of global ischemia. Both enalaprilat and losartan improved recovery of the contractile state at the end of the reperfusion period. Enalaprilat abolished the 'hyperfunction' phase in early reperfusion. *p < 0.05 vs. stunned group.

stunning, it did not abolish the hypercontractile phase at the beginning of reperfusion.

Passive diastolic stiffness was assessed by evaluating LVEDP at constant volume in the three groups and the results are depicted in Fig. 2. LVEDP is expressed as a percent of the preischemic value. LVEDP increased during the reperfusion period in all three groups. However after 30 min of reperfusion LVEDP rose significantly less in the groups treated with enalaprilat and losartan (399 ± 36% vs. 215 ± 22% and 302 ± 29%, respectively). Although both drugs protected the heart from the increase in diastolic stiffness, the protective effect was more pronounced with enalaprilat.

Figure 3 shows 2 indices of isovolumic relaxation in the 3 groups: τ (upper panel) and +P/–P (lower panel). τ and +P/–P increased transiently in the stunned group (from 29.7 ± 2.3 msec to 54.0 ± 3.3 msec and from 1.39 ± 0.07 to 2.01 ± 0.18, respectively) but returned to preischemic values by the end of the reperfusion period. This transient elevation in both

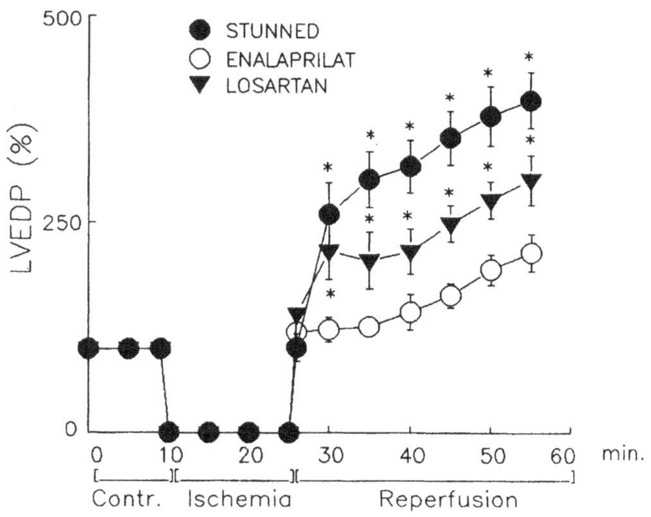

Fig. 2. Time course for the recovery of LVEDP in 3 groups of isolated rabbit hearts following 15 min of global ischemia. Although both compounds attenuate the increase in stiffness, the protective effect in the enalaprilat group is greater than that in the losartan group. *p < 0.05 vs. stunned group.

parameters was not modified by the infusion of either drug. Thus, the administration of enalaprilat and losartan showed a dissociated effect on the diastolic alterations of myocardial stunning. Whereas diastolic stiffness was attenuated by the 2 compounds, the relaxation rate was not improved.

Discussion

We investigated the role of Ang in the development of myocardial injury during global ischemia and reperfusion by comparing the effects of treatment with either an ACE inhibitor or an Ang antagonist on the systolic and diastolic dysfunction of myocardial stunning. Addition of enalaprilat to the perfusate before ischemia resulted in significant improvement of the contractile state. This finding is in agreement with those of several published studies in which an ACE inhibitor resulted in reduced myocardial damage after ischemia in the isolated perfused heart [9, 21, 22]. Additionally, the early 'hyperfunction' phase of systole observed at the beginning of reperfusion was attenuated by enalaprilat. Previous work [23] has shown that this early hypercontratile phase was abolished by verapamil, suggesting that calcium influx is involved in this early phenomenon. In our study we also examined the effects of enalaprilat on indices of both early and late diastolic function. Whereas some experimental studies have examined abnormalities of diastolic stiffness in myocardial stunning and the effects of pharmacological protection, no prior studies have examined indices relating to all phases of diastole [7, 13, 14]. It has been

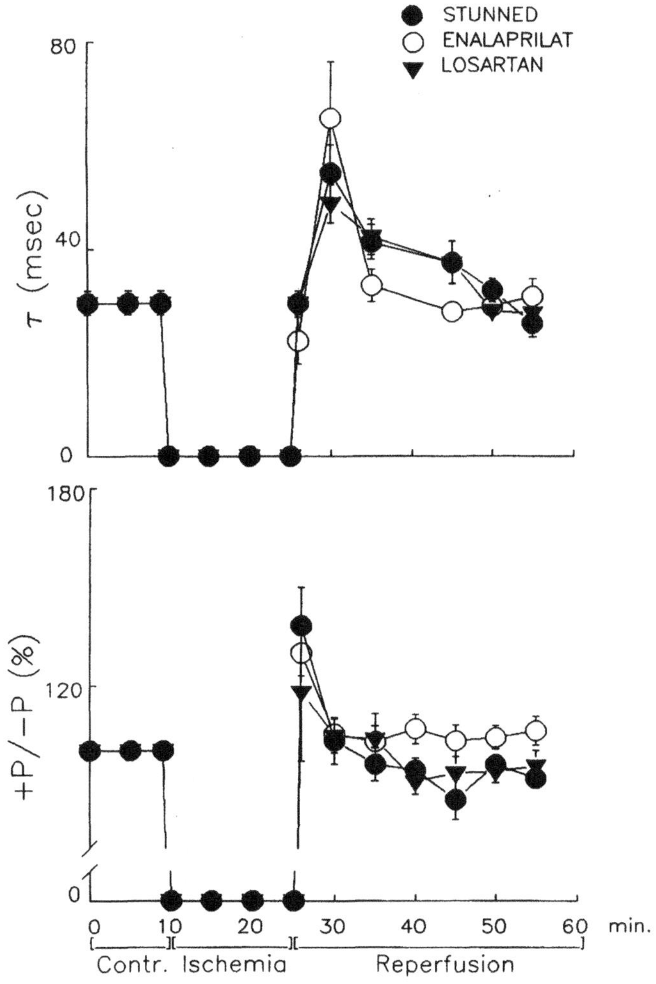

Fig. 3. Time course for the recovery of 2 relaxation indices, τ (upper panel) and +P/–P ratio (lower panel) in 3 groups of isolated rabbit hearts following 15 min of global ischemia. Myocardial stunning decreased the relaxation rate in early reperfusion. Neither enalaprilat nor losartan modified the impairment of early relaxation.

shown that in the isolated heart reperfusion following ischemia induces a monotonic increase in diastolic stiffness and biphasic change in relaxation characterized by an early transient impairment followed by a faster relaxation when stunning is established [24]. Interestingly, enalaprilat attenuated the stiffness alterations from the beginning of reperfusion, but did not modify the early phase of the reperfusion isovolumic relaxation impairment.

We also documented the effect of losartan, an Ang antagonist, when it was administered before ischemia. Hearts from treated animals demonstrated significantly improved systolic and diastolic function, including diastolic stiffness. However, like enalaprilat, losartan failed to improve the isovolumic relaxation rate. Reports of the effects of Ang antagonists on ischemic myocardium are relatively few. In one early study in rats, functional recovery after myocardial infarction did not improve [25]. In another study a greater degree of vasodilation in the coronary vasculature was demonstrated in dogs after treatment *in vivo* with losartan than ACE inhibitor treatment [26]. Our data on systolic function agree with the results from an earlier study in which chronic treatment with an Ang type 1 antagonist improved the contractile state of stunned myocardium [5]. On the other hand, a recent report showed markedly impaired recovery of systolic function after 30 min of ischemia and 30 min of reperfusion [27]. However, all of these studies failed to evaluate alterations in diastolic function. Whereas the cardioprotective effects of ACE inhibitors have been attributed, at least in part, to their effect on bradykinin [28], our results are consistent with the suggestion that a blockade of the effects of Ang alone may be the dominant cardioprotective mechanism of ACE inhibition, since treatment with an AT_1 selective Ang receptor antagonist afforded protection similar to that seen with enalaprilat. Some involvement of bradykinin in cardioprotective activity cannot be ruled out, however, because evaluation of a specific antagonist of bradykinin was not included in this study. A previous investigation [26] of the functional role of Ang AT_2 receptors in endothelial cells and in isolated rat heart indicates that some protective effects of AT_1 may be mediated indirectly by kinins. The latter observations indicate that in endothelial cells in which AT_1 receptors are blocked, endogenously released Ang would act on AT_2 receptors, causing enhanced synthesis and release of bradykinin with possible effects on ischemic myocardium. Thus, a potential limitation of this study is that we did not evaluate if bradykinin might be associated with protective effects on myocardial stunning, a hypothesis that deserves future study.

In summary, these data strongly suggest that myocardial protection resulting from Ang antagonist treatment *in vitro* results from a direct blockade of the local cardiac renin-angiotensin system. The protective effect includes not only improvement of the contractile state but also salutary diminution of stiffness.

References

1. Gavras H: Angiotensin converting enzyme inhibition and its impact on cardiovascular disease. Circulation 81: 381–388, 1990
2. Greenberg B, Quinones MA, Koilpillai C, Limacher M, Shindler D, Benedict C, Shelton B: Effects of long-term enalapril therapy on cardiac structure and function in patients with left ventricular dysfunction: Results of the SOLVD echocardiography substudy. Circulation 91: 2573–2581, 1995
3. Santos RAS, Brum JM, Brosnihan KB, Ferrario CM: The renin-angiotensin system during acute myocardial ischemia in dogs. Hypertension 15(Supl I): I121–I127, 1990
4. Wiemer G, Schölkens BA, Wagner A, Heitsch H, Linz W: The possible role of angiotensin II subtype AT_2 receptors in endothelial

cells and isolated ischemic rat hearts. J Hypertension 11: S234–S235, 1993

5. Yoshiyama M, Kim S, Yamagishi H, Omura T, Tani T, Yanagi S, Toda I, Teragaki M, Akioka K, Takeuchi K, Takeda T: Cardioprotective effect of the angiotensin II type I receptor antagonist TCV-116 on ischemia-reperfusion injury. Am Heart J 128: 1–6, 1994
6. Schölkens BA, Linz W, König W: Effects of the Angiotensin converting enzyme inhibitor, ramipril, in isolated ischaemic rat heart are abolished by a bradykinin antagonist. J Hypertension 6 (Suppl 4): S25–S28, 1988
7. Cargnoni A, Ferrari R, Curello S, Ceconi C, Boraso A, Visioli O: Protection of the ischemic myocardium by the Converting-Enzyme inhibitor Zofenopril: Insight into its mechanism of action. J Cardiovasc Pharmacol 20: 694–704, 1992
8. Cargnoni A, Boraso A, Scotti C, Ghirardelli N, Benigno M, Bernocchi P, Pedersini P, Ferrari R: Effect of Angiotensin converting enzyme inhibition with Quinaprilat on the ischaemic and reperfused myocardium. J Mol Cell Cardiol 26: 69–86, 1994
9. Li K, Chen X: Protective effects of Captopril and Enalapril on myocardial ischemia and reperfusion damage of rat. J Mol Cell Cardiol 19: 909–915, 1987
10. Linz W, Schölkens B: Influence of local converting enzyme inhibition on angiotensin and Bradykinin effects in ischemic rat hearts. J Cardiovasc Pharmacol 10(Supl 7): S75–S82, 1987
11. Przyklenk K, Kloner RA: Angiotensin converting enzyme inhibitors improve contractile function of stunned myocardium by different mechanisms of action. Am Heart J 121: 1319–1330, 1991
12. Przyklenk K, Kloner RA: Acute effects of Hydralazine and Enalapril on contractile function of postischemic 'stunned' myocardium. Am J Cardiol 60: 934–936, 1987
13. Eng C, Weil B, Sonnenblick EH: Effect of Captopril on postischemic myocardial expansion. J Cardiovasc Pharmacol 21: 560–566, 1993
14. Eberli FR, Apstein CS, Ngoy S, Lorell BH: Exacerbation of left ventricular ischemic diastolic dysfunction by pressure-overload hypertrophy. Modification by specific inhibition of cardiac Angiotensin converting enzyme. Circ Res 70: 931–943, 1992
15. Haber HL, Powers ER, Gimple LW, Wu CC, Subbiam K, Johnson WH, Feldman MD: Intracoronary Angiotensin-Converting enzyme inhibitor improves diastolic function in patients with Hypertensive left ventricular hypertrophy. Circulation 89: 2616–2625, 1994
16. Gelpi RJ, Mosca SM, Cingolani HE: The effects of Bay K 8644 on diastolic function in the dog heart. J Moll Cell Cardiol 22: 1285–1296, 1990
17. Grassi de Gende AO, Perez Alzueta AD, Cingolani HE: Effect of isoproterenol on the relation between maximal rate of contraction and maximal rate of relaxation. Am J Physiol 233: H404–H409, 1997
18. Abe Y, Kitada Y, Narimatsu A: Effects of a calcium-sensitizing positive inotropic agent MCI-154 and its combinade use with Enalapril on postischemic contractile dysfunction of dogs hearts. J Cardiovasc Pharmacol 26: 653–659, 1995
19. Werrmann JG, Cohen SM: Use of losartan to examine the role of the cardiac renin-angiotensin system in myocardial dysfunction during ischemia and reperfusion. J Cardiovasc Pharmacol 27: 177–182, 1996
20. Zhuo J, Song K, Abdelrahman A, Mendelsohn FA: Blockade by intravenous losartan of AT_1 angiotensin II receptors in rat brain, kidney, and adrenals demonstrated by in vitro autoradiography. Clin Exp Pharmacol Physiol 21: 557–567, 1994
21. Liang CS, Gavras H, Black J, Sherman LG, Hood WB: Renin-Angiotensin system inhibition in acute myocardial infarction in dogs. Effects on systemic hemodynamics, myocardial blood flow, segmental myocardial function, and infarct size. Circulation 66: 1249–1255, 1982
22. Sharpe N, Smith H, Murphy J, Greaves S, Hart H, Gamble G: Early prevention of left ventricular dysfunction after myocardial infarction with angiotensin-converting enzyme inhibition. Lancet 337: 872–876, 1991
23. Yoshikawa T, Akaishi M, Ikeda F, Handa S, Nakamura Y: Postischemic hypercontraction is enhanced in ischaemically injured canine myocardium. Cardiovasc Res 26: 337–341, 1992
24. Mosca SM, Carriquiriborde M, Cingolani HE: Biphasic changes in relaxation following reperfusion after myocardial ischemia. Moll Cell Biochem 160/161: 123–128, 1996
25. Smits JFM, Krimpen CV, Schoemaker RG, Cleutjens JPM, Daemen MJAP: Angiotensin II receptor blockade after myocardial infarction in rats: Effects on hemodynamics, myocardial DNA synthesis, and interstitial collagen content. J Cardiovasc Pharmacol 20: 772–778, 1992
26. Sudhir K, MacGregor JS, Gupta M, Barbant SD, Redberg R, Yock PG, Chatterjee K: Effect of selective angiotensin II receptor antagonism and angiotensin converting enzyme-inhibition on the coronary vasculature *in vivo*, intravascular 2-dimensional and Doppler ultrasound studies. Circulation 87: 931–938, 1993
27. Ford WR, Clanachan AS, Bodh IJ: Opposite effects of Angiotensin AT_1 and AT_2 receptor antagonists on recovery of mechanical fucntion after ischemia-reperfusion in isolated working rat hearts. Circulation 94: 3087–3089, 1996
28. Zhu P, Zaugg CE, Simper D, Hornstein P, Allegrini PR, Buser PT: Bradykinin improves postischemic recovery in the rat heart: Role of high energy phosphates, nitric oxide, and prostacyclin. Cardiovasc Res 29: 658–663, 1995

Molecular and Cellular Biochemistry **186:** 123–129, 1998.
© 1998 *Kluwer Academic Publishers.*

Is stunning prevented by ischemic preconditioning?

Susana M. Mosca,* Ricardo J. Gelpi,* Jose Milei, Graciela Fernández Alonso and Horacio E. Cingolani*

Centro de Investigaciones Cardiovasculares Universidad Nacional de La Plata, La Plata, Argentina

Abstract

In a model of global ischemia in the isolated perfused rat heart, a 20 min ischemic period followed by 30 min of reperfusion induces a decrease in isovolumic developed pressure (LVDP) and $+dP/dt_{max}$ to 61 ± 6% and 61 ± 7% of baseline, respectively. Left ventricular end-diastolic pressure (LVEDP) increases to 36 ± 4 mmHg at the end of the reperfusion period. No significant necrotic area as assessed by triphenyltetrazolium chloride (TTC) was detected at the end of the reperfusion period. By an immunohistochemical method using antiactin monoclonal antibodies 10.8 ± 1.9% of unstained cells were detected in the stunned hearts and 10.3 ± 1.2% in control hearts. Preceding the ischemic episode with a cycle of 5 min of ischemia followed by 10 min of reperfusion (ischemic preconditioning) protected contractile function. LVDP and $+dP/dt_{max}$ now stabilized at 89 ± 5% and 94 ± 5% of baseline respectively. LVEDP was 20 ± 2 mmHg at the end of the reperfusion period. The protection of contractile dysfunction after 20 min of ischemia was achieved also by early reperfusion of low Ca^{2+}-low pH perfusate. With this intervention LVDP stabilized at 87 ± 5% of baseline. LVEDP was 12 ± 2 mmHg at the end of the reperfusion period. A positive inotropic intervention induced by a modified postextrasystolic potentiation protocol at the end of the reperfusion period increases LVDP to levels higher than baseline in the stunned hearts. However, these values were less than those obtained in control hearts. Ischemic preconditioning significantly increased the maximal inotropic response. Therefore, ischemic preconditioning diminishes the contractile dysfunction of early stunning. (Mol Cell Biochem **186:** 123–129, 1998)

Key words: stunning, ischemic preconditioning

Introduction

Following a brief ischemic insult myocardial contractility decreases for hours, days or even weeks. This reversible form of dysfunction is known as stunning and has been the subject of extensive investigation [1–7]. Ischemic preconditioning induces protection when the ischemic period is preceded by a short cycle(s) of ischemia and reperfusion and it is known to reduce myocardial necrosis and arrhythmias [8–12]. However, the protective effect of preconditioning on myocardial stunning has not been sufficiently characterized. Preconditioning has been shown to improve post-ischemic recovery of function in *in vitro* models of global ischemia [5, 7, 12–14]. However, it remains uncertain whether preconditioning is capable of reducing stunning or merely protects by decreasing irreversible damage of myocardial cells. Furthermore, the possibility that the first preconditioning ischemic period induces protection by decreasing contractility as a result of some degree of stunning has also been controversial [15–18].

One way to determine whether this preparation represents a true model of stunning is to insure that no irreversible damage occurs during the ischemic period. The absence of irreversible damage can be determined either directly by studying the tissue or indirectly by assessing the response of the heart to inotropic stimulation.

The goal of this study is to analyze the characteristics of the mechanical protection mediated by ischemic preconditioning in a rat model of global ischemia.

Address for offprints: H.E. Cingolani, Centro de Investigaciones Cardiovasculares Universidad Nacional de La Plata, 60 y 120, 1900 La Plata, Argentina
*Members of Consejo Nacional de Investigaciones Científicas y Técnicas (CONICET). Centro de Investigaciones Cardiovasculares, Universidad Nacional de La Plata, 60 y 120, 1900 La Plata, Argentina

Materials and methods

Isolated heart preparation

Rats were anesthetized with an intraperitoneal injection of sodium pentobarbital (60 mg/kg body wt). The heart was rapidly excised and perfused by the non-recirculating Langendorff technique with Ringer's solution containing (in mM): 118 NaCl, 5.9 KCl, 1.2 $MgSO_4$, 1.35 $CaCl_2$ and 11.1 dextrose. The buffer was saturated with a mixture of 95%O_2–5%CO_2, had a pH 7.4, and was maintained at 37°C. The conductive tissue in the atrial septum was damaged with a fine needle to achieve atrioventricular block, and the right ventricle was paced at 280 ± 10 beats/min. A latex balloon tied to the end of a polyethylene tube was passed into the left ventricle through the mitral valve; the opposite end of the tube was then connected to a Statham P23XL pressure transducer. The balloon was filled with aqueous solution to give an end-diastolic pressure of 8–12 mmHg, and this volume was unchanged for the remainder of the experiment. Coronary perfusion pressure was monitored at the point of cannulation of the aorta and adjusted to approximately 60–70 mmHg. Coronary flow, controlled with a peristaltic pump, was 11 ± 2 ml/min. Left ventricular pressure (LVP) and its first derivative (dP/dt) were recorded with a direct writing recorder. By subtracting end-diastolic pressure (LVEDP) from the peak systolic pressure, left ventricular developed pressure (LVDP) was obtained. During ischemia the time of onset of contracture, defined as the time when diastolic pressure increased by 2 mmHg, was measured.

Experimental protocols

After 10 min during which pressure development was allowed to stabilize, hearts were divided into four experimental groups:

Group I (control): Hearts were perfused aerobically for 80 min.
Group II (MS): Hearts were submitted to 20 min of normothermic global ischemia and 30 min of reperfusion. Global ischemia was created by cross-clamping the perfusate inflow line and the heart was placed in a saline bath held at 37°C.
Group III (IP): Ischemic preconditioning was induced by only one cycle of 5 min of ischemia and 10 min of reperfusion. Thereafter hearts underwent 20 min of global ischemia followed by 30 min of reperfusion as in Group II.
Group IV (Low Ca^{2+}, low pH): As in Group II, hearts were subjected to 20 min of normothermic ischemia. However, during the first 5 min of reperfusion the perfusing solution was changed to contain only 0.1 mM Ca^{2+} at pH 6.7. During the second 5-min period the perfusate was again changed to contain 0.3 mM Ca^{2+} [19]. For the remaining 20 min of the reperfusion period hearts were perfused with the control solution.

Potentiated contractions

In all groups a modified postextrasystolic potentiation (PSP) test was used to achieve maximal developed pressure (M_xDP). To evoke postextrasystolic potentiated contractions (PC) a computer-controlled automated routine was used. This procedure was originally applied by Urthaler *et al.* [20] on papillary muscles. These authors demonstrated that the peak tension obtained by PSP was usually much higher than tetanized papillary muscle tension. We adapted this procedure for the isovolumic heart. The greatest PC was obtained with a train of 10 stimuli (5 msec stimulus duration) delivered at a frequency of 10–15 Hz followed by a pause of 1 sec (Fig. 1).

Measurement of infarct area

After the reperfusion period, six rat hearts of Groups I and II were perfused with 1% triphenyltetrazolium chloride (TTC) in phosphate buffer (pH = 7.4) for 5 min and immersed in this solution for another 5 min. With this technique viable tissue stains brick red and dead tissue is left unstained, allowing the differentiation of infarcted tissue [21]. At the end of the incubation period hearts were perfused with 10% formalin in saline for 3–4 min and immersed for 48 h in this solution which not only fixes the tissue but also enhances the color contrast. Then, hearts were sliced into 1 mm thick transverse

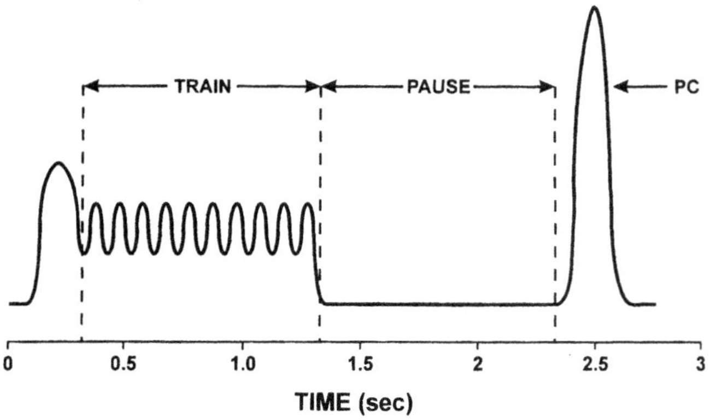

Fig. 1. Scheme showing the modified postextrasystolic potentiation (PC) achieved by application of a train of 10 stimuli followed by a 1 sec pause in isovolumic rat hearts.

Fig. 2. Photograph of rat heart slices stained with triphenyltetrazolium chloride at the end of a 30-min reperfusion period following 20 min of global ischemia. Unstained areas were not detected, indicating absence of necrosis. Arrow: indicates the location of suture for pacing electrode that was unstained.

sections along the long axis from apex to base. Slices were immediately photographed and used for macroscopic determination of damaged myocardium. Then, the specimens were routinely processed for light microscopy by embedding in paraffin. Histological sections (4 μm thick) were serially cut, stained with hematoxylin-eosin and processed for immunohistochemical analysis for which HHF35 antiactin monoclonal antibodies (Biogenex, San Ramon, Ca) and avidin-peroxidase complex (ABC) as detection system were used [22, 23]. Endogenous peroxidase activity was blocked with methanol containing 0.03% H_2O_2. Peroxidase reactions were developed with 3,3'-diaminobenzidine. For the morphometric study an integrated computerized image analysis system was utilized (Vidas, Kontron electronics, Munich, Germany). Scanning windows were set for each specimen to quantify brown staining representing actin reactivity. All available fields were scanned for each section providing an average of 56 fields per heart (range 37–79). With this method the diminution of actin reactivity could be used for the easy detection of very mild degrees of degeneration of cardiac muscle fibers [22].

Data analysis

Data are given as means ± S.E. Statistical differences were analyzed using analysis of variance for repeated measurements followed by Newman–Keuls test for *post-hoc* comparison of values obtained during reperfusion in the different protocols.

Results

The perfused rat heart injured by a 20 min period of global ischemia followed by 30 min of reperfusion exhibits recovery of LVDP to 61 ± 6% of the baseline value and of +dP/dt$_{max}$ to 61 ± 7%. LVEDP was 11 ± 0.5 mmHg in the control period and increased to 36 ± 4 mmHg after 30 min of reperfusion. In hearts stained at the end of the reperfusion period with TTC (Fig. 2), no significant unstained zone was detected. Since it is known that TTC staining underestimates the area of small infarcts, hearts were also studied by immunohistochemical detection of actin staining. Figure 3 shows the percentage of unstained cells in experimental groups I and II using anti-actin monoclonal antibodies. About 10% (10.8 ± 1.9%) of myocardial cells were unstained in the stunned hearts. However, control hearts without any ischemic period other than that which is unavoidable during the isolation procedure showed a similar percentage of unstained cells (10.3 ± 1.2%). Within the limitations of the methods used we can conclude that the stunned hearts did not demonstrate any more irreversible damage than non-ischemic control hearts. The present data are in agreement with previous studies in which no significant irreversible damage has been detected in hearts of different species [24, 25] after 15–20 min of myocardial ischemia followed by reperfusion. The greater loss of actin staining than TTC staining after an ischemic period has been described [22]. Previous results have indicated that the immunohistochemical reaction of actin filaments can be used for detection of very mild degrees of degeneration of cardiac muscle fibers [22].

Figure 4 shows the beneficial effect of one episode of 5 min of global ischemia on contractile behaviour following prolonged ischemia. LVDP and +dP/dt$_{max}$ values before any intervention were 88 ± 3 mmHg and 1657 ± 68 mmHg/sec, respectively. At the end of the first reperfusion period (10 min) both parameters returned to baseline values (87 ± 4 mmHg and 1610 ± 75 mmHg/sec). When flow was restarted after the 20 min ischemic period LVDP and +dP/dt$_{max}$ stabilized

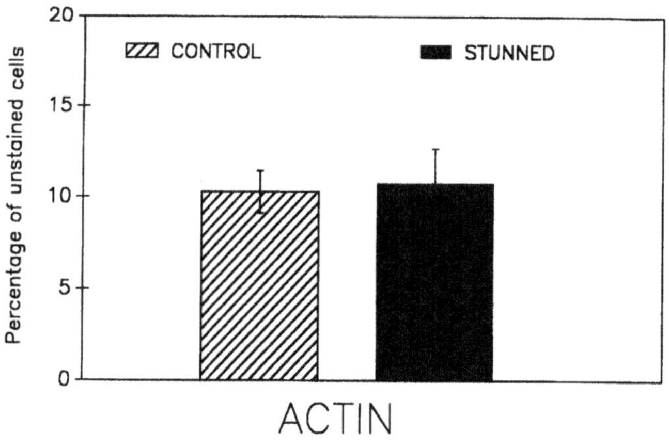

Fig. 3. The percentage of damaged myocardium in control and stunned hearts detected with antiactin monoclonal antibodies. No significant differences were observed between the two groups.

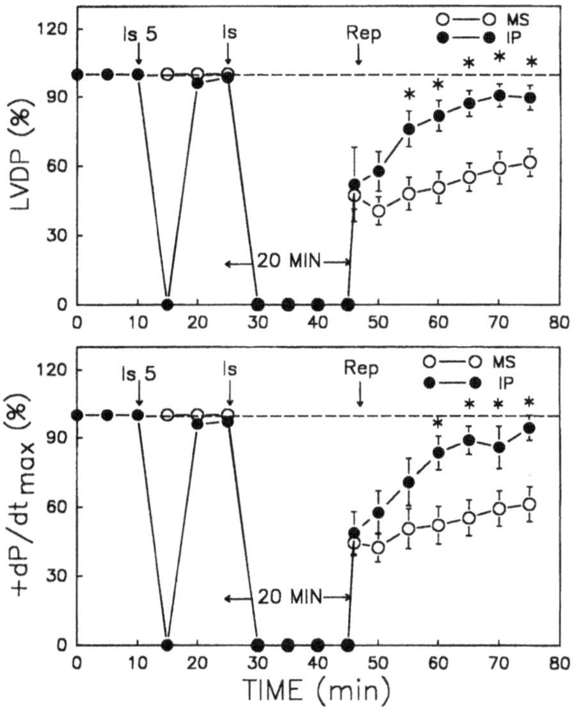

Fig. 4. Changes in left ventricular developed pressure (LVDP) and maximal rate of rise of LV pressure (+dP/dt$_{max}$) in stunned (MS) and preconditioned (IP) rat hearts. Note the protective effect of IP on the systolic dysfunction of stunned myocardium. *$p < 0.05$

at higher values than those obtained in stunned hearts. At the end of the reperfusion period LVDP and +dP/dt$_{max}$ in the preconditioned hearts were 89 ± 5% (76 ± 4 mmHg) and 94 ± 5% (1535 ± 142 mmHg) of baseline values, respectively.

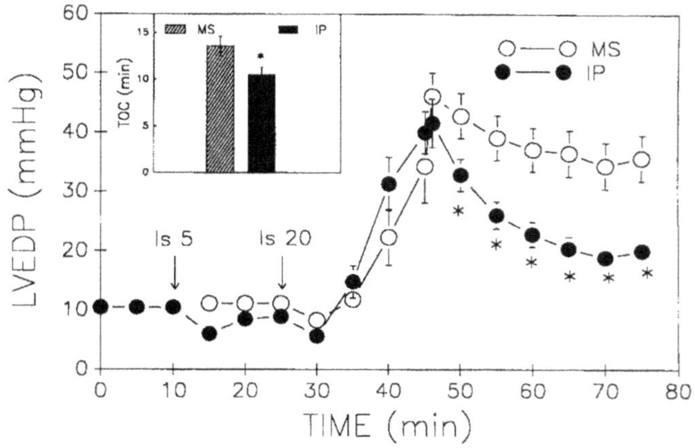

Fig. 5. Changes in left ventricular end-diastolic pressure (LVEDP) during preischemic, ischemic and reperfusion periods in stunned (MS) and preconditioned (IP) rat hearts. Note that IP attenuated the increase of LVEDP seen in MS during reperfusion. Inset shows the shorter time of onset of contracture (TOC) following initiation of ischemia in preconditioned than stunned hearts. *$p < 0.05$

Figure 5 shows the effect of IP on the increase in diastolic chamber stiffness induced by the 20 min ischemic period. The beneficial effects were detected during reperfusion. After 30 min of reperfusion LVEDP was 20 ± 2 mmHg, a value significantly different than the one obtained in stunned hearts (36 ± 4 mmHg). The time of onset of contracture (TOC) during the ischemic period however was shorter in the preconditioned hearts (10.5 ± 0.8 min vs. 13.5 ± 1.1 min in stunned hearts). These data are in agreement with previous reports [14, 26].

To gain further insight into the mechanisms involved in the contractile dysfunction induced by the ischemic insult a protocol of early reperfusion with a low Ca^{2+}-low pH perfusate (first 10 min of reperfusion) was carried out. The results are shown in Fig. 6. At the end of the 30 min reperfusion period LVDP reached 87 ± 5% of baseline values. Thus, the recovery of myocardial contractility was improved when compared with the stunned hearts of group II. The improvement in the recovery of contractile function induced by altering the perfusate early in reperfusion is consistent with the idea that the contractile impairment detected after reperfusion was not entirely the result of the ischemic period, and perhaps also a result of events occuring during reperfusion. Similar results have recently been reported [19].

In the four experimental groups, contractile reserve was explored with a modified PSP test. This maneuver was done to estimate M_xDP similar to the maximal developed force assessed in papillary muscle [20]. With this inotropic intervention, values of M_xDP higher than baseline were obtained in the four experimental groups. In the stunned hearts M_xDP reached 113 ± 9 mmHg. In the preconditioned group the maximal inotropic response achieved was 142 ± 5 mmHg and in the group protected by low Ca^{2+}-low pH during reperfusion the value of M_xDP obtained was 171 ± 14 mmHg. In control hearts without any intervention M_xDP was 164 ± 7 mmHg. Therefore, the M_xDP in stunned hearts was lower than that obtained in the other 3 groups (Fig. 7).

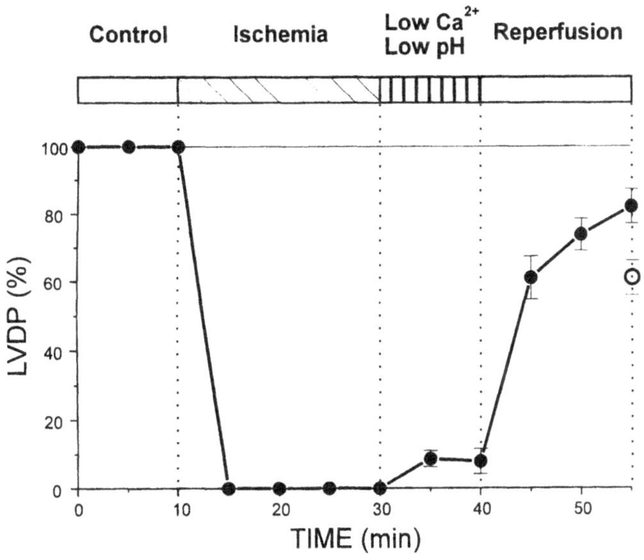

Fig. 6. Changes in left ventricular developed pressure (LVDP) in rat hearts perfused early during reperfusion with low Ca^{2+}-low pH solution. To facilitate comparison, the value of LVDP of group II at the end of the reperfusion period (open circle) is depicted. Note the protective effect of this intervention on the systolic dysfunction of myocardial stunning.

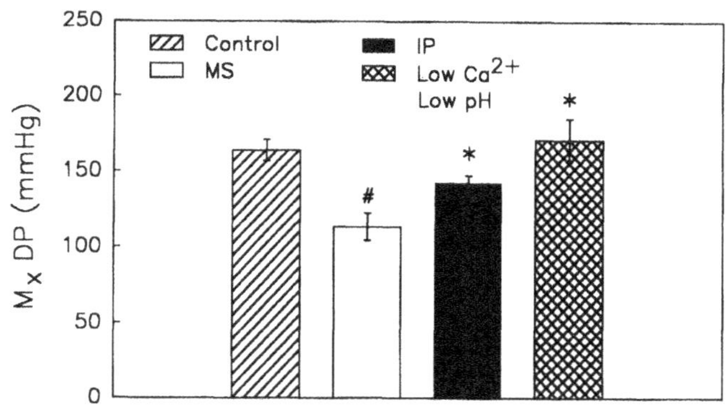

Fig. 7. Values of maximal developed pressure (M_xDP) following postextrasystolic potentiation at the end of the reperfusion period in the four experimental groups [control, myocardial stunning (MS), ischemic preconditioning (IP) and low Ca^{2+}-low pH]. Note that values are significantly less in stunned than control hearts and were improved by IP and reperfusion with low Ca^{2+}-low pH solution. *$p < 0.05$ with respect to MS. #$p < 0.05$ with respect to the other 3 groups.

Discussion

Since our experiments were performed in an *in vitro* model of global ischemia, the discussion will be restricted to possible protective mechanisms involved with this preparation during the first window of protection. Our data are confirmatory of several studies showing that in a model of global ischemia, ischemic preconditioning improves the recovery of contractility following a period of ischemia-reperfusion [13, 14, 28, 29]. In addition our data show that the protection can be obtained without an initial depression of the contractile function (stunning). These results are in agreement with previous studies [16–18]. However, is ischemic preconditioning preserving contractile function by protection from stunning, from cell death or both? In other words, is the isolated heart subjected to 15–20 min of ischemia followed by reperfusion a good model of stunning or only a model of necrosis? Many experimental studies have used this model of stunning [5, 7, 12–14, 24, 25, 30–32]. However, not many have assessed the possible extent of irreversible damage after the ischemic period [24, 25, 30–32].

We were unable to detect irreversible damage after 20 min of ischemia by TTC staining. However, it is known that the TTC technique underestimates the necrotic area. The possibility of necrosis not detected by TTC stainning due to the short reperfusion period (30 min) used in our study should also be considered. For these considerations we should be cautious in the interpretation of the TTC data. With monoclonal antiactin antibodies no differences in the percentage of unstained cells were detected between control hearts and those exposed to the ischemic period of 20 min.

If we accept these data and those of others [24, 25, 31] showing no significant irreversible damage in this model of global ischemia, we can conclude that ischemic preconditioning is protecting from stunning in this model. The protection obtained in our experiments by early reperfusion with low Ca^{2+}-low pH is in agreement with the position that the contractile dysfunction after 20 min of ischemia in the rat can be defined as reperfusion injury and not the result of necrosis. However, the possibility of low Ca^{2+}-low pH reperfusion altering the rate of recovery from reversible ischemic injury should also be considered.

The myocardial dysfunction of stunning is characterized by depressed contractile performance. The force of contraction at any given $[Ca^{2+}]_i$ is lower in stunned muscles [27] than in controls. The production of cytotoxic, oxygen-derived free radicals [33–36] has been implicated in the pathogenesis of postischemic stunning. Ca^{2+} overload also has been proposed to be the trigger of events leading to stunning [37, 38]. Increased Ca^{2+} influx can augment free radical generation, while free radical generation can increase Ca^{2+} influx [39, 40]. The mechanism by which the myofilament response to Ca^{2+} is reduced is not yet known. Recently it has been proposed that myocardial stunning is caused by a partial degradation of troponin I (TnI) as a result of the activation of a Ca^{2+}-dependent protease during reperfusion [19]. The partial degradation of TnI was prevented by low Ca^{2+}-low pH during reperfusion. Whether or not this partial degradation of TnI can be blunted by ischemic preconditioning is not known at the present time.

References

1. Heyndrickx G, Millard RW, McRitchie RJ, Maroko PR, Vatner SF: Regional myocardial function and electrophysiologic alterations after brief coronary artery occlusions in conscious dogs. J Clin Invest 56: 978–985, 1975
2. Braunwald E, Kloner R: The stunned myocardium: prolonged, postischemic ventricular dysfunction. Circulation 66: 1146–1149, 1982
3. Thaulow E, Guth B, Heusch G, Ross JJr: Characteristics of regional myocardial stunning after exercise in dogs with chronic coronary stenosis. Am J Physiol 257: H113–H119, 1989
4. Charlat M, O'Neill P, Hartley C, Roberts R, Bolli R: Prolonged abnormalities of left ventricular diastolic wall thinning in the 'stunned' myocardium in conscious dogs: time course and relation to systolic function. J Am Coll Cardiol 13: 185–194, 1989
5. Haster SB, Becker LC, Ambrosio G, Jacobus WE: Reduced aerobic metabolic eficency in globally 'stunned' heart. J Moll Cell Cardiol 21: 419–426, 1989
6. Lawrence W, Maugham L, Kass D: Mechanism of global functional recovery despite sustained postischemic regional stunning. Circulation 85: 816–827, 1992
7. Krause SM: Effect of global myocardial stunning on Ca^{2+}-sensitive myofibrilar ATPase activity and creatine kinetics. Am J Physiol 259: H813–H819, 1990
8. Murry CE, Jennings RB, Reimer KA: Preconditioning with ischemia: a delay of lethal cell injury in ischemic myocardium. Circulation 74: 1124–1136, 1986
9. Downey JM, Liu GS, Thornton JD: Adenosine and the anti-infarct effects of preconditioning. Cardiovasc Res 27: 3–8, 1993
10. Shiki K, Hearse DJ: Preconditioning of ischemic myocardium: reperfusion-induced arrhythmias. Am J Physiol 253: H1470–H1476, 1987
11. Parratt JR, Vegh A: Pronounced antiarrhythmic effects of ischemic preconditioning. Cardioscience 5: 9–18, 1994
12. Cave AC, Apstein CS: Polymyxin B, a protein kinase C inhibitor, abolishes preconditioning-induced protection against contractile dysfunction in the isolated blood perfused rat heart. J Mol Cell Cardiol 28: 977–987, 1996
13. Cave AC, Collis CS, Downey JM, Hearse DJ: Improved functional recovery by ischaemic preconditioning is not mediated by adenosine in the globally ischaemic, isolated rat heart. Cardiovasc Res 27: 663–668, 1993
14. Cave AC: Preconditioning induced protection against post-ischaemic contractile dysfunction: characteristics and mechanisms. J Mol Cell Cardiol 27: 969–979, 1995
15. Cargnoni A, Ceconi C, Bernocchi P, Pasini E, Curello S, Ferrari R: Is stunning an important component of preconditioning? J Mol Cell Cardiol 28: 2323–2331, 1996
16. Moolman JA, Genade S, Winterbach R, Harper IS, Williams K, Lochner A: Preconditioning with a single short episode of global ischemia in the isolated working rat heart: effect on structure, mechanical function and energy metabolism for various durations of sustained global ischemia. Cardiovasc Drugs Ther 9: 103–115, 1995
17. Asimakis GK: Myocardial glycogen depletion can not explain the cardioprotective effects of ischemic preconditioning in the rat heart. J Mol Cell Cardiol 28: 563–570, 1996
18. Cave AC, Hearse DJ: Ischemic preconditioning and contractile function: studies with normothermic and hypothermic global ischemia. J Mol Cell Cardiol 24: 1113–1123, 1992
19. Gao WD, Atar D, Liu Y, Perez NG, Murphy AM, Marban E: Role of troponin I proteolysis in the pathogenesis of stunned myocardium. Circ Res 80: 393–399, 1997
20. Urthaler F, Walker AA, Reeves DNS, Hefner LL: Maximal twitch tension in intact lenght-clamped ferret papillary muscles evoked by modified postextrasystolic potentiation. Circ Res 62: 65–74, 1988
21. Fishbein MC, Meerbaum S, Rit J, Lando U, Kanmatsuse K, Mercier JC, Corday E, Ganz W: Early phase acute myocardial infarct size quantification: validation of the triphenyltetrazolium chloride tissue enzyme staining technique. Am Heart J 101: 593–600, 1981
22. Hashimoto S, Nishida S, Himura S, Takahashi M, Enomoto Y, Sakatani K: Immunohistological reaction of myocardial fibers with actin antiserum in autopsy cases of myocardial infarction. Histol Histopath 1: 241–249, 1986
23. Hsu SM, Raine L, Farger H: Use of avidin-biotin-peroxidase complex (ABC) in immunoperoxidase techniques: a comparison between ABC and unlabeled antibody (PAP) procedures. J Histochem Cytochem 29: 577–580, 1981
24. Cingolani HE, Mosca SM, Moreyra AE: Diastolic dysfunction in stunning: its prevention. The failing heart. Eds: Dhalla NS, Beamish RE, Nagano M, Raven Press, New York, pp. 67–78, 1995

25. Kusuoka H, Porterfield JK, Weissman HF, Weisfeld M, Marban E: Pathophysiology and pathogenesis of stunned myocardium. J Clin Invest 79: 950–961, 1987
26. de Albuquerque CP, Gerstenblith G, Weiss RG: Importance of metabolic inhibition and cellular pH in mediating preconditioning contractile and metabolic effects in rat hearts. Circ Res 74: 139–150, 1994
27. Gao WD, Atar D, Backx PH, Marban E: Relationship between intracellular calcium and contractile force in stunned myocardium. Direct evidence for decreased myofilament Ca^{2+} responsiveness and altered diastolic function in intact ventricular muscle. Circ Res 76: 1036–1048, 1995
28. Asimakis GK, Inners-McBride K, Medellin G, Conti VR: Ischemic preconditioning attenuates acidosis and post-ischemic dysfuntion in isolated rat heart. Am J Physiol 263: H887–H894, 1992
29. Lasley RD, Anderson GM, Mentzer RM: Ischemic and hypoxic preconditioning enhance post-ischemic recovery of function in the rat heart. Cardiovasc Res 27: 565–570, 1993
30. Jenkins DP, Pugsley WB, Yellon DM: Ischaemic preconditioning in a model of global ischemia: infarct size limitation, but not reduction of stunning. J Mol Cell Cardiol 27: 1623–1632, 1995
31. Nayler WG, Elz JS, Buckley DJ: The stunned myocardium: effect of electrical and mechanical arrest and osmolarity. Am J Physiol 254 (Heart Circ Physiol 23): H60–H69, 1988
32. Pucheu S, Coudray Ch, Tresallet N, Favier A, de Leiris J: Effect of dietary antioxidant trace element supply on cardiac tolerance to ischemia-reperfusion in the rat. J Mol Cell Cardiol 27: 2303–2314, 1995
33. Downey JM: Free radicals and their involvement during long-term myocardial ischemia and reperfusion. Annu Rev Physiol 52: 487–504, 1990
34. Garlick PP, Davies MJ, Hearse DJ, Slater TF: Direct detection of free radicals in the reperfused rat heart using electron spin resonance spectrocopy. Circ Res 61: 757–760, 1987
35. Bolli R: Oxygen-derived free radicals and postischemic myocardial dysfunction ("stunned myocardium"). J Am Coll Cardiol 12: 239–249, 1988
36. Gao WD, Liu Y, Marban E: Selective effects of oxygen free radicals on excitation-contraction coupling in ventricular muscle: implications for the mechanism of stunned myocardium. Circulation 94(suppl I): I–185 (Abstract), 1996
37. Kusuoka H, Jacobus WE, Marban E: Calcium oscillations in digitalis-induced ventricular fibrillation: Pathogenetic role and metabolic consequences in isolated ferret hearts. Circ Res 62: 609–619, 1988
38. Tani M: Mechanisms of Ca^{2+} overload in reperfused ischemic myocardium. Ann Rev Physiol 52: 543–559, 1990
39. Correti M, Koretsune Y, Kusuoka H, Chacko VP, Zweier J, Marban E: Glycolitic inhibition and calcium overload as consequences of exogenously generated free radicals in rabbits hearts. J Clin Invest 88: 1014–1025, 1991
40. Josephson RA, Silverman HS, Lakatta EG, Swtern MD, Zweier JL: Study of the mechanisms of hydrogen peroxide and hydroxil free radical induced cellular injury and calcium overload in cardiac myocytes. J Biol Chem 266: 2354–2361, 1991

Molecular and Cellular Biochemistry **186**: 131–137, 1998.
© 1998 *Kluwer Academic Publishers.*

Physiological and biochemical adrenergic regulation of the stunned myocardium

Dorothy E. Vatner and Stephen F. Vatner
Cardiovascular and Pulmonary Research Institute, Allegheny University of the Health Sciences, Pittsburgh, PA, USA

Abstract

A dual approach was employed to study β-adrenergic receptor signal transduction in post ischemic (stunned) myocardium, examining physiological interventions in awake, chronically instrumented pigs and biochemical, cellular mechanisms in sarcolemmal preparations from the stunned hearts using the contralateral non-ischemic zone as a control. Ten min of coronary artery occlusion (CAO) and 30 min coronary artery reperfusion (CAR) resulted in depressed posterior wall-thickening (myocardial stunning). Isoproterenol increased transmural wall thickening more in stunned myocardium than in non-ischemic myocardium. In contrast, the responses of wall thickening to forskolin, actually decreased during stunning compared with control. NKH 477, a water soluble forskolin derivative, that does not activate cardiac nerves, increased wall thickening in non-ischemic tissue similarly to the effects on stunned myocardium. Increasing cardiac neural tone reflexly with inferior venal caval occlusion (IVCO) elicited similar results to forskolin, i.e., stunned myocardium responded with less of an increase in wall thickening as compared with non-ischemic myocardium. β-adrenergic receptor density, as determined with ^{125}I-cyanopindolol binding, was significantly increased in stunned subendocardium and subepicardium compared with respective values in non-ischemic myocardium. There were no differences in the response of adenylyl cyclase to isoproterenol in stunned and non-ischemic myocardium. The enhanced responsiveness of the β-adrenergic receptor to isoproterenol stimulation in stunned myocardium corresponded to the increase in β-adrenergic receptor density. The combination of enhanced responses to isoproterenol, and decreased responses to forskolin and to IVCO and preserved responsiveness to NKH 477, suggest that stunned myocardium is characterized by transient sympathetic neural stunning. The enhanced sensitivity to β-adrenergic receptor stimulation has important clinical implications, both in terms of therapy of stunned myocardium and detection of stunned and /or hibernating myocardium, i.e., low dose dobutamine echocardiography. (Mol Cell Biochem **186**: 131–137, 1998)

Key words: myocardial stunning, β-adrenergic receptor, adenylyl cyclase, coronary artery occlusion, coronary artery reperfusion

Introduction

β-adrenergic receptor signal transduction is fundamental to regulation of cardiac contractility. It is most important in mediating increases in cardiac inotropy, e.g., in response to stress. Conversely, myocardial ischemia is one of the most potent negative inotropic interventions. An interesting clinical conundrum is whether the positive effects of β-adrenergic receptor stimulation can offset and even overcome the cardiac depressant effects of ischemia. The answer to this question has been complicated by three major factors. First, many paradigms of ischemia are sufficiently prolonged, thereby resulting in infarction. The admixture of infarcted tissue with ischemic but viable myocardium, makes the interpretation of results from β-adrenergic receptor stimulation complex. Secondly, in paradigms of ischemia with total coronary artery occlusion (CAO) the attendant increases in myocardial oxygen demand induced by β-adrenergic receptor stimulation cannot be met with commensurate increases in coronary blood flow, thereby prohibiting the expected increase in cardiac function. Thirdly, ischemia is thought to alter β-adrenergic receptor signal transduction.

Our primary goal was to examine potential alterations in β-adrenergic receptor signaling in stunned myocardium, i.e., after coronary artery reperfusion (CAR) had intervened. This avoids the above mentioned complications due to persistent

Address for offprints: D.E. Vatner, Cardiovascular and Pulmonary Research Institute, Allegheny University of the Health Sciences, Allegheny Campus, 15th Floor, South Tower, 320 East North Avenue, Pittsburgh, PA 15212, USA

CAO or to the effects of attendant infarction, since in stunned myocardium the coronary artery is patent and necrosis is absent. This is the ideal setting to examine β-adrenergic receptor regulation in post-ischemic tissue. A secondary goal was to examine the extent to which β-adrenergic receptor stimulation can overcome the myocardial depressant effects inherent to stunned myocardium.

We have recently utilized a porcine model of 'stunned myocardium', i.e., ultimately reversible post-ischemic myocardial dysfunction, following a 10 min period of CAO, to examine effects of β-adrenergic receptor stimulation [1, 2]. By employing this model, complications due to necrosis and persistent CAO are avoided. Our experimental design involved a dual approach, examining physiological interventions in awake, chronically instrumented pigs, and biochemical/cellular mechanisms in sarcolemmal membrane preparations from the stunned hearts using the contralateral non-ischemic zone as a control. β-adrenergic receptor activation was carried out in conscious pigs, using interventions that stimulated the β-adrenergic receptor directly (isoproterenol), or stimulated the adenylyl cyclase catalytic unit directly (forskolin, NKH 477). Since forskolin also stimulates cardiac nerves *in vivo* [3], it was important to stimulate β-adrenergic receptors via the neural pathway as well. This was achieved with baroreflex unloading, *in vivo*, which causes a reflex increase in cardiac neural tone. NKH 477 is a water soluble forskolin derivative, which unlike forskolin, does not activate cardiac nerves, because of its water solubility [3].

There are several advantages of the porcine model. Pigs have poorly developed coronary collateral circulation in the heart [4, 5], and accordingly, CAO results in more intense transmural myocardial ischemia in pigs than in dogs, who have well-developed coronary collaterals. Thus, it is likely that the stunning would be more uniform transmurally in pigs [6]. In addition, the conscious pig is an excellent model for examining interventions that might alter myocardial stunning because the extent of stunning is similar with brief CAO when carried out several days apart [6]. The current results are based primarily on recent studies, carried out in Dr. D. Vatner's laboratory [1].

Physiological studies

Regional myocardial blood flow and wall thickness

Blood flow was determined with isotopically labeled microspheres (15± 2 μm in diameter, New England Nuclear, Boston, MA, USA). The type of labeled microsphere was chosen randomly. One to two million micropheres were injected through a left atrial catheter. A reference sample of arterial blood was taken from a catheter in the descending thoracic aorta. Withdrawal of blood began 15 sec prior to microsphere injection and continued for 90 sec after the microsphere injection was completed. After sacrifice, tissue samples from the non-ischemic and stunned zones were taken for blood flow analysis as shown in Fig. 1, blood flow fell during 10 min CAO in the subendocardium and sub-epicardium, and recovered after 30 min CAR. In the ischemic (stunned) zone, the blood flow was slightly but not significantly lower during CAR.

Wall thickness crystals were implanted to measure wall thickness in the posterior LV free wall. The sub-endocardial crystals were introduced obliquely, so the myocardium between the crystals would not be impaired by injury or fibrosis. Ten min of CAO reduced wall thickening completely, such that paradoxical wall thinning was observed. After CAR there was recovery of systolic wall thickening, but it remained depressed even after 30 min CAR (Fig. 1), i.e., typical stunning was observed.

Effects of adrenergic stimulation

Isoproterenol (0.2 μg/kg) increased transmural wall thickening more $p < 0.05$, in stunned myocardium (+1.40 ± 0.16 mm) than non-ischemic myocardium (+0.71 ± 0.22 mm) (Figs 2 and 3). In contrast to isoproterenol, the responses of wall thickening to forskolin actually decreased, $p < 0.05$, during stunning (+0.54 ± 0.10 mm, $p < 0.05$) compared with control (+0.78 ± 0.11 mm) (Figs 2 and 3). As shown in Figs 2 and 3, NKH 477 increased wall thickening in non-ischemic tissue (+0.95 ± 0.25 mm) similarly to the effects on stunned myocardium (+1.01 ± 0.24 mm). Increasing cardiac neural tone reflexly with IVCO elicited similar results to forskolin, i.e., stunned myocardium responded with less, $p < 0.05$, increase in wall thickening (+0.53 ± 0.05 mm) as compared with non-ischemic myocardium (+0.95 ± 0.07mm). In summary, the response to isoproterenol in stunned myocardium was unique because the increased response was significantly different, $p < 0.05$, from the responses observed with forskolin, NKH 477 and IVCO. Furthermore, forskolin and IVCO actually elicited decreases in wall thickness in stunned myocardium compared with responses in non-ischemic tissue.

Biochemical studies

β--Adrenergic receptor antagonist binding studies

Using the conscious pig model of 10 min CAO followed by 30 min CAR, the animals were killed, hearts were removed and the area at risk was divided into subendocardium and subepicardium. The tissues were homogenized, centrifuged at 14,000 xg for 20 min. The pellet was resuspended in Tris buffer. ^{125}I-cyanopindolol (^{125}I-cyp) 0.01 to 1.0 nM was used with 10 μg of membrane protein. The samples were incubated for 60 min, and filtered on glass fiber filters. The filters were counted in a gamma counter. The data was computer modeled with the Ligand program [7].

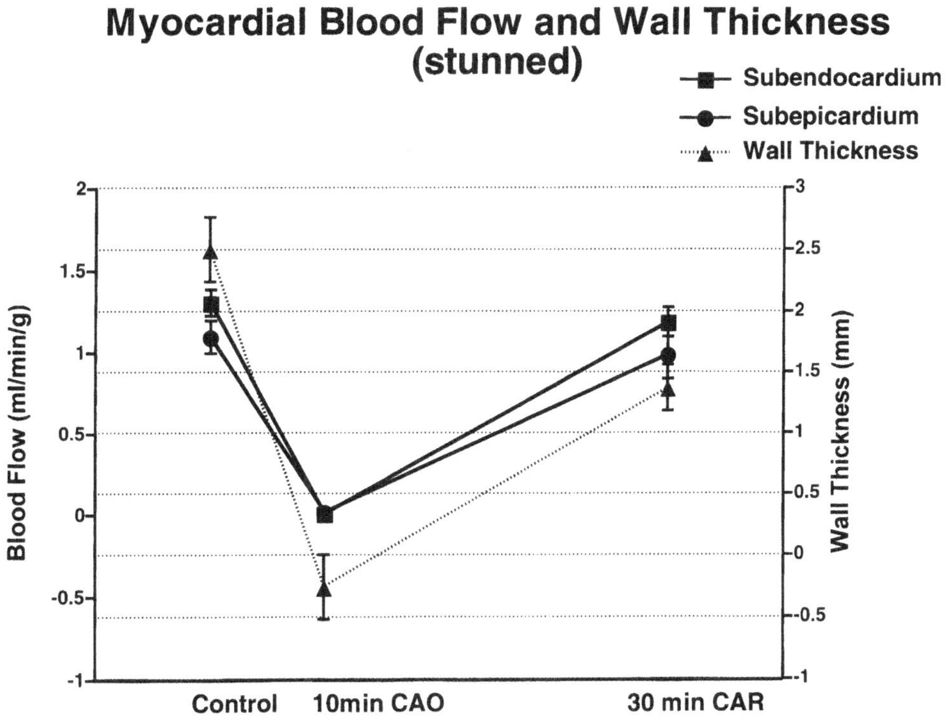

Fig. 1. Comparison of effects of 10 min of CAO and 30 min of CAR on subendocardial (■) and subepicardial blood flow (●) and changes in wall thickness (▲) in the pig are shown. The scale for blood flow is on the left and that for wall thickness on the right. At 30 min CAR wall thickness remained depressed (stunning), while blood flow had almost recovered. Redrawn with permission from Sato *et al.* (1997) (Reference 1).

β-adrenergic receptor density, as determined with ^{125}I-cyp binding, was significantly increased in stunned subendocardium and subepicardium compared with respective values in non-ischemic myocardium (Fig. 4). These data are consistent with the enhanced physiological responses to isoproterenol observed, *in vivo*. The affinity (K_D) for ^{125}I-cyp was not significantly different between the stunned myocardium (subendocardium 41 ± 3 pM, subepicardium 38 ± 5 pM) and non-ischemic myocardium (subendocardium 42 ± 6 pM, subepicardium 46 ± 6 pM).

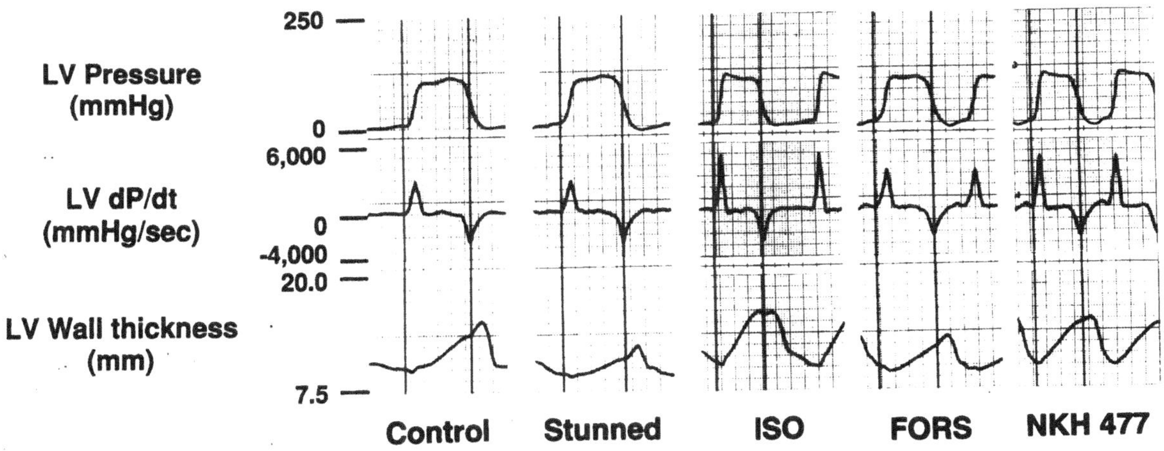

Fig. 2. Representative recording of measurements of LV pressure and dP/dt, and LV wall thickness before coronary artery occlusion (control), during stunning, and the response of isoproterenol (ISO, 0.2 μg/kg), forskolin (FORS, 100 nmol/kg) and NHK 477 (30 μg/kg) during myocardial stunning in a conscious pig. Reproduced with permission from Sato *et al.* (1997) (Reference 1).

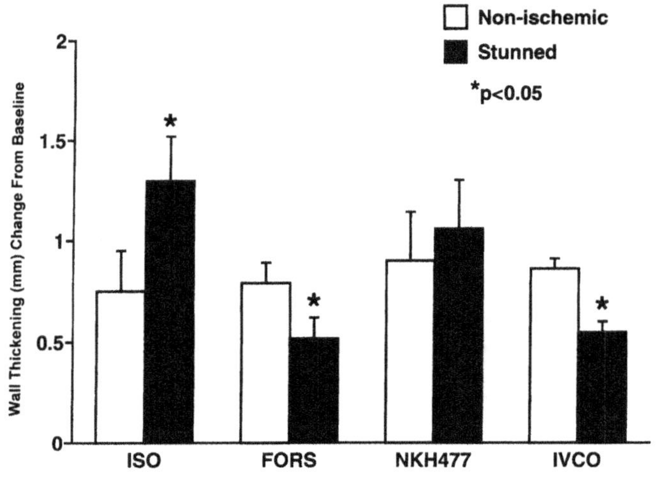

Fig. 3. The average ± S.E. wall thickening responses to ISO (isoproterenol), FORS (forskolin), NKH 477 and IVCO (inferior vena caval occlusion) are shown for control (□) and stunned myocardium (■). ISO increased wall thickening significantly more during stunning than before stunning. In contrast, FORS increased wall thickening significantly less during stunning than control. The responses to NKH 477 were not changed in stunned myocardium compared to control. The responses to IVCO were depressed significantly during stunning compared to control. The asterisks denote significant differences, $p < 0.05$, before and during stunning. Reproduced with permission from Sato *et al.* (1997) (Reference 1).

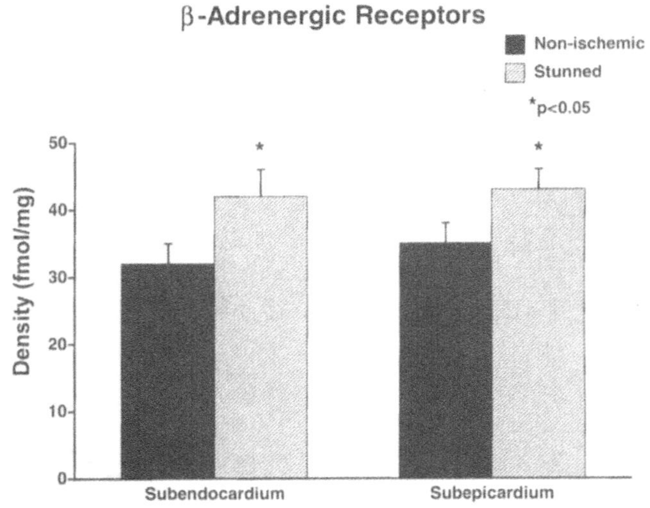

Fig. 4. β-adrenergic receptor density was measured using ^{125}I-cyp in cardiac sarcolemmal membrane preparations from subendocardium and subepicardium of pigs with 10 min CAO followed by 20 min of CAR. In both subendocardium and subepicardium the stunned region showed a significant increase in β-adrenergic receptors as compared to the non-ischemic region of the same heart (n = 10). Redrawn with permission from Sato *et al.* (1997) (Reference 1).

β-Adrenergic receptor agonist binding studies
Agonist binding competition curves were assayed using 15 μg of membrane protein, 0.07 nM ^{125}I-cyp, and 22 concentrations of isoproterenol (10^{-9} to 10^{-4}M). Assays were incubated for 45 min at 37°C, rapidly filtered through glass fiber filters, and counted in a gamma counter. The binding data was analyzed with the Ligand program [7]. The data fit best to a two-site model, with high and low affinity sites. As shown in Fig. 5, the proportion of receptors from each site showed no significant differences between non-ischemic and stunned myocardium.

Adenylyl cyclase activity
Isoproterenol-stimulated adenylyl cyclase activity stimulation curves for non-ischemic and stunned endocardium and epicardium are shown in Fig. 6. There were no differences in the responses of adenylyl cyclase to isoproterenol in stunned and non-ischemic myocardium. These data are consistent with preserved inotropic responses to NKH 477, *in vivo*, but not the depressed responses to forskolin observed, *in vivo*.

Discussion

One major finding was that the inotropic response to humoral β-adrenergic receptor stimulation was actually enhanced in stunned myocardium [1, 2]. Several other studies of stunned myocardium have shown preserved responsiveness to β-adrenergic receptor stimulation with isoproterenol [8–10], epinephrine [11, 12], norepinephrine [13], and dobutamine [14–18]. Also, there is evidence that stimulation of β-

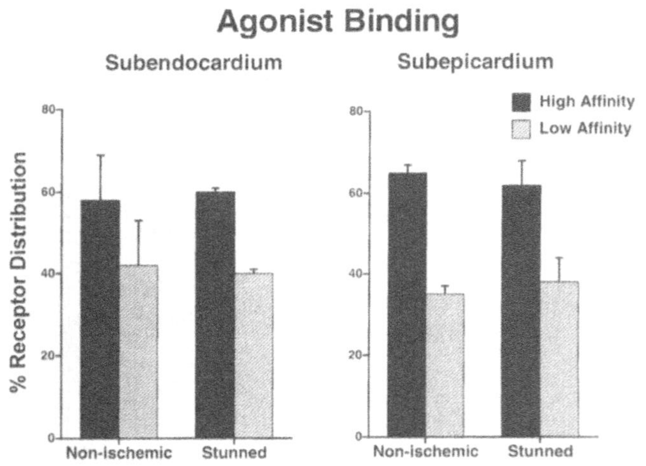

Fig. 5. β-adrenergic receptor agonist binding using ^{125}I-cyp and a series of concentrations of isoproterenol were used with subendocardial and subepicardial sarcolemmal membrane preparations from non-ischemic and stunned regions. There was no difference in the distribution of high and low affinity receptors. Redrawn with permission from Sato *et al.* (1997) (Reference 1).

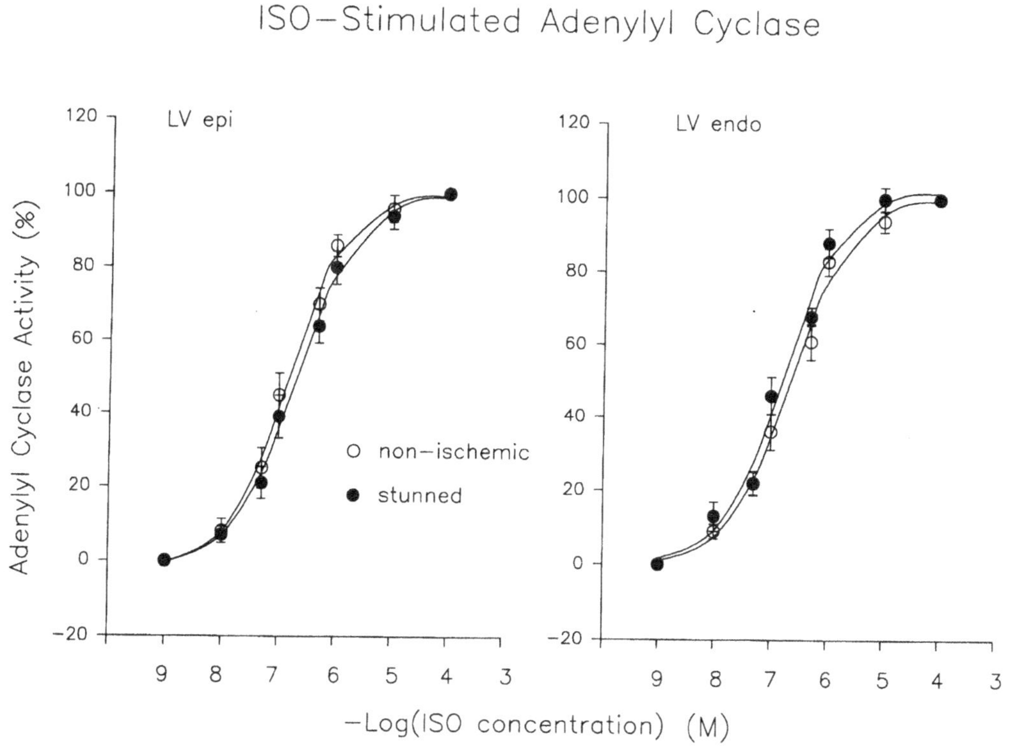

Fig. 6. Isoproterenol (ISO) stimulated adenylyl cyclase activities in non-ischemic (O) and stunned (●) myocardium are shown transmurally. There were no differences between non-ischemic and stunned myocardium in both subendocardium (endo) and subepicardium (epi). Reproduced with permission from Sato *et al.* (1997) (Reference 1).

adrenergic receptors actually can reverse stunning [9, 10, 19]. Our observation demonstrating enhanced sensitivity to β-adrenergic receptor stimulation, has potentially important clinical implications. Specifically, better understanding the role of β-adrenergic receptors in the stunned myocardium is clinically important because of the common use of dobutamine echocardiography [20] to identify viable myocardium, which is dysfunctional at baseline, potentially due to chronic stunning, but demonstrates increased contractility with β-adrenergic receptor stimulation. It is possible that the use of dobutamine echocardiography [20] is based in part on the enhanced response of stunned myocardium to β-adrenergic receptor stimulation, as shown in our experiments.

There has been considerable debate regarding the effects of myocardial ischemia on adenylyl cyclase activity. Studies with longer periods of ischemia, i.e., greater than 20 min, where some of the effects are irreversible, generally demonstrated a decrease in adenylyl cyclase activity, more significant in the subendocardium [21–23]. Conversely, studies in isolated rat hearts with very brief episodes of ischemia demonstrate the opposite, i.e., sensitization of adenylyl cyclase activity [24–27]. Little is known about the effects of stimulating adenylyl cyclase directly in stunned myocardium, which was one of the major goals of our investigation [1]. Interestingly, forskolin, which stimulates the catalytic unit of adenylyl cyclase elicited a depressed physiological response in stunned myocardium. This occurred in the face of no change in adenylyl cyclase activity *in vitro*. We have recently shown that forskolin elicits a neural component *in vivo* as well as its direct action on adenylyl cyclase [3]. It is also known that brief periods of CAO, elicit stunning of cardiac nerves, as well as myocardium [28–32]. Therefore, we hypothesized that it was the neural component of forskolin's action, which was depressed, and accounted for the reduced responsiveness to forskolin in stunned myocardium in conscious pigs.

To test the hypothesis that the neural component of forskolin was depressed, but that the direct effect adenylyl cyclase was preserved, we administered NKH 477, a cyclase activator, which is water soluble, and accordingly, does not affect autonomic efferent activity [3]. Indeed, we have shown the effects of NKH477 are similar in the presence and absence of ganglionic blockade, whereas the effects of forskolin were depressed after ganglionic blockade. The responses to NKH 477 were not depressed in stunned myocardium, consistent with the *in vitro* experiments demonstrating preserved adenylyl cyclase activity. Interestingly, reflex inotropic responses to IVCO, which activates cardiac nerves reflexly, were significantly diminished. These experiments support the concept that stunned myocardium is not characterized by altered adenylyl cyclase activity, but responses to stimulating sympathetic nerves are depressed due to an effect of transient ischemia on cardiac nerves, i.e., neural stunning, as demonstrated by other studies of brief periods of myocardial ischemia followed by CAR (stunning) [28–32].

The enhanced responsiveness of the β-adrenergic receptor to isoproterenol stimulation in stunned myocardium corresponded to the increase in β-adrenergic receptor density. However, the response to isoproterenol stimulated adenylyl cyclase activity was similar to the non-ischemic region even though β-adrenergic receptor density was increased. This apparent discrepancy between enhanced physiological responsiveness to isoproterenol, appears to be inconsistent with the *in vitro* dose response curve for isoproterenol-stimulated adenylyl cyclase activity, which demonstrated no change in stunned myocardium (Fig. 6). Therefore, there is not a simple explanation of increased β-adrenergic receptor density transducing increased function through increased adenylyl cyclase activity. Other explanations must be explored. One possibility is that stunned myocardium is characterized by transient sympathetic neural stunning. Indeed, brief CAO has been shown to result in efferent sympathetic denervation [29]. This action could result in upregulation of β-adrenergic receptors, which could explain the enhanced physiological response to isoproterenol, whereas the transient denervation was responsible for the depressed neurally mediated responses. Another possibility is that the stunned myocardium has increased amounts of β-adrenergic receptors and adenylyl cyclase in the light vesicle fraction which is not analyzed in the crude membrane fraction used in this study [33]. This is unlikely, since the amount of β-adrenergic receptors and adenylyl cyclase activity in this discarded fraction is very small compared to the crude membrane fraction that is studied [33]. It is also possible that the increased β-adrenergic receptors are not coupled to adenylyl cyclase because of impaired β-adrenergic receptor cycling from the plasma membrane to the cytosol [34]. However, the finding that responsiveness to NKH 477 was also significantly less than the responses to isoproterenol, suggesting that β-adrenergic receptor stimulation *in vivo* resulted in unique enhanced activity. In view of the fact that adenylyl cyclase activity was not enhanced, it is also possible that there is a direct activation of calcium channels with isoproterenol, bypassing cyclic AMP generation. However this mechanism, whether or not involving G-proteins, remains controversial [35, 36]. Finally, it is possible that biochemical assays are not sensitive enough to fully reflect the rapid changes characteristic of *in vivo* activity [37].

In summary, the results of enhanced responses to isoproterenol, and decreased responses to forskolin and to IVCO and preserved responsiveness to NKH 477, suggest that stunned myocardium is characterized by transient sympathetic neural stunning. This transient reduced neural function may result in upregulation of β-adrenergic receptors and enhanced isoproterenol responsiveness without an effect on adenylyl cyclase. The enhanced sensitivity to β-adrenergic receptor stimulation has important clinical implications, both in terms of therapy of stunned myocardium and detection of stunned and/or hibernating myocardium, i.e., low dose dobutamine echocardiography [20].

References

1. Sato S, Sato N, Kudej RK, Uechi M, Asai K, Shen Y-T, Ishikawa Y, Vatner SF, Vatner DE: β-Adrenergic receptor signalling in stunned myocardium of conscious pigs. J Mol Cell Cardiol 29: 1387–1400, 1997
2. Vatner DE, Sato S, Vatner SF, Stasser RH: Sympathetic signal transduction in myocardial ischemia. In: GR Heyndrickx, SF Vatner, WC Wijns, (eds). Stunning, Hibernation and Preconditioning: Clinical Pathophysiology of Myocardial Ischemia. Lippincott-Raven, Philadelphia, New York, 1997, pp 31–48
3. Iwase M, Ishikawa Y, Shen Y-T, Shannon RP, Sato N, Ganguly PK, Eki T, Vatner DE, Vatner SF: Neurally mediated cardiac effects of forskolin in conscious dogs. Am J Physiol 271(Heart Circ Physiol 40): H1473–H1482, 1996
4. Eckstein RW: Coronary interarterial anastomoses in young pigs, and mongrel dogs. Circ Res 2: 460–465, 1954
5. Patterson RE, Kirk ES: Analysis of coronary collateral structure, function, and ischemic border zones in pigs. Am J Physiol 244 (Heart Circ Physiol 13): H23–H31, 1983
6. Shen Y-T, Vatner SF: Differences in myocardial stunning following coronary artery occulsion in conscious dogs, pigs and baboons. Am J Physiol 270: H1312–H1322, 1996
7. Munson RJ, Rodbard D: LIGAND: A versatile computerized approach for characterization of ligand-binding systems. Anal Biochem 107: 220–239, 1980
8. Ambrosio G, Jacobus WE, Bergman CA, Weisman HF, Becker LC: Preserved high energy phosphate metabolic reserve in globally 'stunned' hearts despite reduction of basal ATP content and contractility. J Mol Cell Cardiol 19: 953–964, 1987
9. Bolli R, Zhu W-X, Myers MD, Hartley CJ, Roberts R: Beta-adrenergic stimulation reverses postischemic myocardial dysfunction without producing subsequent functional deterioration. Am J Cardiol 56: 964–968, 1985
10. Chiu WC, Kedem J, Scholz PM, Weiss HR: Regional asynchrony of segmental contraction may explain the 'oxygen consumption paradox' in stunned myocardium. Basic Res Cardiol 89: 149–162, 1994
11. Becker LC, Levine JH, DiPaula AF, Guarnieri T, Aversano T: Reversal of dysfunction in postischemic stunned myocardium by epinephrine and postextrasystolic potentiation. J Am Coll Cardiol 7: 580–589, 1986
12. Gorge G, Papageorgiou I, Lerch R: Epinephrine-stimulated contractile and metabolic reserve in postischemic rat myocardium. Basic Res Cardiol 85: 595–605, 1990
13. Ciuffo AA, Ouyang P, Becker LC, Levin L, Weisfeldt ML: Reduction of sympathetic inotropic response after ischemia in dogs: Contributor to stunned myocardium. J Clin Invest 75: 1504–1509, 1985
14. Fan D, Soei LK, Sassen LMA, Krams R, Verdouw PD: Mechanical efficiency of stunned myocardium is modulated by increased afterload dependency. Cardiovasc Res 29: 428–437, 1995

15. Hashimoto T, Buxton DB, Krivokapick J, Hansen HW, Phelps MD, Schelbert HR: Responses of blood flow, oxygen consumption, and contractile function to inotropic stimulation in stunned canine myocardium. Am Heart J 126: 347–351, 1994
16. Krams R, Soei LK, McFalls EO, Prins EAW, Sassen LMA, Verdouw PD: End-systolic pressure length relations of stunned right and left ventricles after inotropic stimulation. Am J Physiol 265: H2099–H2109, 1993
17. Watanabe S, Buffington CW: Electromechanical association in regionally stunned swine myocardium. Anesthesiology 79: 1266–1277, 1993
18. Zhou Z, Laskey RD, Hegge JO, Bunger R, Mentzer RM: Myocardial stunning: A therapeutic conundrum. J Thorac Cardiovas Surg 110: 1391–1401, 1995
19. Mercier JC, Lando U, Kanmatsuse K, Ninomiya K, Meerbaum S, Fishbein MD, Swan HJC, Ganz W: Divergent effects of inotropic stimulation on the ischemic and severely depressed reperfused myocardium. Circulation 66: 397–400, 1982
20. La Canna G, Alfieri O, Giubbini R, Gargano M, Ferrari R, Visioli O: Echocardiography during infusion of dobutamine for identification of reversible dysfunction in patients with chronic coronary artery disease. J Am Coll Cardiol 23: 617–626, 1994
21. Susanni EE, Manders WT, Knight DR, Vatner DE, Vatner SF, Homcy CJ: One hour of myocardial ischemia decreases the activity of the stimulatory guanine-nucleotide regulatory protein Gs. Circ Res 65(4): 1145–1150, 1989
22. Vatner DE, Young MA, Knight DR, Vatner SF: Beta-receptors and adenylate cyclase: Comparison of nonischemic, ischemic, and postmortem tissue. Am J Physiol 258: H140–H144, 1990
23. Vatner DE, Kiuchi K, Manders WT, Vatner SF: Effects of coronary arterial reperfusion on β-adrenergic receptor-adenylyl cyclase coupling. Am J Physiol 264: H196–H204, 1993
24. Strasser RH, Marquetant R, Kubler W: Adrenergic receptors and sensitization of adenylyl cyclase in acute myocardial ischemia. Circulation 82(Suppl P): 23–29, 1990
25. Strasser RH, Marquetant R: Supersensitivity of the adenylyl cyclase system in acute myocardial ischemia: Evaluation of three independent mechanisms. Basic Res Cardiol 85(Suppl 1): 67–78, 1990
26. Strasser RH, Krimmer J, Braun-Dullaeus R, Marquetant R, Kubler W: Dual sensitization of the adrenergic system in early myocardial ischemia: Independent regulation of the β-adrenergic receptors and adenylyl cyclase. J Mol Cell Cardiol 22(12): 1405–1423, 1990
27. Strasser RH, Marquetant R, Kubler W: Independent sensitization of beta-adrenoceptors and adenylate cyclase in acute myocardial ischaemia. British J Clin Pharmacol 30(Suppl 1): 27S–35S, 1990
28. Gutterman DD, Morgan DA, Miller FJ: Effect of brief myocardial ischemia on sympathetic coronary vasoconstriction. Circ Res 71(4): 960–969, 1992
29. Hageman GR, Gantenberg NS: Attenuation of baroreflex changes in cardiac sympathetic efferent activities during acute myocardial ischemia. Am Heart J 126: 347–351, 1993
30. Inoue H, Zipes DP: Time course of denervation of efferent sympathetic and vagal nerves after occlusion of the coronary artery in the canine heart. Circ Res 62(6): 1111–1120, 1988
31. Neely BH, Hageman GR: Differential cardiac sympathetic activity during acute myocardial ischemia. Am J Physiol 258: H1534–H1541, 1990
32. Ninomiya I, Matsukawa K, Honda T, Nishiura N, Shirai M: Cardiac sympathetic nerve activity and heart rate during coronary occlusion in awake cats. Am J Physiol 251: H528–H537, 1986
33. Vatner DE: Characterization of the subfractions from a purified sarcolemma preparation of canine left ventricle. J Mol Cell Cardiol 22: 1349–1357, 1990
34. Maisel AS, Motulsky HJ, Insel PA: Externalization of β-adrenergic receptors promoted by myocardial ischemia. Science 230: 183–186, 1985
35. Hartzell HC, Mery P-F, Fischmeister R, Szabo G: Sympathetic regulation of cardiac calcium current is due exclusively to cAMP-dependent phosphorylation. Nature 351: 573–576, 1991
36. Yatani A, Codina J, Imoto Y, Reeves JP, Birnbaumer L, Brown AM: A G protein directly regulates mammalian cardiac calcium channels. Science 238: 1288–1292, 1987
37. Iwase M, Bishop SP, Uechi M, Vatner DE, Shannon RP, Kudej RK, Wight DC, Wagner TE, Ishikawa Y, Homcy CJ, Vatner SF: Adverse effects of chronic endogenous sympathetic drive induced by cardiac G_s alpha overexpression. Circ Res 78(4): 517–524, 1996

Molecular and Cellular Biochemistry **186:** 139–145, 1998.
© 1998 *Kluwer Academic Publishers.*

Ischemic preconditioning attenuates apoptotic cell death associated with ischemia/reperfusion

Nilanjana Maulik,[1] Tetsuya Yoshida,[1] Richard M. Engelman,[2] David Deaton,[2] Joseph E. Flack III,[2] John A. Rousou[2] and Dipak K. Das[1]

[1]*Cardiovascular Division, Department of Surgery, University of Connecticut School of Medicine, Farmington, CT;* [2]*Department of Surgery, Baystate Medical Center, Springfield, MA, USA*

Abstract

Apoptosis or programmed cell death is a genetically controlled response for cells to commit suicide and is associated with DNA fragmentation or laddering. The common inducers of apoptosis include oxygen free radicals/oxidative stress and Ca^{2+} which are also implicated in the pathogenesis of myocardial ischemic reperfusion injury. To examine whether ischemic reperfusion injury is mediated by apoptotic cell death, isolated perfused rat hearts were subjected to 15, 30 or 60 min of ischemia as well as 15 min of ischemia followed by 30, 60, 90 or 120 min of reperfusion. At the end of each experiment, the heart was processed for the evaluation of apoptosis and DNA laddering. Apoptosis was studied by visualizing the apoptotic cardiomyocytes by direct fluorescence detection of digoxigenin-labeled genomic DNA using APOPTAG® *in situ* apoptosis detection kit. DNA laddering was evaluated by subjecting the DNA obtained from the hearts to 1.8% agarose gel electrophoresis and photographed under UV illumination. The results of our study revealed apoptotic cells only in the 90 and 120 min reperfused hearts as demonstrated by the intense fluorescence of the immunostained digoxigenin-labeled genomic DNA when observed under fluorescence microscopy. None of the ischemic hearts showed any evidence of apoptosis. These results were corroborated with the findings of DNA fragmentation which showed increased ladders of DNA bands in the same reperfused hearts representing integer multiples of the internucleosomal DNA length (about 180 bp). The presence of apoptotic cells and DNA fragmentation in the myocardium were completely abolished by subjecting the myocardium to repeated short-term ischemia and reperfusion which also reduced the ischemic reperfusion injury as evidenced by better recovery of left ventricular performance in the preconditioned myocardium. The results of this study indicate that reperfusion of ischemic heart, but not ischemia, induces apoptotic cell death and DNA fragmentation which can be inhibited by myocardial adaptation to ischemia. (Mol Cell Biochem **186:** 139–145, 1998)

Key words: apoptosis, DNA fragmentation, ischemia/reperfusion, ischemic preconditioning, myocardial adaptation

Introduction

Despite the complexity in understanding the mechanism of reperfusion, the importance of reperfusion injury in the biological system has become increasingly recognized and appreciated in recent years. It is generally believed that several interrelated factors are responsible for reperfusion injury; of these three are considered to be major: (i) free radical generation, (ii) intracellular Ca^{2+} overloading, and (iii) loss of membrane phospholipids. A number of recent studies have proved the importance of phospholipids in the pathophysiology of reperfusion injury and in fact it has been suggested that the two important features of reperfusion injury, free radical generation and Ca^{2+} overloading, may be secondary to phospholipid breakdown [1, 2]. Generation of free radicals have been demonstrated directly using ESR and HPLC technology [3] and indirectly by measuring oxidative stress, lipid peroxidation, and malonaldehyde formation [4]. Numerous reports demonstrate the efficacy of antioxidants and free radical scavengers in moderating ischemia/reperfusion injury [5].

Apoptosis is a form of programmed cell death that can be

Address for offprints: N. Maulik, Department of Surgery, University of Connecticut, School of Medicine, Farmington, CT 06030-1110, USA

induced in susceptible cells by a wide variety of normal physiological stimuli as well as by deleterious environmental conditions and cytotoxic agents [6]. The significance of apoptosis is increasingly recognized both in physiological regulation and in pathological conditions. It is now widely accepted that apoptosis is a gene-directed process and can be seen as part of the repertoire available to the cell to respond to external and internal stimuli [7, 8]. Only very recently, important information about the intracellular molecular signals involved in stimulation and suppression of apoptosis in mammalian cells has emerged. Several different second messenger systems are likely to be associated with the induction of apoptosis, and the final response varies with the signals being received [7, 8].

Recent studies support the notion that myocardial adaptation to ischemia reduces ischemic reperfusion injury and myocardial infarction [9, 10]. The objective of this study was 2-fold: first, to examine whether myocardial ischemia/reperfusion induces apoptotic cell death; and second, to determine whether ischemic adaptation reduces/inhibits apoptosis. Our results indicate that reperfusion of ischemic myocardium induces apoptosis in concert with DNA fragmentation which can be inhibited by myocardial adaptation to ischemia.

Materials and methods

Isolated rat heart preparation

Fifty-four Sprague-Dawley rats weighing about 300 g were anesthetized with pentobarbital (80 mg/kg, i.p.). After intravenous administration of heparin (500 IU/kg), the chests were opened, and the hearts were rapidly excised and mounted on a non-recirculating Langendorff perfusion apparatus [11]. Retrograde perfusion was established at a pressure of 100 cm H_2O with an oxygenated normothermic Krebs-Henseleit bicarbonate (KHB) buffer containing (in mM): 118.0 NaCl, 24.0 $NaHCO_3$, 4.7 KCL, 1.2 KH_2PO_4, 1.2 $MgSO_4$, 1.7 $CaCl_2$, and 10.0 glucose. The KHB buffer had been previously equilibrated with 95% O_2/5% CO_2, pH 7.4 at 37°C. The hearts were randomly divided into three major groups : (i) 18 isolated perfused rat hearts subdivided into 3 groups of 6 hearts were subjected to 15, 30 or 60 min of ischemia; (ii) 24 hearts were subjected to 15 min of ischemia followed by either 30, 60, 90 or 120 min of reperfusion (n = 6 in each group); and (iii) 6 hearts were preconditioned by subjecting them to 5 min of ischemia and 10 min of reperfusion, repeated four times (PC), followed by 15 min of ischemia and 120 min of reperfusion. At the end of each experiment, hearts were processed for the evaluation of apoptosis, DNA laddering and gene expression.

Another set of experiments involving 12 rats was carried out to evaluate the effects of ischemic preconditioning on myocardial performance. For this, the Langendorff preparation was switched to the working mode following the preconditioning protocol (n = 6) [11]. Control experiments were performed without subjecting the hearts to preconditioning (n = 6). Aortic flow was measured by a calibrated rotameter. Coronary flow rate was measured by a timed collection of the coronary perfusate that dripped from the heart. After a 10-min aerobic perfusion of the heart, the aortic inflow line was clamped at a point close to the origin of the aortic cannula. Reperfusion was initiated by unclamping the aortic line. Before ischemia and during reperfusion, heart rate (HR), coronary flow (CF) and aortic flow (AF) rates were registered. Left ventricular developed pressure (LVDP), which was defined as the difference between LV systolic and end-diastolic pressure (EDP), and the first derivative of LVDP (LVdp/dt), were also recorded.

Evaluation of apoptosis

Apoptotic cell death was evaluated using the Tunel method [12] and APOPTAG® detection kit (Oncor, Gaithersburg, MD). In brief, control, ischemic, and reperfused heart tissues were immediately put in 10% formalin and fixed in an automatic tissue fixing machine. The tissues were carefully embedded in molten paraffin in metallic blocks, covered with flexible plastic moulds and kept under freezing plates to allow the paraffin to solidify. The metallic containers were removed and tissues became embedded in paraffin on the plastic moulds. Ten sections were made from each sample in order to make a clear judgement for each experiment. Prior to analyzing tissues for apoptosis, tissue sections were deparaffinized with xylene and washed in succession with different concentrations of ethanol (absolute, 95%, 70%). Tissues were then a) treated with proteinase k for 15 min at room temperature, b) blotted carefully to remove excess liquids, c) treated with equilibrium buffer applied directly on the specimens, and d) placed in a humidified chamber for 5 min at room temperature. Specimens were then treated with terminal deoxynucleotidyl transferase (TdT) treatment at 37°C in a humidified chamber. After 1 h, cover slips were removed, and the specimens were placed in a Coplin jar containing stop/wash buffer (supplied in the kit) for 10 min at room temperature. Fifty two μl of working strength anti digoxigenin-fluorescein was added to the slides which were incubated for 30 min at room temperature, washed in PBS, and counterstained with propidium iodide/antifade (supplied with the kit) applied to the slide. Apoptotic cells were visualized by direct fluorescence detection of digoxigenin-labeled genomic DNA by epifluorescence using standard fluorescein excitation and emission filters with an Axiovert

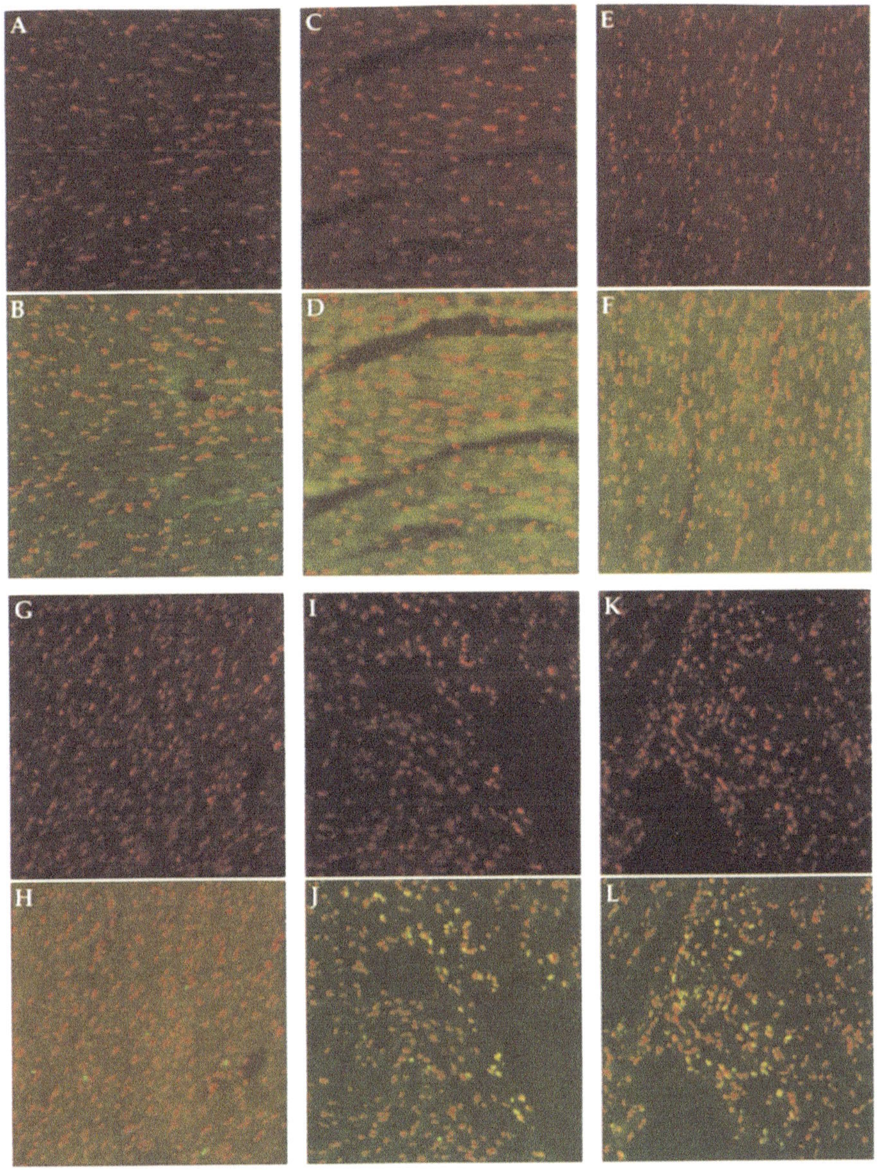

Fig. 1. Photomicrograph showing immunohistochemical staining of extended DNA in ischemic hearts. Residues of digoxigenin-nucleotide were catalytically added to the DNA by TdT enzyme for nick end labeling (TUNEL). TUNEL immunostaining identified multiple apoptotic cells in ischemic/reperfused heart tissue sections which were shown as very bright glowing yellow-green cells under the microscope. Corresponding propidium iodide (PI) staining is shown in red (A, C, E, G, I, K). PI and fluorescein channels are superimposed to show apoptotic cells as glowing yellow-green cells under the microscope (B, D, F, H, J, L) on green background. All normal cells are stained in orange when superimposed. A and B: – Control; C and D: 15 min ischemia; E and F: 30 min ischemia; G and H: 15 min ischemia followed by 60 min reperfusion; I and J: 15 min ischemia followed by 90 min reperfusion; K and L: 15 min ischemia followed by 120 min reperfusion.

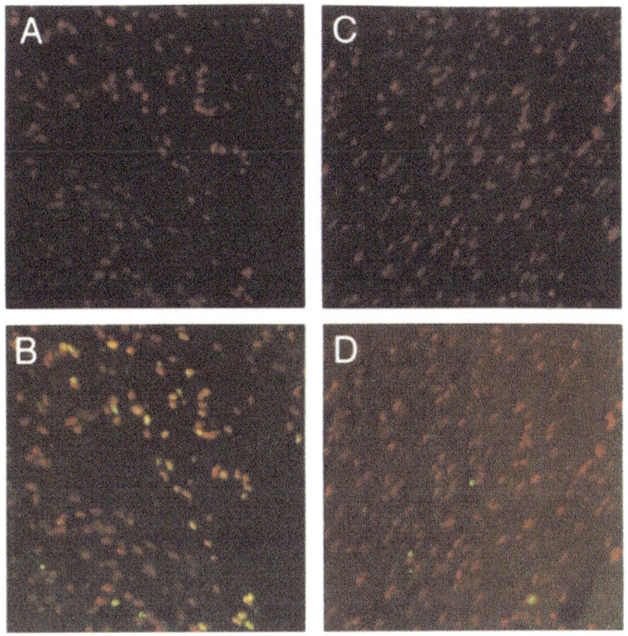

Fig. 2. Effects of preconditioning on the apoptotic cell death. Immunostaining was performed as described under the legend of Fig. 1. The results for propidium iodide (PI) staining are shown in red (A, C). PI and fluorescein channels are superimposed to show apoptotic cells as glowing yellow-green cells under the microscope (B,D) on green background. All normal cells are stained in orange when superimposed. A and B: 15 min ischemia followed by 120 min of reperfusion; C and D: PC followed by 15 min of ischemia and 120 min of reperfusion.

100 TV microscope. This method is based on the appearance of new 3′-OH DNA ends generated by DNA fragmentation and typically localized in morphologically identifiable nuclei and *apoptotic* bodies. In contrast, normal nuclei, which have relatively insignificant numbers of DNA 3′-OH ends, are not stained with this reagent.

DNA fragmentation

Apoptosis is best characterized biochemically by the cleavage of genomic DNA into nucleosomal fragments of 180 bp or multiples thereof that are readily detected as a DNA ladder by gel electrophoresis. DNA was isolated by standard techniques [13]. To perform DNA laddering, 10 μg of DNA obtained from control, ischemic, and reperfused myocardium was subjected to electrophoresis on 1.8% agarose gel containing 0.5 μg/ml ethidium bromide, and photographed under UV illumination.

Statistical analysis

For statistical analysis, a two-way analysis of variance (ANOVA) followed by Scheffe's test was first carried out using Primer Computer Program (McGraw-Hill, 1988) to test for any differences between groups. If differences were established, the values were compared using Student's *t*-test for unpaired data. The values were expressed as mean ± S.E.M. The results were considered significant if p was less than 0.05.

Results

Induction of apoptosis by ischemia/reperfusion

We were unable to detect apoptotic cells in the control and in the ischemic hearts. Even 60 min of ischemia could not induce apoptosis as evidenced under propidium iodide filter or under dual filter (superimposing propidium iodide on fluorescence filter) exposure (not shown) Thirty min of ischemia causes irreversible cell injury and necrosis in the rat heart. Surprisingly, there was no sign of fragmented nuclear DNA in these biopsies. Digoxigenin-labeled apoptotic cells were identified only in the reperfused hearts. The extent of apoptosis increased with the progression of reperfusion time. Apoptotic cell death was first evidenced after 90 min of reperfusion (Fig. 1). The number of apoptotic cells increased after 90 and 120 min of reperfusion as evidenced from the immunohistochemical staining of the extended DNA in these hearts. The number of apoptotic cells was sharply reduced in the preconditioned hearts (Fig. 2).

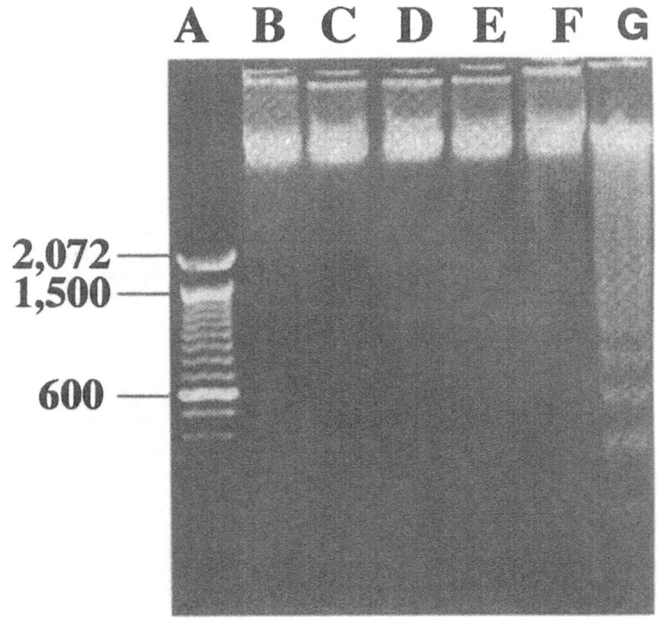

Fig. 3. Effects of ischemia and reperfusion on DNA fragmentation. DNA was isolated from the hearts and was subjected to 1.8% agarose gel electrophoresis. 100 bp DNA ladder was used as molecular weight marker (lane A). B: baseline; C: 15 min ischemia; D: 15 min ischemia followed by 30 min reperfusion; E: 15 min ischemia followed by 60 min reperfusion; F: 15 min ischemia followed by 90 min reperfusion; G: 15 min ischemia followed by 120 min reperfusion.

DNA fragmentation induced by ischemia/reperfusion

Like apoptotic cell death, even a prolonged period of ischemia did not cause DNA laddering. DNA fragmentation became obvious after 90 min of reperfusion following 15 min of ischemia (not very clear in the figure) (Fig. 3). Hearts reperfused for 120 min demonstrated clearer DNA fragmentation as compared to hearts reperfused for 90 min. The amount of DNA fragmentation was dramatically reduced in the preconditioned hearts (Fig. 4).

Effects of preconditioning on myocardial performance

As shown in Table 1, preconditioning did not alter the heart rate during ischemia and reperfusion. Aortic flow was drastically reduced during reperfusion in both groups. However, the degree of reduction was significantly smaller in the preconditioned group. Coronary flow was slightly but significantly lower after ischemia/reperfusion, and improvement was noticed in the preconditioning group compared to the control group. For both groups, developed pressure and its first derivative were lower during reperfusion compared to baseline. As depicted in Table 1, these values were

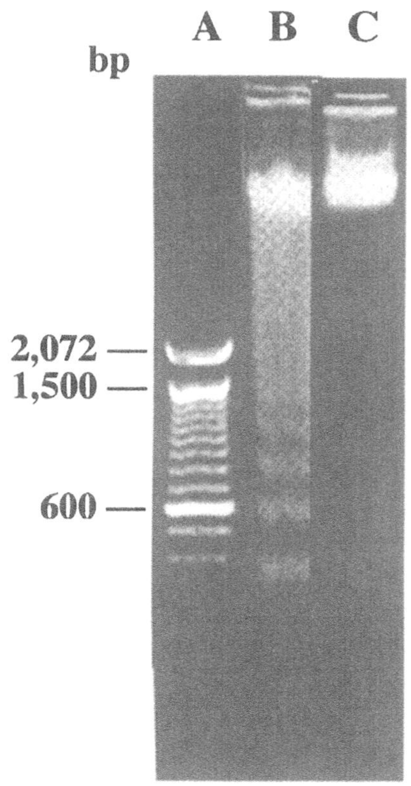

Fig. 4. Effects of preconditioning on DNA fragmentation. DNA was isolated from the hearts and was subjected to 1.8% agarose gel electrophoresis. 100 bp DNA ladder was used as molecular weight marker (lane A). B: 15 min ischemia followed by 120 min reperfusion; C: PC followed by 15 min ischemia and 120 min reperfusion.

Table 1. Effects of ischemic preconditioning on myocardial function

	Control		Preconditioned	
	Baseline	I/R	Baseline	I/R
Heart rate	305 ± 3.2	295 ± 4.7	294 ± 4.9	285 ± 9.0
DP	73 ± 1.5	35.4 ± 1.8†	74 ± 3.1	45.4 ± 1.4*†
dp/dtmax	2983 ± 74	1520 ± 63†	3064 ± 111	2055 ± 89*†
AF	43.5 ± 0.9	14.4 ± 0.7†	44.0 ± 0.9	24.3 ± 1.2*†
CF	24.1 ± 0.6	17.1 ± 0.4†	24.3 ± 0.4	19.6 ± 0.5*†
AP	142.5 ± 5.6	105 ± 3.7†	147.2 ± 4.9	125.0 ± 4.6*

$^{\dagger}p < 0.05$ compared to baseline $^{*}p < 0.05$ compared to control. DP: Developed pressure; dp/dtmax: Maximum first derivative of developed pressure; AF: Aortic flow; CF: Coronary flow; AP: Aortic pressure; I/R: measurement made after 15 min of ischemia followed by 60 min of reperfusion.

significantly higher in the preconditioned hearts ($p < 0.05$). Aortic pressure was also lower during reperfusion for all the hearts, but it was significantly higher for the preconditioned heart group ($p < 0.05$).

Discussion

Although cardiomyocyte death and infarction associated with ischemia/reperfusion injury are traditionally believed to be induced via necrosis, which is a clear-cut mechanism of accidental cell death, its precise mechanism remains unclear. Recent studies have implicated apoptotic cell death in ischemic brain [15] and ischemic liver [16]. Reperfusion of ischemic renal tissues was found to be associated with apoptotic cell death [17]. Cardiomyocytes exposed to hypoxia revealed apoptotic cell death as evidenced by DNA fragmentation in conjunction with the expression of Fas mRNA [18]. More recently, evidence was furnished in support of reperfusion injury-mediated apoptosis in cardiomyocytes [19]. Based on pathologic evaluation, these investigators concluded that apoptosis might be a specific feature of reperfusion injury in cardiac myocytes. In another study, apoptotic and necrotic myocyte cell deaths associated with ischemia/reperfusion were shown to be independent contributing variables of infarct size in rats [20]. A separate study has shown apoptosis to be a feature of human vascular pathology, including restenotic lesions and, to a lesser extent, atherosclerotic lesions, suggesting that apoptosis may modulate the cellularity of lesions that produce human vascular obstruction [21]. In this report, we have demonstrated that apoptosis is not induced by ischemia. Even 60 min of ischemia could not potentiate apoptotic cell deaths. Apoptotic cells become apparent after reperfusion, and the maximum number of such cells become visible after 120 min of reperfusion. In concert, DNA fragmentation also occurs in the ischemic reperfused myocardium. However, these results cannot rule out the possibility that more prolonged ischemia could result in apoptotic cell death. In fact, we have recently demonstrated that the signal for apoptosis is initiated during ischemia [22]. Thus, it is possible that more severe ischemia could lead to apoptotic cell death.

Collective evidence has led to the notion that ischemic preconditioning can reduce the incidence of ventricular fibrillation [23] and moderate ischemic reperfusion injury [24, 25]. The precise mechanism of preconditioning remains speculative. Ischemic preconditioning has been found to be accompanied by the preservation of high energy phosphates, and reduction of oxidative stress developed during the reperfusion of ischemic myocardium [26, 27]. Preconditioning can reduce free radical production by preserving myocardial ATP since the breakdown products of ATP, hypoxanthine and xanthine may potentiate free radical production in the ischemic reperfused myocardium [28]. Since, free radicals play an important role in apoptosis, we speculated that preconditioning might reduce apoptotic cell death during post-ischemic reperfusion. The results of our study clearly demonstrate that preconditioning reduced apoptosis and DNA fragmentation, simultaneously ameliorating ischemic/reperfusion injury.

Considerable evidence exists that reperfusion of ischemic myocardium leads to a specific type of cellular injury generally known as reperfusion injury. Although the precise mechanism of reperfusion injury remains under considerable debate, three interrelated factors are believed to play a role in the pathophysiology of reperfusion injury. These are generation of oxygen-derived free radicals, intracellular Ca^{2+} overloading and loss of membrane phospholipids, all of which are associated with the reperfusion injury [1]. Interestingly, the same three factors have been shown to be involved in apoptotic cell death . The hallmarks of apoptosis are increased cytosolic Ca^{2+} content, oxygen free radicals, and redistribution of membrane phospholipids [6,7]. Taken together, this suggests an interesting possibility that at least a part of reperfusion injury may be mediated by apoptosis. We have shown that reperfusion, but not ischemia, triggers apoptosis and DNA fragmentation. Indeed, free radicals are generated during the reperfusion of ischemic myocardium, and not during ischemia, suggesting that free radicals may be the common link for both apoptosis and reperfusion injury. Preconditioning of heart by repeated ischemia and reperfusion inhibited apoptotic cell dealth and DNA laddering presumably by reducing oxidative stress.

Acknowledgments

This study was supported by NIH HL 34360, NIH HL 22559, NIH HL 33889 as well as a Grant-in-Aid from the American Heart Association.

References

1. Das DK, Engelman RM: Phospholipids in myocardial reperfusion injury. In: DK Das (ed). Pathophysiology of Reperfusion Injury, CRC Press, Florida, 1992, pp 149–179
2. Otani H, Engelman RM, Rousou JA, Breyer RH, Das DK: Enhanced prostaglandin synthesis due to phospholipase breakdown in ischemic-reperfused myocardium. Control of its production by a phospholipase inhibitor or free radical scavengers. J. Mol Cell Cardiol 18: 953–959, 1986
3. Tosaki A, Bagchi D, Hellegouarch A, Pali T, Cordis GA, Das DK: Comparisons of ESR and HPLC methods for the detection of hydroxy radicals in ischemic/reperfused hearts. A relationship between the genesis of oxygen-free radicals and reperfusion-induced arrhythmias. Biochem Pharmacol 45: 961–969, 1993
4. Cordis GA, Maulik N, Das DK: Detection of oxidative stress in heart by estimating the dinitrophenylhydrazine derivative of malonaldehyde. J Mol Cell Cardiol 27: 1645–1653, 1995
5. Das DK, Maulik N: Evaluation of antioxidant effectiveness in ischemia reperfusion tissue injury methods. Meth Enzymol 233: 601–610, 1994
6. McConkey DJ, Orrenius S, Jondal M. Cellular signaling in programmed cell death (apoptosis). Immunol Today 11: 120–121, 1990
7. Korsmeyer SJ: Regulators of cell death. TIG 11: 101–105, 1995
8. Richter C: Pro-oxidants and mitochondrial Ca^{2+}: Their relationship to apoptosis and oncogenesis. FEBS Lett 325: 104–107, 1993
9. Banerjee A, Locke-Winter C, Rogers KB, Mitchell MB, Bensard DD, Brew EC, Cairns CB, Harken AH: Transient ischemia preconditions against subsequent cardiac ischemia reperfusion injury by an alpha-1 adrenergic mechanism. Circ Res 73: 656–670, 1993
10. Flack J, Kimura Y, Engelman RM, Das DK: Preconditioning the heart by repeated stunning improves myocardial salvage. Circulation 84 (suppl III): 369–374, 1991
11. Engelman DT, Watanabe M, Engelman RM, Rousou JA, Kisin E, Kagan VE, Maulik N, Das DK: Hypoxic preconditioning preserves antioxidant reserve in the working rat heart. Cardiovasc Res 29: 133–140, 1995
12. Schmitz GG,Walter T, Seible R, Kessler C: Nonradioactive labeling of oligonucleotides in vitro with the Hapten Digoxigenin by tailing with terminal transferase. Anal Biochem 192: 222–231, 1991
13. Bortner CD, Oldenburg NBE, Cidlowski JA: The role of DNA in fragmentation in apoptosis. Trends Cell Biol 5: 21–26, 1995
14. Das DK, Engelman RM, Kimura Y: Molecular adaptation of cellular defences following preconditioning of the heart by repeated ischemia. Cardiovasc. Res 27: 578–584, 1993
15. Tominaga T, Kure S, Narisawa K, Yoshimoto T: Endonuclease activation following focal ischemic injury in the rat brain. Brain Res 608: 21–26, 1993
16. Fukuda K, Kojiro M, Chiu JF: Demonstration of extensive chromatin cleavage in transplanted Morris Hepatoma 7777 tissue: Apoptosis or necrosis. Am J Pathol 142: 935–946, 1993
17. Schumer M, Colombel MC, Sawczuk IS, Gobe G, Connor J, O'Toole KM, Wise GJ, Buttyan R: Morphologic, biochemical and molecular evidence of apoptosis during the reperfusion phase after brief periods of renal ischemia. Am J. Pathol 140: 831–838, 1992
18. Tanaka M, Ito H, Adachi S, Akimoto H, Nishikawa T, Kasajima T, Marumo F, Hiroe M: Hypoxia induces apoptosis with enhanced expression of Fas antigen messenger RNA in cultured neonatal rat cardiomyocytes. Circ Res 75: 426–433, 1994
19. Gottlieb RA, Burleson KO, Kloner RA, Babior BM, Engler RL: Reperfusion injury induces apoptosis in rabbit cardiomyocytes. J. Clin Invest. 94: 1621–1628, 1994
20. Kajstura J, Cheng W, Reiss K, Clark WA, Sonnenblick EH, Krajewski S, Reed JC, Olivetti G, Anversa P: Apoptotic and necrotic myocyte cell deaths are independent contributing variables of infarct size in rats. Lab Invest 74: 86–107, 1996
21. Isner JM, Kearney M, Bortman S, Passeri J: Apoptosis in human atherosclerosis and restenosis. Circulation 91: 2703–2711, 1995.
22. Maulik N, Kagan VE, Tyurin VA, Das DK: Redistribution of membrane phosphatidylethanolamine and phosphatidylserine precedes reperfusion–induced apoptosis in heart. Am J. Physiol (Heart and Circ. Physiology) 43: H242–H248, 1998
23. Tosaki A, Cordis GA, Szerdahelyi P, Engelman RM, Das DK: Effects of preconditioning on reperfusion arrhythmias, myocardial functions, formation of free radicals, and ion shifts in isolated ischemic/reperfused rat hearts. J. Cardiovasc Pharmacol 23: 365–373, 1994
24. Cohen MV, Downey JM: Preconditioning during ischemia: Basic mechanisms and potential clinical applications. Cardiol Res. 3: 137–149, 1995
25. Rao V, Ikonomidis JS, Weisel RD, Cohen G: Preconditioning to improve myocardial protection. Annals N.Y. Acad Sci. 793: 338–354, 1996.
26. Engelman D, Watanabe M, Engelman RM, Rousou JA, Kisin E, Kagan VE, Maulik N, Das DK: Hypoxic preconditioning preserves antioxidant reserve in the working rat heart. Cardiovasc Res 29: 133–140, 1995

27. Reimer KA: The slowing of ischemic energy demand in preconditioned myocardium. Annals N.Y. Acad Sci. 793: 13–26, 1996
28. Das DK, Engelman RM, Rousou JA, Breyer RH, Otani H, Lemeshow S: Pathophysiology of superoxide radical as potential mediator of ischemic and reperfusion injury in pig heart. Basic Res Cardiol 81: 155–166, 1986

Molecular and Cellular Biochemistry **186**: 147–158, 1998.
© 1998 *Kluwer Academic Publishers.*

The extracellular matrix in hibernating myocardium – a significant factor causing structural defects and cardiac dysfunction

A. Elsässer,[1] M. Schlepper,[2] R. Zimmermann,[2] K.-D. Müller,[2] R. Strasser,[2] W.-P. Klövekorn[2] and J. Schaper[3]

[1]*Albert-Ludwigs-University Freiburg, Department of Cardiology;* [2]*Kerckhoff-Clinic, Bad Nauheim;* [3]*Max-Planck-Institute, Department of Exp. Cardiology, Bad Nauheim, Germany*

Abstract

Recently, we described chronic intracellular degeneration accompanied by fibrosis as typical structural features of hibernating myocardium and we concluded that cellular degeneration as a sign of the incomplete adaptation to the reduced blood flow is characteristic of hibernation. This study has been extended by analyzing the composition of the extracellular matrix proteins of the diseased myocardium. Areas of hibernating myocardium were identified in 38 patients by angiography, multigated radionuclide ventriculography, thallium scintigraphy with reinjection and low-dose dobutamine echocardiography. These areas were biopsied at cardiac surgery and were studied by electron microscopic and immunofluorescence techniques. Electron microscopy showed an enlarged extracellular space containing numerous particles of cellular debris, macrophages, fibroblasts, homogeneous matrix material and collagen fibrils. The basement membrane of the cardiomyocytes was thickened by an augmentation of laminin, fibronectin and collagen VI, but these proteins also were present in the matrix itself. Collagen fibrils were numerous and macrophages (CD68) and fibroblasts (vimentin) were increased. *In situ* hybridization showed an increase in mRNA for laminin, fibronectin and collagen. This observation is consistent with the conclusion that fibrotic scar formation was occurring continuously. It is postulated that fibrosis is the consequence of myocyte loss due to chronic underperfusion in the hibernating tissue. This will further injure myocytes so that a vicious cycle is established that leads to progressive loss of structural integrity and functional capacity. Since these changes are progressive, revascularization should be performed at the earliest time point possible in patients with areas of hibernating myocardium. (Mol Cell Biochem **186**: 147–158, 1998)

Key words: hibernating myocardium, degeneration, fibrosis, laminin, fibronectin, collagen

Introduction

Myocardium that shows persistently impaired left ventricular function at rest due to a reduced coronary blood flow that can be partially or completely restored to normal following revascularization has been termed 'hibernating myocardium' [1, 2]. This definition implies that the myocardial structure is well preserved under the conditions of hibernation and that functional disturbances are temporary and can be reversed by myocardial revascularization. However, the recovery of contractile function of hibernating myocardium frequently is significantly delayed or may never be completely restored after revascularization by coronary bypass surgery. In the course of a recent extensive investigation on the structural alterations in hibernating myocardium, severe degeneration of the cardiomyocytes accompanied by cardiac fibrosis was described [3] and the hypothesis was put forward that hibernation represents an incomplete adaptation to ischemia that eventually results in chronic tissue destruction. The previous report was concerned mainly with intracellular

Address for offprints: J. Schaper, Max-Planck-Institute, Department of Exp. Cardiology, Beneke strasse 2, D-61231 Bad Nauheim, Germany

changes, while in this paper we will describe more extensively the changes occurring in the interstitial space. This space is filled with extracellular matrix that consists of cellular components such as fibroblasts and macrophages, structures of the microvasculature, and various, mostly fibrillar proteins such as the different collagens, proteoglycans, fibronectin and laminin.

An increase in interstitial tissue, i.e. fibrosis, in hibernating myocardium has been reported by Ausma *et al.* and Vanoverschelde and was also observed by our group. In several studies, the Belgian group [4–7] reported that the degree of fibrosis determines the ability of the tissue to contract, i.e. regional ejection fraction is increasingly reduced with a higher content of fibrosis; our studies have confirmed this finding. Fibrosis was interpreted by our group to be the result of the loss of myocytes and has to be regarded as 'replacement or reparative' fibrosis. 'Reactive' fibrosis is assumed to exist by others [5, 8] . The composition of the extracellular matrix, however, has not been investigated in detail until now and will therefore be described in the present study.

Material and methods

Patients and study protocol

Thirty-eight selected patients with coronary heart disease (37 men and one woman, mean age 61 ± 7 years) were included in the present study (Table 1). In all patients, angiography showed reduced left ventricular function and multigated radionuclide ventriculography showed that global and regional ejection fractions were decreased. A reduced rate of perfusion and persisting tissue viability were determined by thallium-201 scintigraphy with reinjection. Echocardiography with low dose dobutamine indicated a reduced but, upon pharmacologic stimulation, still preserved functional capacity. The indication for coronary bypass surgery was established and during open-heart surgery transmural Tru-cut needle biopsies were removed from the center of the area previously diagnosed as hibernating. The degree of res–titution of perfusion and of functional recovery was determined in all patients by echocardiography at 10–14 days and by all methods employed preoperatively 3 months after the operation (Fig. 1).

Table 1. Clinical characteristics

number of patients		38
sex	male	37
	female	1
age range (years)		45–72
symptoms (number of patients)		
angina pectoris	CCS I	0
	CCS II	19
	CCS III	6
	CCS IV	4
dyspnoea	NYHA I	0
	NYHA II	10
	NYHA III	8
	NYHA IV	4
angina pectoris and dyspnoea		14
mean duration of symptoms (months)		14.4 ± 18.5
myocardial infarction (number of patients)		25
thrombolytic therapy		5
ECG - Q wave		21

CSS – classification of angina pectoris by the Canadian Society of Cardiology; NYHA – Classification of dyspnoea of the New York Heart Association. The hibernating area was in a vessel distribution different from that of myocardial infarction.

Informed written consent was obtained from each patient for every investigation as well as the approval of the local hospital review board.

Clinical methods

Dobutamine 2D echocardiography

A Hewlett-Packard wide-angle phased-array imaging system (Sonos 1500/2,5 MHz transducer) was used for transthoracic echocardiography at rest and during the intravenous infusion of dobutamine at a rate of 5 and 10 mg/kg BW/min infused over 10 min using a standard protocol. Images of standard views were acquired.

Echocardiographic analysis. The left ventricle was divided into 16 segments according to the recommendations of the American Society of Echocardiography and regional function of the left ventricle was defined as 1: normal, 2: slightly hypokinetic, 3: severely hypokinetic, 4: akinetic and 5: dyskinetic. A wall motion score index of the global left ventricle and of the hibernating area was calculated at baseline and after dobutamine infusion for each patient by the sum of the score of segments divided by the number of segments evaluated. When regional function improved in at least two adjacent abnormal segments by a factor 1 during dobutamine infusion at 10–14 days and/or 3 months after revascularization, the area was diagnosed as hibernating myocardium.

Thallium-201 scintigraphy

A stress-rest-reinjection protocol was used. Thallium-201 at a dose of 111 MBq was injected intravenously at the peak of bicycle exercise which had started at 25 W with increments of 25 W every 2 min. Myocardial SPECT acquisition was performed using a conventional circular-field

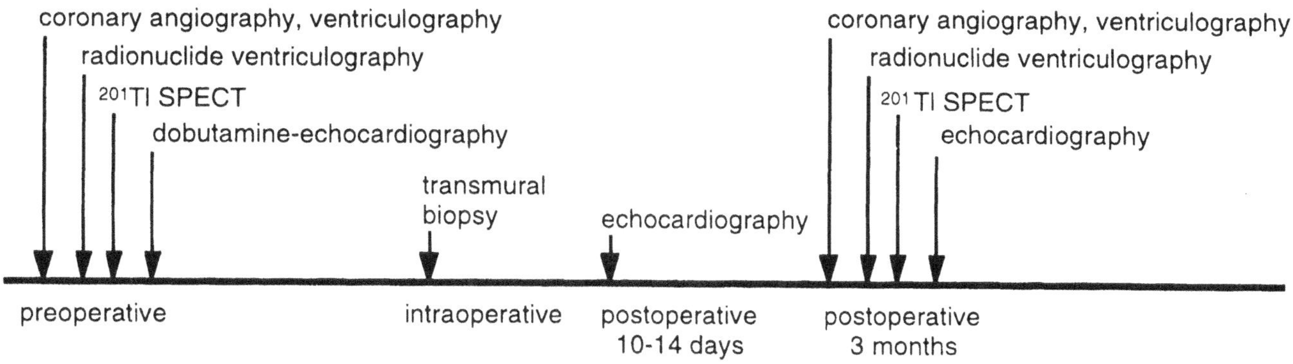

Fig. 1. Study protocol.

single-head rotating digital gamma camera (Sopha, DS7) immediately after the exercise. After redistribution imaging, an additional thallium dose of 37 MBq was administered at rest and reinjection images were acquired 30 min thereafter.

Qualitative and quantitative analysis. Qualitative analysis was done in 13 sectors and was based on the visual interpretation of regional tracer activity in the stress images in contrast to the reinjection images. Quantitative determination of thallium-201 activity was carried out in 3 short-axis tomograms (basal, 2 midventricular) and one vertical long-axis view. These values were compared to the database of a reference population of the same sex (probability < 5% for significant coronary artery disease in 46 women and 38 men investigated in this hospital).

Thallium-201 uptake on stress images below 2 S.D. of the mean in this reference population was classified as abnormal. An increase of thallium uptake of 15% or more measured at reinjection in comparison to the stress images in the baseline study was considered to represent a reversible myocardial defect. Hibernating myocardium was diagnosed when a defect was reversible by qualitative analysis and/or when the uptake value was in the range of the mean normal value ± 2 S.D. in the baseline study and hypokinetic or akinetic wall motion was found at rest in baseline radionuclide ventriculography.

Radionuclide ventriculography
Left ventricular wall motion at rest and during graded bicycle exercise was assessed by multigated radionuclide ventriculography using a dose of 740 MBq of Technetium-99 m *in vivo* labelled red blood cells and a small-field-of-view gamma camera (LEM, Siemens, Germany).

Quantitative sector ejection fraction (EF) calculation was carried out according to the method described by Stanke [9].

For the definition of hibernating myocardium a postoperative improvement of at least 5% of the left ventricular EF was needed.

Coronary angiography and left ventriculography
Coronary narrowing was determined in all patients by coronary angiography with multiple projections. Biplane left ventriculography was performed (DCI, Philips).

Comparison. All left ventricular segments were grouped into 3 vascular territories: (1) the anterior and anterolateral wall as perfusion area of the left anterior descending coronary artery; (2) the lateral left ventricular wall as perfusion area of the left circumflex artery; and (3) the inferior and posterior wall as perfusion area of the right coronary artery.

Statistical analysis

Data are presented as mean values ± 1 S.D. The Friedman and the Dunn test (echocardiography), the paired *t*-test (radionuclide ventriculography) and the Wilcoxon-test (thallium-201 uptake) were used for the evaluation of differences between the different time points of the study.

A p value of < 0.05 was considered to represent a significant difference.

Tissue preparation

Two transmural biopsies were removed from each patient and either immediately frozen in liquid nitrogen for immunohistochemistry or immersed in 3% glutaraldehyde buffered with 0.1 M Na cacodylate (at pH 7.4, 440 mosmol) for electron microscopy.

Electron microscopy

Small tissue samples were embedded in Epon following routine procedures. Semithin sections were stained with

PAS for glycogen and evaluated in the light microscope. Ultrathin sections were stained with uranyl acetate and lead citrate and viewed and photographed in a Philips CM 10 electron microscope.

Histology and immunohistochemistry

The tissue samples were mounted with Tissue Tek (OCT compound, Miles Inc.); cryosections were prepared, air dried and fixed with acetone at –20°C for 10 min. The following primary antibodies were used: collagen I and III (Bioscience), collagen VI (Telios/Biomol), fibronectin (ICN Biomedicals), laminin and vimentin (Dianova) and CD68 as marker for macrophages (Dako). Incubation with the first antibody for 60 min at room temperature was followed by treatment with the biotinylated second antibody (Amersham) for 60 min. The last incubation was carried out with fluoroisothiocyanate (FITC)-linked streptavidin (Amersham). Nuclei were stained with actinomycin D (Molecular Probes) diluted 1:100. Frequent rinsing with phosphate buffer (PBS) was done between all steps. The sections were covered with Mowiol (Hoechst), cover slipped and viewed in a Leica Aristoplan microscope with fluorescence equipment or in a confocal laser microscope (Leica). Representative sections were photographed on professional Kodak Ektachrome 100 HC film. All reproductions were made from 35 mm slides.

In situ *hybridization*

In situ hybridization was carried out for laminin, fibronectin and collagen I $\alpha 1$. The probe for fibronectin was a 3.5 kb Hind III – Xba I or a 360 bp Sac I – Bgl II fragment of clone pRCabFN1 (generous gift of Dr. K. Boheler, London). The probe for laminin was a 1.5 kb Hind III-fragment of human laminin B2 (ATCC 65028, Rockville, USA), subcloned for this purpose in pBluescript SK +. The probe for collagen I was a 1.5 kb EcoR1-subclone of human collagen $\alpha 1$ (I) (ATTC, Hf677). Sense and antisense RNA-probes labeled with 100 μCi of [a-^{35S}] UTP (1000 Ci/mmol) were tested by conventional Northern blot analysis for specificity and stored in 0.1 M dithiothreitol (DTT) at –80°C until further use.

Cryostat sections (4 μm) were placed on glass slides coated with 3-aminopropyltriethoxysilane (Sigma) and fixed with 4% paraformaldehyde for 10 min. After washing, the sections were dehydrated with graded ethanol, dried, and directly used according to Simmons [10].

Each section was hybridized overnight in a humidified chamber at 50°C with hybridization buffer including 2.5 × 10^6 cpm of denatured antisense or sense cRNA probe. Following several washes with increasing stringency [11] dehydrated slides were exposed to Kodak NTB-2 emulsion for up to 2–4 weeks. Microscopic analysis was done after counterstaining with 0.1% toluidine blue.

Controls

Small left ventricular biopsies from patients undergoing correction of an atrial septal defect were used as normal human myocardium.

For immunohistochemistry omission of the first antibody was used to check the specificity of the reaction.

For *in situ* hybridization, incubation with the sense probe was carried out. The latter should not react with the tissue RNA and the autoradiogram should, therefore, be negative.

Results

The structural changes observed by both electron microscopy and immunohistochemistry correlated well with the severity of the reduction of function and the extent of postoperative recovery of the hibernating area, i.e. patients with more severe functional disturbances showed more fibrotic material and disappearance of myocytes than those with slight or moderate dysfunction [3].

Dobutamine echocardiography

The mean values of the wall motion score index for hibernating myocardium are indicated in Table 2. Segmental function was significantly improved at 3 months.

Thallium-201 scintigraphy

Qualitatively, visual analysis showed 346 of 494 analyzed segments (70%) to be abnormal post-stress. On 4 h images and after reinjection 80 segments (23%) showed complete

Table 2. Clinical data before and after coronary bypass surgery

	Preoperative	Postoperative	
		10–14 days	3 months
Dobutamine echocardiography (wall motion score)	3.4 ± 0.4	2.9 ± 0.6*	1.8 ± 0.6*
Radionuclide ventriculography (EF %)	23.6 ± 12.7	n.d.	48.0 ± 16.8**
Thallium-201 uptake (%)	40.5 ± 9.9	n.d.	66.5 ± 11.4**

Values are means ± S.D. *p < 0.01; **p < 0.0001.

redistribution, 128 (37%) incomplete redistribution, and 138 (40%) a fixed defect which was also present in the baseline study.

The effect of revascularization was assessed by comparing pre- and post-revascularization scans. Of 138 segments with a fixed defect before operation, 12 showed an increase of thallium-201 uptake. Of 128 segments with a partially reversible defect, 105 had an enhanced thallium-201 uptake postoperatively. All 80 segments with complete redistribution showed a better thallium-201 uptake 3 months after revascularization. Ischemia was absent in the postoperative SPECT images.

Quantitative evaluation of thallium-201 uptake as an indicator of perfusion showed a significant increase in the hibernating region (Table 2).

Radionuclide ventriculography

Regional LV-EF at rest increased at 3 months after revascularization (Table 2).

Success of revascularization

Aortocoronary bypass surgery with 3.4 ± 0.98 venous and internal mammary artery bypass grafts per patient were performed. Three months after revascularization all hibernating areas were revascularized.

Electron microscopy

The myocytes showed different degrees of degeneration that involved myofilaments as well as the cytoskeleton, the mitochondria and nuclei [3]. The interstitial space contained numerous fibroblasts and macrophages together with elements of the microvasculature. Cellular debris was obvious and small cellular particles either could be identified as remnants of myocytes or were not recognizable because of advanced destruction. The extracellular matrix was homogeneously electron-dense indicating the presence of proteinous material. Collagen fibrils and elastic fibers were evident (Fig. 2). The areas filled with fibrotic material varied in extent. PAS staining for glycogen was positive in myocytes independent of the degree of alterations.

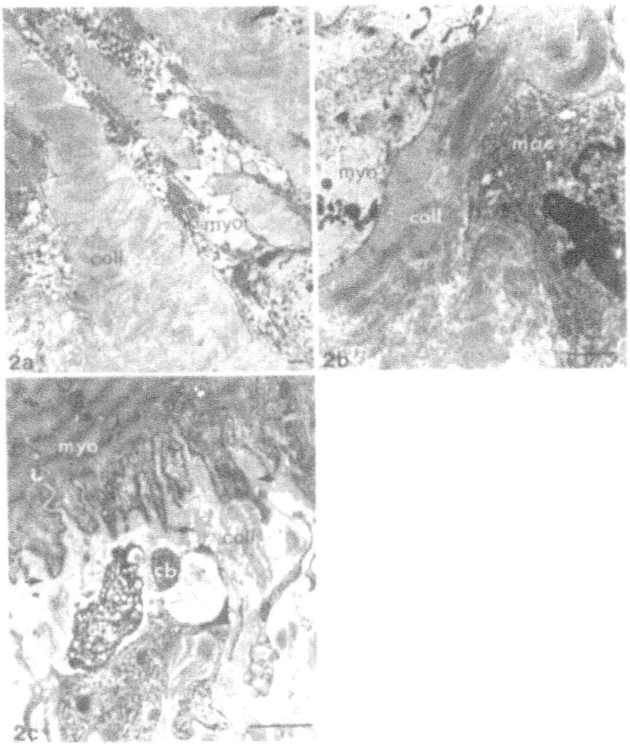

Fig. 2. Electron microscopy shows a degenerated myocyte (myo) surrounded by wavy collagen fibrils (coll) in 2a. 2b shows a degenerated myocyte (myo) on the left, and collagen fibrils (coll) and a macrophage (mac) in the center and right. 2c. A myocyte (myo) with many extensions is situated at the top, cellular debris (cb), collagen fibrils (coll) and a macrophage (mac) are also visible. Note the thickened basement membrane (small arrows). Bars are 2 μm.

Immunohistochemistry

Fibronectin

Fibronectin is present in normal tissue in small amounts. It constitutes the homogeneous matrix in the extracellular space surrounding the cellular and fibrillar structures. Fibronectin is a major component of the basement membrane and was observed around myocytes and endothelial cells but not around macrophages or fibroblasts. Fibronectin also was noticed as fine lines within the myocytes in a pseudo-intracellular localization. This is caused by the labeling of the basement membrane within the T-tubular system and this is a part of the extracellular space.

In hibernating myocardium the amount of fibronectin was increased and it filled the entire extracellular space. The enlargement of the fibrotic area in patients with more severe functional defects was paralleled by an increase in the area staining positively for fibronectin (Fig. 3).

Laminin

Laminin is another component of the basement membrane and therefore it is seen as a fine line surrounding all myocytes and endothelial as well as smooth muscle cells

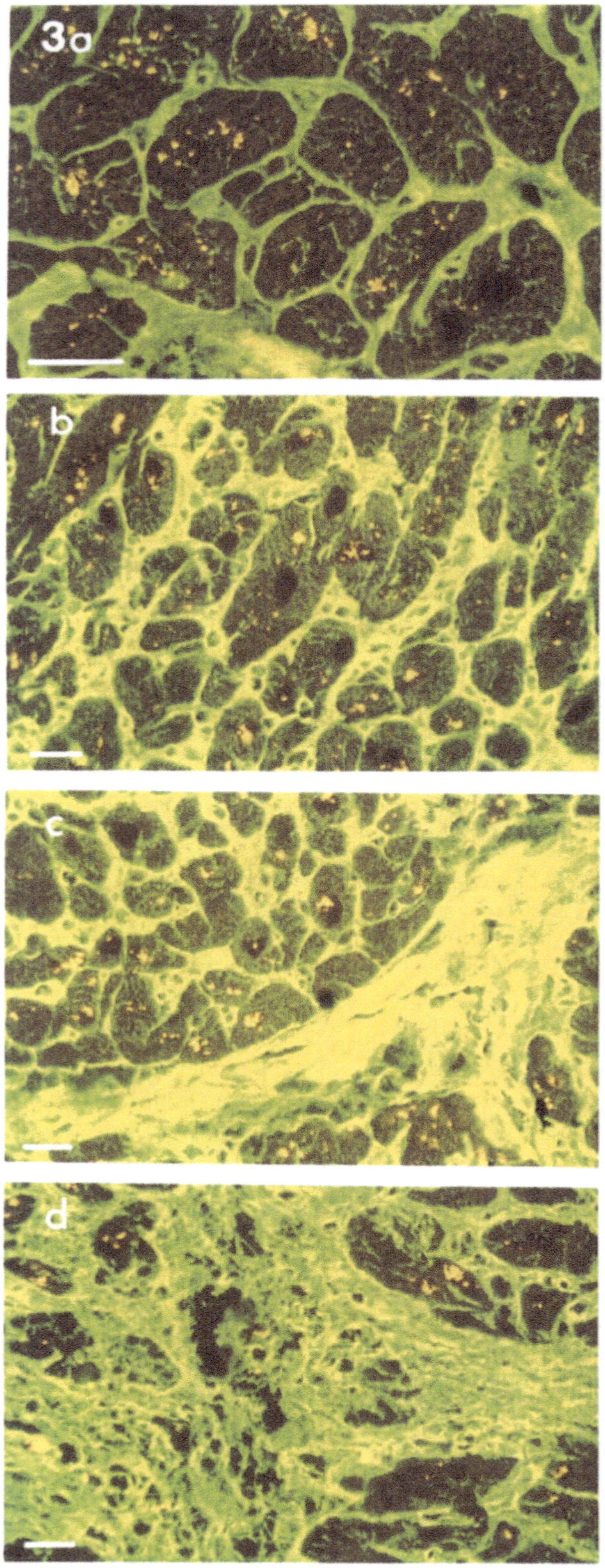
3a
b
c
d

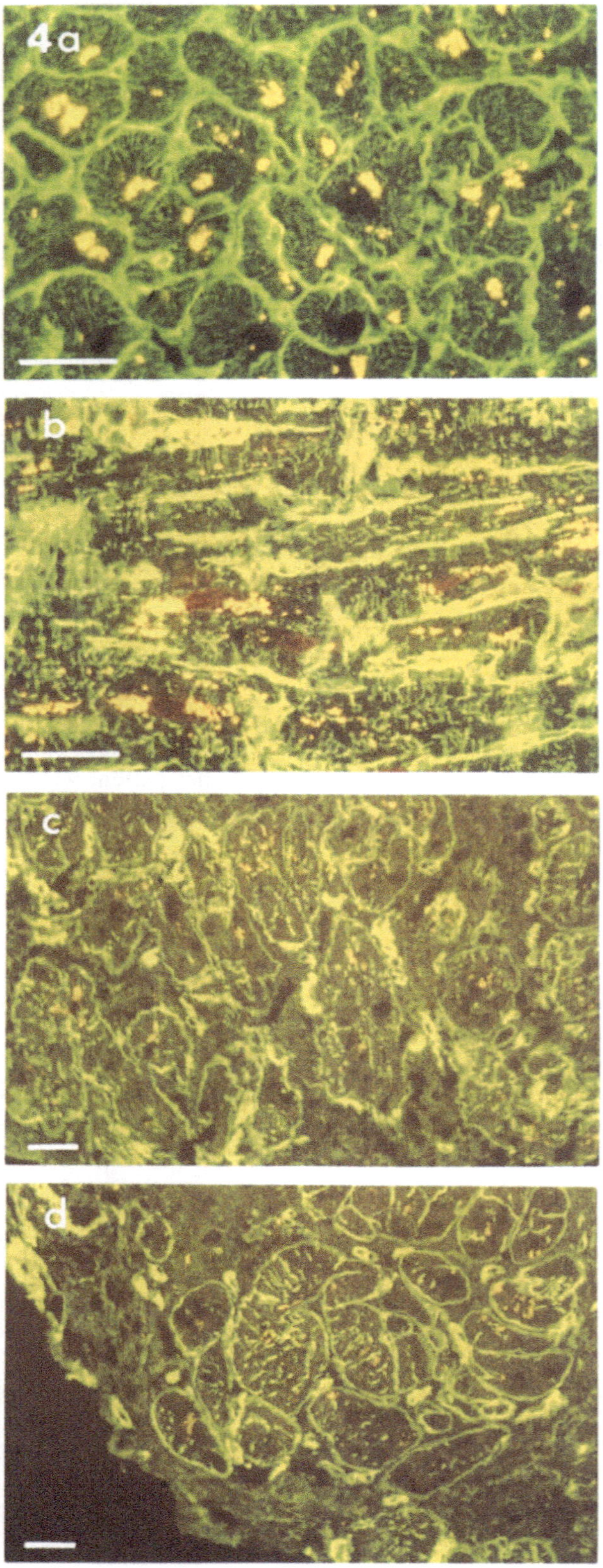
4a
b
c
d

(Fig. 4a). It also stains the basal membrane part within the T-tubuli causing fine transverse lines running from the sarcolemma into the myocytes. This material is increased in the extracellular space of the T-tubular system (Fig. 4b). This was especially pronounced in a few patients with hibernating myocardium in which the entire cell was apparently filled with laminin.
In hibernating myocardium the degenerating cardiomyocytes and small cellular particles were clearly delineated by laminin labeling, and the fibrotic areas contained an increased amount of this material (Fig. 4).

Collagen I, III and VI
These types of collagen were found in normal myocardium within the extracellular space. Thick wavy collagen bundles were present in the perimysium, whereas the amount of collagen fibrils was rather limited between myocytes.

In hibernating myocardium the amount of collagen was increased similar to the augmentation of fibronectin, and numerous myocytes, usually exhibiting signs of degeneration, were separated from each other by collagen fibrils. Collagen I (Fig. 5) showed a distribution similar to that of collagen VI (Fig. 6).

Vimentin
Vimentin was observed in all mesenchymal cells in the interstitium, i.e. in endothelial and smooth muscle cells as well as in fibroblasts, and it can, therefore, be used as an indicator of the cellularity of any tissue. In normal myocardium, the number of vimentin-positive cells was small, but it was significantly increased in hibernating myocardium. In the latter, a focal accumulation of fibroblasts could be observed, especially in areas of fibrosis (Fig. 7).

Labeling for macrophages (CD68)
Macrophages were rarely observed in normal myocardium and their number was only slightly increased in hibernating cardiac tissue. These cells were either situated close to smaller arteries or veins, or they were observed between myocytes. Usually, they are rather large and protrude with long slender extensions between the cells (Fig. 8).

In situ *hybridization*

In normal and diseased tissue, the mRNA for fibronectin was localized in different cell types including myocytes, fibroblasts and endothelial cells. The intensity of labeling was increased in hibernating myocardium as compared to normal tissue. This was especially the case in areas of focal fibrosis (Fig. 9).

Labeling for the mRNA of laminin was present over myocytes, fibroblasts and endothelial cells in normal and diseased tissue and the intensity of staining of fibroblasts was slightly stronger in diseased myocardium (Fig. 10).

In situ hybridization for collagen I $\alpha1$ showed an increase of the mRNA in both types of myocardium (data not shown).

Discussion

The following definition of hibernating myocardium was used in this investigation: myocardium was hibernating if it was underperfused but still viable and exhibited left ventricular dysfunction at rest diagnosed by Tl-201 scintigraphy and radionuclide ventriculography. Furthermore, the same region exhibited functional improvement during dobutamine infusion preoperatively as well as recovery of regional function 3 months after coronary bypass surgery as measured by all preoperative methods. All patients included in this study fulfilled these clinical criteria [12–14].

Degeneration of myocytes involving all cellular components and the occurrence of apoptosis previously have been described as the typical structural alterations observed in hibernating myocardium. Glycogen storage in cellular areas with nonspecified cytoplasm, i.e. cytoplasm free of contractile elements, was another typical feature [15, 16]. It was concluded that the adaptation to the lack of oxygen in hibernating myocardium is incomplete and that the final outcome of this disease process is loss of myocytes followed by scavenging by macrophages [3].

In the present study the structural composition of the fibrotic tissue has been examined in more detail and further conclusions regarding the pathomechanism of the process of 'hibernation' can be inferred.

Laminin and fibronectin, major proteins of the basal lamina that surround all myocytes, were both increased. Both proteins possess their specific receptors, the integrins, in the sarcolemma through which the interaction between the intracellular milieu and the extracellular matrix is mediated. The connection via integrins is not only

Fig. 3. Tissue sections prepared for immunohistochemistry. With increasing fibrosis the areas positive for fibronectin increase. With severe damage in d only a few myocytes are still present. Specific fluorescence is green, lipofuscin appears as yellow dots, nuclei are red. All bars are 20 μm. a – normal; b – slight; c – moderate; d – severe structural alteration.

Fig. 4. Tissue sections prepared for immunohistochemistry. Laminin labeling is fine around normal myocytes in a; it is abundantly present in the T-tubules in b and it surrounds the remaining myocytes as a thick layer in c and d. Specific fluorescence is green, lipofuscin appears as yellow dots, nuclei are red. All bars are 20 μm. a – normal; b – slight; c – moderate; d – severe structural alteration.

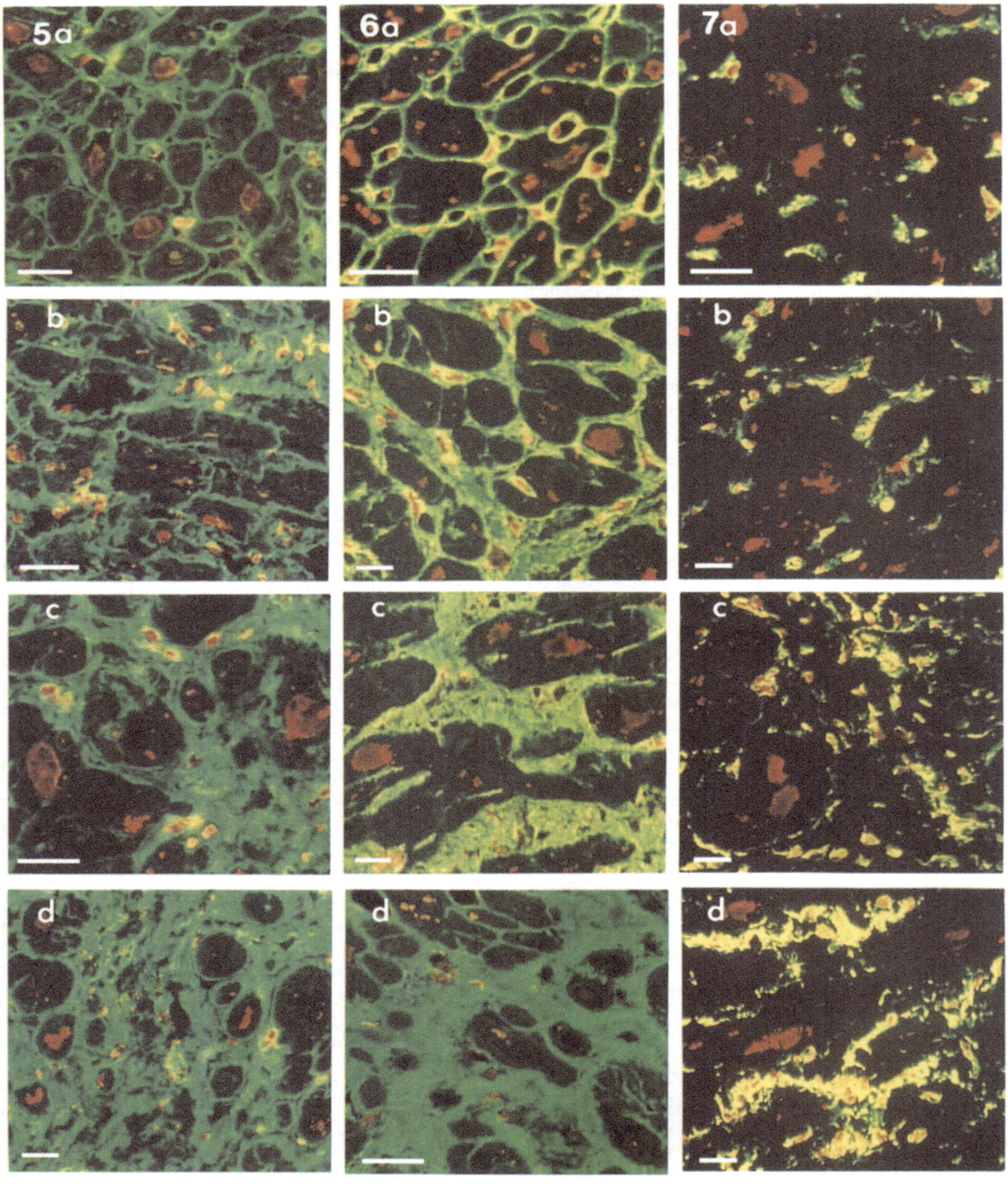
5a
6a
7a
b
b
b
c
c
c
d
d
d

important for the mechanical coupling of the cells and the extracellular matrix but also for signal transduction to the nucleus, thereby influencing transcription and gene expression [17–19].

Laminin is a large glycoprotein consisting of 3 polypeptide chains, A, B1, and B2, that are interconnected by disulfide links. In the basement membrane the cruciform laminin molecules form an elastic but solid network and determine largely the stability of the cells they are encasing [20–22]. Fibronectins are dimeric glycoproteins that possess binding domains for fibrin and fibrinogen, gelatin, heparin, collagen, and proteoglycans that play a role in the formation of the basement membrane and of the extracellular matrix and are important in would healing and in blood coagulation [17, 23, 24].

It was obvious that the proteins fibronectin and laminin were increased in hibernating myocardium. The mRNAs for fibronectin and laminin were only slightly increased as compared to normal tissue. This difference may be related to the long biologic half-life of both proteins. Thus accumulation by lack of breakdown will be significant. The fact that the mRNA was present at all indicates a continuous production of both proteins and an enlargement of the fibrotic area. Both myocytes of normal size and small remnants of degenerated myocytes were surrounded by a thickened basal lamina, a finding which suggests that this structure is resistant to noxious stimuli. On the other hand, the thickening of the basement membrane may have negative effects on the myocytes by limiting or altering signal transduction and isolating the cells from oxygen and substrate perfusion. Since the basement membrane acts as a highly selective filter that has a regulatory influence on cellular polarity and differentiation as well as on the metabolic properties of the cells, many more cellular functions may be disturbed by an increase in laminin and fibronectin. Furthermore, the interaction between the myocytes and the extracellular matrix will be disturbed which may further damage the myocytes [25, 26].

The different collagen types consist of helical polypeptide chains. A balance between collagen synthesis and degradation is present in normal tissue. Collagen I is the major structural protein that forms thick fibrils with typical cross striations. Collagen III forms a network connecting collagen I fibrils and collagen VI is a globular molecule [27–30]. The collagens form the interstitial scaffold of the myocardium [31]. They insert into the basal lamina of myocytes and endothelial cells, ensure the stability of the tissue structure and determine to a major degree the ventricular geometry. Because of the inextensibility of the collagen fibrils, the ventricular shape is maintained during diastole, whereas the collagens are force transducers during systole [32, 33].

The disproportionate increase in the mass of collagen fibrils observed in hibernating myocardium results in increased stiffness and therefore a loss of elasticity of the ventricular myocardium. The consequence is first a loss of the diastolic and later also of the systolic ventricular function, i.e. a loss of diastolic and systolic compliance.

The increase in collagen is most probably due to a disturbed balance between synthesis and degradation of this material. Fibroblasts are the most important cell type for the synthesis of collagen as well as for the production of collagenases and of metalloproteinases, the enzymes degrading collagen [34–36]. The increased number of fibroblasts in hibernating myocardium observed by vimentin staining indicates that many more cells are available for the production of collagen and other matrix components. The collagen-producing activity of fibroblasts most probably is increased by a direct mechanical effect because of the abnormal hemodynamic situation, but activation of the renin-angiotensin system resulting in growth factor release also may play a role [37–39].

Since an increased degree of fibrosis was observed concomitantly with degeneration and loss of myocytes, the term 'replacement or reparative fibrosis' rather than 'reactive' fibrosis [5] seems appropriate for hibernating myocardium. The number of macrophages observed in hibernating myocardium was increased slightly as compared to normal tissue, but signs of inflammation such as accumulation of leukocytes or expression of cellular adhesion molecules in endothelial cells were not observed in this chronically underperfused tissue (unpublished observation). Macrophages often contained clusters of lipofuscin indicating phagocytosis of cellular debris originating from myocytes.

The amount of fibronectin, laminin, and the collagens was increased both within the basement membrane and in the entire extracellular space. As a consequence, the degeneration of myocytes will continue and more pronounced accumulation

Fig. 5. Tissue sections prepared for immunohistochemistry. Collagen I is present in greater amounts with increasing loss of myocytes and the appearance of reparative fibrosis. Note the isolated myocytes in d. Specific fluorescence is green, lipofuscin appears as yellow dots, nuclei are red. All bars are 20 μm. a – normal; b – slight; c – moderate; d – severe structural alteration.

Fig. 6. Tissue sections prepared for immunohistochemistry. Collagen VI is present around myocytes in normal tissue but it increases and is present as homogeneous or fibrillar mass in the extracellular matrix with increasing fibrosis. Specific fluorescence is green, lipofuscin appears as yellow dots, nuclei are red. All bars are 20 μm. a – normal; b – slight; c – moderate; d – severe structural alteration.

Fig. 7. Tissue sections prepared for immunohistochemistry. Vimentin labeled fibroblasts increase in number in more severe fibrosis. Specific fluorescence is green, lipofuscin appears as yellow dots, nuclei are red. All bars are 20 μm. a – normal; b – slight; c – moderate; d – severe structural alteration.

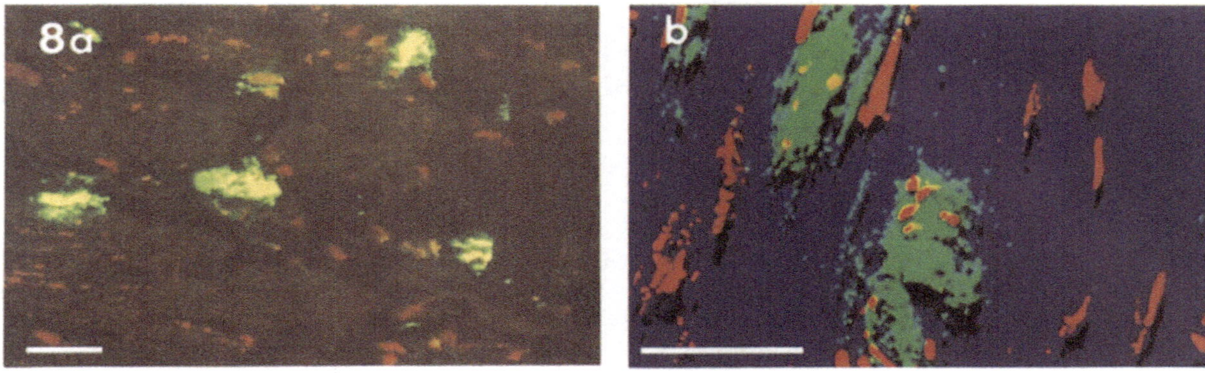

Fig. 8. Tissue sections prepared for immunohistochemistry. Staining for CD68 reveals the presence of large macrophages in hibernating myocardium. b was taken with the confocal microscope. Specific fluorescence is green, lipofuscin appears as yellow dots, nuclei are red. All bars are 20 μm.

of extracellular matrix material will occur. In this manner, a vicious cycle may be established whereby degeneration of myocytes leads to thickening of the basement membrane and an increased deposition of matrix proteins which then in turn may further damage still existing myocytes. The myocytes will be more and more separated from each other. They will cease to contract and finally will die by apoptosis. Deposition of collagen by the increased number of vimentin-positive fibroblasts will further aggravate this development.

We would like to postulate that hibernating myocardium at first represents an almost complete adaptation to the lack of oxygen with preservation of the myocardial structure. Under these conditions, a new stable balance between supply and demand is established. However, with longer durations of myocardial underperfusion this delicate situation can be disturbed easily and immediately by an increased oxygen demand or by the progression of coronary artery stenosis. As a consequence, the degree of adaptation no longer will be sufficient. Myocyte degeneration will set in and fibrosis will develop. Sequestration of cellular particles into the extracellular space, a phenomenon frequently observed, may stimulate macrophages and fibroblasts to an increased phagocytotic activity. These cells thus stimulated will also start to

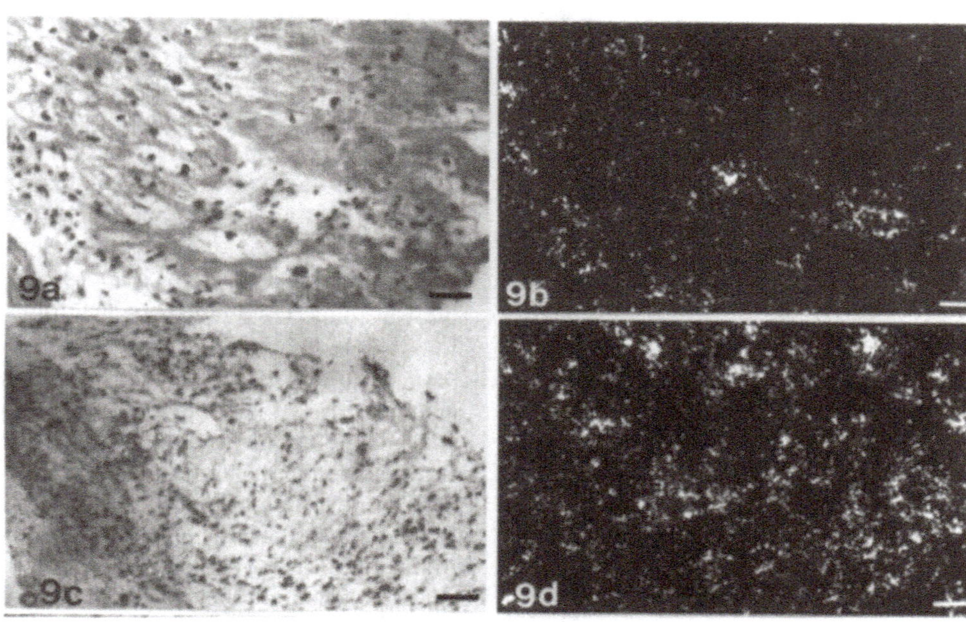

Fig. 9. In situ hybridization for fibronectin. a and b show tissue with slight alterations and labeling of only a few structures. In c and d the tissue is severely damaged and the labeling is increased. All bars are 20 μm.

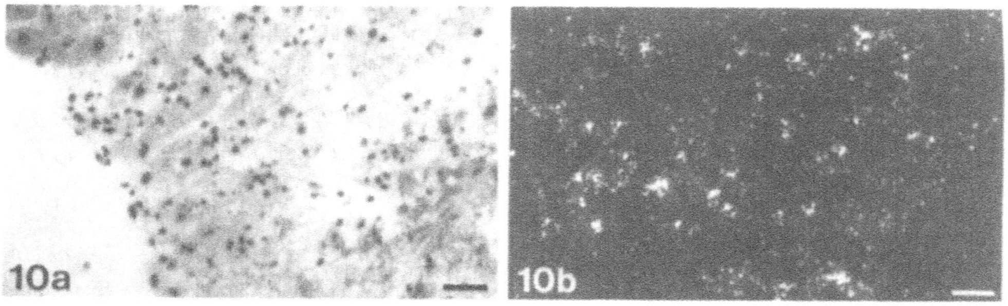

Fig. 10. In situ hybridization for laminin in severely altered myocardium shows that the mRNA is produced by interstitial cells and probably also by myocytes. All bars are 20 μm.

synthesize extracellular proteins at an accelerated rate, thereby worsening the structural deterioration.

In this way, a new cycle is started that will produce more cellular degeneration until apoptosis occurs. Atrophy and death of myocytes are followed by replacement fibrosis that will further compromise myocyte function.

A summary of our hypothesis regarding the role of reparative fibrosis and the intracellular degeneration in hibernating myocardium and their pathophysiological consequences is illustrated in Fig. 11.

In contrast to the events in acute myocardial infarction where myocytes die acutely and in great numbers, we believe that in hibernating myocardium the slowly developing cellular degeneration and final cell death produce the mixture of myocytes and scar tissue typical of hibernating myocardium. However, when revascularization has not been done, the final outcome of this process will be a large area of fibrosis indistinguishable from that of postinfarct scars. The final conclusion of this study is the necessity to interrupt the vicious cycle of continuous myocyte degeneration and development of fibrosis in hibernating myocardium by early revascularization.

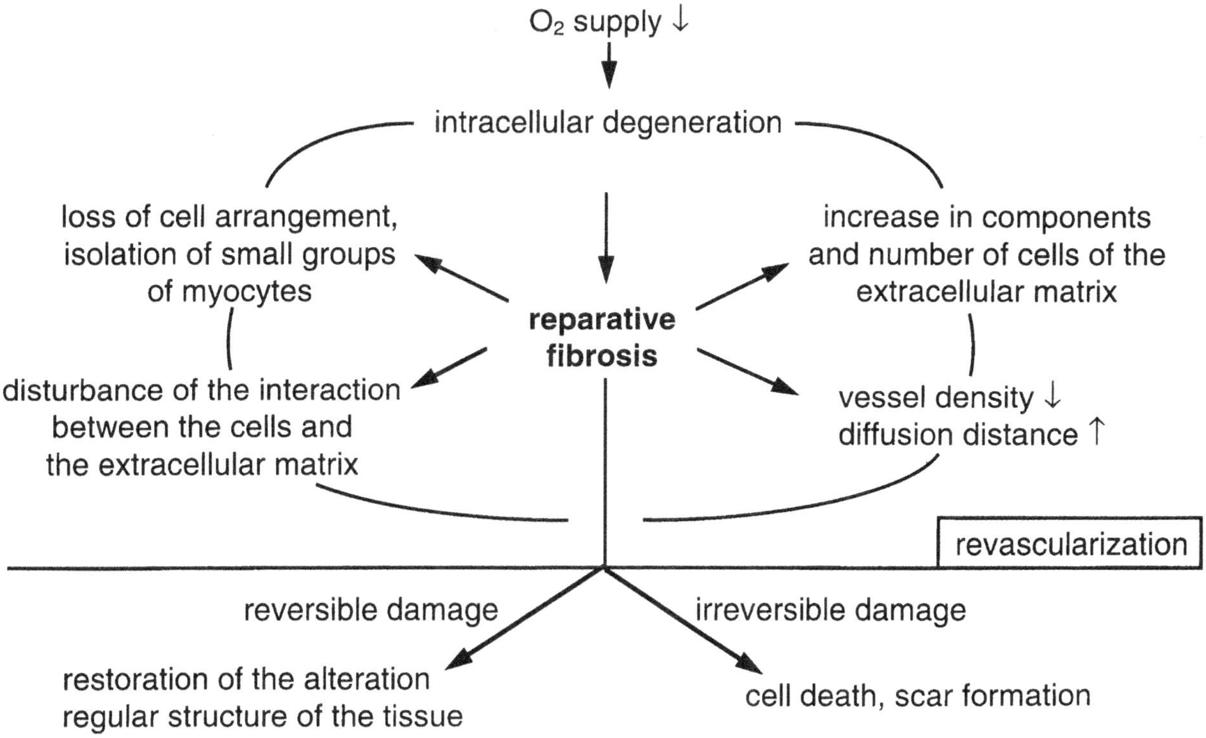

Fig. 11. Hypothesis regarding role of intracellular degeneration and reparative fibrosis in myocardial hibernation.

References

1. Rahimtoola SH: The hibernating myocardium. Am Heart J 117: 211–221, 1989
2. Rahimtoola SH: The hibernating myocardium in ischaemia and congestive heart failure. Eur Heart J 14: 212-220, 1993
3. Elsässer A, Schlepper M, Klövekorn W-P, Cai WJ, Zimmermann R, Müller K-D, Strasser R, Kostin S, Gagel C, Münkel B, Schaper W, Schaper J: Hibernating myocardium: An incomplete adaptation to ischemia. Circulation 96: 2920–2931, 1997
4. Vanoverschelde JL, Wijns W, Deprè C, Essamri B, Heyndrickx GR, Borgers M, Bol A, Melin JA: Mechanisms of chronic regional postischemic dysfunction in humans – new insights from the study of noninfarcted collateral-dependent myocardium. Circulation 87: 1513–1523, 1993
5. Ausma J, Cleutjens J, Thone F, Flameng W, Ramaekers F, Borgers M: Chronic hibernating myocardium: interstitial changes. Mol Cell Biochem 147: 35–42, 1995
6. Ausma J, Schaart G, Thone F, Shivalkar B, Flameng W, Depre C, Vanoverschelde J, Ramaekers F, Borgers M: Chronic ischemic viable myocardium in man: aspects of dedifferentiation. Cardiovasc Pathol 4: 29–37, 1995
7. Borgers M, Thone F, Wouters L, Ausma J, Shivalkar B, Flameng W: Structural correlates of regional myocardial dysfunction in patients with critical coronary artery stenosis: chronic hibernation? Cardiovasc Pathol 2: 237–245, 1993
8. Maes A, Flameng W, Borgers M, Ausma J, Bormans G, Vandewerf F, Deroo M, Mortelmans L: Regional myocardial blood flow before and after revascularization and ultrastructural findings in patients with chronic coronary artery disease. Eur J Nucl Med 22: 1299–1305, 1995
9. Standke R, Hör G, Maul FD: Fully automated sectorial equilibrium radionuclide ventriculography. Eur J Nucl Med 8: 77–83, 1983
10. Simmons DM, Arriza JL, Swanson JW: A complete protocol for *in situ* hybridization of messenger RNAs in brain and other tissues with radiolabeled single-stranded RNA probes. J Histotechnol 12: 169–181, 1989
11. Kluge A, Zimmermann R, Münkel B, Mohri M, Schaper J, Schaper W: Insulin-like growth factor I is involved in inflammation linked angiogenic processes after microembolization in porcine heart. Cardiovasc Res 29: 407–415, 1995
12. Afridi I, Kleiman NS, Raizner AE, Zoghbi WA: Dobutamine echocardiography in myocardial hibernation. Optimal dose and accuracy in predicting recovery of ventricular function after coronary angioplasty. Circulation 91: 663–670, 1995
13. Dilsizian V, Perrone-Filardi P, Arrighi JA, Bacharach SL, Quyyumi AA, Freedman NM, Bonow RO: Concordance and discordance between stress-redistribution-reinjection and rest-redistribution thallium imaging for assessing viable myocardium. Comparison with metabolic activity by positron emission tomography. Circulation 88: 941–952, 1993
14. Dilsizian V, Bonow RO: Current diagnostic techniques of assessing myocardial viability. Circulation 87: 1–20, 1993
15. Camici P, Ferranini E, Opie LH: Myocardial metabolism in ischemic heart disease: basic principles and application to imaging by positron emission tomography. Prog Cardiovasc Dis 32: 217–238, 1989
16. Opie LH, Camici PG: Myocardial blood flow, deoxyglucose uptake, and myocyte viability in ischemia. J Nucl Med 33: 1353–1356, 1992
17. Albelda SM, Buck CA: Integrins and other cell adhesion molecules. FASEB J 4: 2868–2880, 1990
18. Geiger B: Intermediate filaments: Looking for a function. Nature 329: 392–393, 1987
19. Simpson D, Decker R: The cardiac cytoskeleton: functional integration of the contractile apparatus. Heart Failure Dec 1992/Jan 1993: 239–254, 1992
20. Engel J, Odermatt E, Engel A, Madri J, Furthmayr H, Rohde H, Timpl R: Shapes, domain organizations and flexibility of laminin and fibronectin. J Mol Biol 150: 97–120, 1981
21. Elliot RW, Barlow D, Hogan BL: Linkage of genes for laminin B1 and B2 subunits on chromosome 1 in mouse. In Vitro Cell Dev Biol 21: 477–484, 1985
22. Paulsson M, Deutzmann R, Timpl R, Dalzappo D, Odermatt E, Engel J: Evidence for coiled-coil alpha-helical regions in the long arm of laminin. EMBO J 4: 309–316, 1985
23. Farhadain F, Contrad F, Corbier A, Barrieux A, Rappaport L, Samuel J: Fibronectin expression during physiological and pathological cardiac growth. J Moll Cell Cardiol 27: 981–990, 1995
24. Hynes RO: Fibronectins. Springer, New York, 1990, pp 1–545
25. Adams JC, Watt FM: Regulation of development and differentiation by the extracellular matrix. Development 117: 1183–1198, 1993
26. Watt FM: The extracellular matrix and cell shape. TIBS 11: 482–485, 1986
27. Weber KT, Sun Y, Katwa LK, Cleutjens JP: Connective tissue: A metabolic entity? J Moll Cell Cardiol 27: 107–120, 1995
28. Weber KT: Cardiac interstitium in health and disease: The fibrillar collagen network. J Am Coll Cardiol 13: 1637–1652, 1989
29. Brilla CG, Maisch B, Weber KT: Myocardial collagen matrix remodelling in arterial hypertension. Eur Heart J 13: 24–32, 1992
30. Bashey RI, Martinez-Hernandez A, Jimenez SA: Isolation, characterization, and localization of cardiac collagen type VI. Associations with other extracellular matrix components. Circ Res 70: 1006–1017, 1992
31. Caulfield JB, Borg TK: The collagen network of the heart. Lab Invest 40: 364–372, 1979
32. Weber KT, Clark W, Janiccki J, Shroff S: Physiologic versus pathologic hypertrophy and the pressure-overloaded myocardium. J Cardiovasc Pharmacol 10: 37–49, 1987
33. Weber KT, Sun Y, Tyagi C, Cleutjens PM: Collagen network of the myocardium: Function, structural remodeling and regulatory mechanisms. J Moll Cell Cardiol 26: 279–292, 1994
34. Brilla CG, Maisch B, Zhou G, Weber KT: Hormonal regulation of cardiac fibroblast function. Eur Heart J 16: 45–50, 1995
35. Chua CC, Chua BH, Zhao ZY, Krebs C, Diglio C, Perrin E: Effects of growth factors on collagen metabolism in cultured human heart fibroblasts. Connect Tissue Res 26: 271–281, 1991
36. Tyagi SC, Ratajaska A, Weber KT: Myocardial matrix metalloproteinase: localisation and activitation. Mol Cell Biochem 126: 49–59, 1993
37. Brilla CG, Janicki JS, Weber KT: Impaired diastolic function and coronary reserve in genetic hypertension. Circ Res 69: 107–115, 1991
38. Butt RP, Laurent GJ, Bishop JE: Mechanical load and polypeptide growth factors stimulate cardiac fibroblast activity. Ann NY Acad Sci 752: 387–393, 1992
39. Weber KT, Brilla CG: Pathological hypertrophy and the cardiac interstitium: Fibrosis and the renin-angiotensin-aldosterone system. Circulation 83: 1849–1865, 1991

Molecular and Cellular Biochemistry **186**: 159–168, 1998.
© 1998 *Kluwer Academic Publishers.*

Dedifferentiated cardiomyocytes from chronic hibernating myocardium are ischemia-tolerant

J. Ausma[1], F. Thoné[2], G.D. Dispersyn,[1] W. Flameng,[3]
J.L. Vanoverschelde,[4] F.C.S. Ramaekers[1] and M. Borgers[1,2]

[1]*Department of Molecular Cell Biology and Genetics, Cardiovascular Research Institute Maastricht, Maastricht University, The Netherlands;* [2]*Department of Morphology, Life Sciences, Janssen Research Foundation, Beerse, Belgium;* [3]*Department of Cardiovascular Surgery, Catholic University of Leuven, Belgium;* [4]*Department of Cardiology, Catholic University of Louvain, Brussels, Belgium*

Abstract

Left ventricular biopsies from 21 patients with clinically diagnosed chronic hibernating myocardium (CHM) were examined by light- and electron microscopy. A mean of 27% of cardiomyocytes were structurally altered and were characterized as hibernating, because of reduced amount of myofibrils and increased glycogen content. Electron microscopy of these cells showed reduction of T-tubules and numerous small mitochondria, but few changes associated with degeneration, acute ischemia or apoptosis. The structural changes found in CHM are reminiscent of dedifferentiation rather than degeneration. The expression patterns of some structural proteins show resemblance with those in embryonic cardiomyocytes.

Histochemically, mitochondrial NADH-oxidase and proton translocating ATPase activities were absent, whereas cytochrome c activity was present. Intracellular calcium distribution indicated normally bound sarcolemmal calcium and absence of excess mitochondrial calcium accumulation. Nuclear chromatin ranged from normal to dispersed with only a few nuclei that were clumped. These results suggest that cardiomyocytes from human CHM hearts are structurally altered, but viable, and lack morphologic and cytochemical characteristics suggestive of apoptosis or acute ischemia. (Mol Cell Biochem **186:** 159–168, 1998)

Key words: ischemia, dedifferentiation, apoptosis, chronic hibernating myocardium

Introduction

In 1985 Rahimtoola [1] described a state of left ventricular dysfunction which occurs during chronic myocardial ischemia. In this condition, the reduced ventricular function could be partially or completely restored to normal if the myocardial oxygen supply/demand was favorably altered by improving blood flow and/or reducing demand. The hibernating response of the heart, namely a diminished function to cope with a reduced myocardial blood flow, has been considered as an act of self-preservation (little blood, little work). Another characteristic of chronic hibernation is that after blood flow has been restored, the recovery of function is slow to very slow (weeks to months) [2].

In recent studies we described the structural alterations in cardiomyocytes from patients with chronic hibernating myocardium (CHM). Typical cellular changes appeared in the cardiomyocytes from CHM that were not degenerative in nature, but were characterized as dedifferentiation [3]. This assumption was supported by the expression and organization pattern of 'early development' markers of the heart muscle, such as α-smooth muscle actin, titin, cardiotin and A-type lamins [4–6].

The dedifferentiation of cardiomyocytes in CHM has been proposed as an adaptation to a limited oxygen supply. Although these dedifferentiated cells lack morphologic hallmarks of cells undergoing apoptosis or necrosis, some doubts remain about the integrity of these cardiomyocytes,

Address for offprints: J. Ausma, Department of Molecular Cell Biology and Genetics, Maastricht University, P.O. Box 616, 6200 MD Maastricht, The Netherlands

i.e., it remains to be established whether or not the hibernating phenotype is a prelude to cell death [7, 8]. With this in mind we assessed the integrity of mitochondria, sarcolemma and nuclei of structurally altered (dedifferentiated) cardiomyocytes.

In this study the nuclear morphology was evaluated in detail by electron microscopy and nuclear chromatin distribution was quantitated by light microscopy. Secondly, the integrity of the mitochondria was assessed by studying the distribution of NADH oxidase, proton translocating ATPase and cytochrome c oxidase activities. Thirdly, studies on the subcellular localization of calcium were performed in order to establish the calcium homeostatic potential of hibernating cardiomyocytes.

Material and methods

Patients

The human cardiac tissue material used in this study consisted of transmural biopsies from 21 patients with severe left ventricular dysfunction. Viability of the myocardium was assessed by positron emission tomography (PET). The detailed description of the patient data regarding anterior wall motion abnormalities, degree of LAD stenosis, flow-metabolic match or mismatch, and functional recovery after coronary bypass surgery has already been presented [9, 10]. During coronary artery bypass surgery, two transmural biopsies were taken from the anterior free wall of the left ventricle. The study was approved by the Ethical Committees for Human Research at the Universities of Leuven and Louvain and all patients have given their informed consent. Left ventricular biopsies derived from 7 donor hearts used for orthotopic transplantation were microscopically examined and served as controls.

Light- and electron-microscopic evaluation

Light and electron microscopy was performed on biopsies which were fixed by immersion in phosphate-buffered 3% glutaraldehyde, postfixed in veronal acetate-buffered 2% osmium tetroxide, dehydrated in graded series of ethanol and embedded in Epon [11].

For electron microscopy, ultrathin sections were cut from each sample, counterstained with uranium acetate and lead citrate, and examined with a Philips CM100 electron microscope.

Two-micron thick Epon sections of the myocardium, stained with either 0.1% toluidine blue or periodic acid-Schiff (PAS), were used for light microscopic morphometry. This staining method permits independent visualization of the contractile elements and glycogen, respectively, and enables the unambiguous determination of the degree of myolysis. We limited our morphologic examinations to areas in which no signs of histological infarction were seen, thus restricting our observations to true hibernating areas [3].

Evaluation of the nuclear morphology

Nuclear morphology was evaluated by light microscopy on two-micron-thick Epon sections stained with 0.1% toluidine blue. Nuclei from cardiomyocytes with more than 20% myolysis were included in the analysis. In a group of 21 patients at least 50 nuclei were evaluated for every myocardial sample. In the light microscope the nuclei were classified into three categories:

> Type 1 nuclei are those showing only a tiny rim of heterochromatin near the nuclear membrane. Some of these nuclei are difficult to distinguish from normal nuclei.
> Type 2 nuclei are bright nuclei with an evenly dispersed heterochromatin. The nuclear margins are hardly visible in this type of nuclei as a result of the absence of clustered heterochromatin.
> Type 3 nuclei contain abundant heterochromatin clumps which are present throughout the nucleoplasm.

Electron microscopy was done to verify the above light microscopic findings.

Enzyme cytochemistry

Cytochemical assays were performed on biopsies fixed for 30 min at 4°C by immersion in 0.1 M sodium cacodylate-buffered 3% glutaraldehyde.

Cytochrome c oxidase (EC 1.9.3.1) activity assay
Frozen sections of 30 μm were incubated for 30–60 min in Seligman's medium [12] [9 ml 0.05M sodium phosphate buffer (pH 7.4), 5 mg 3,3′-diaminobenzidine-4 hydrochloride, 1 ml catalase (20 mg/ml), 10 mg cytochrome c, 0.22 M sucrose].

Proton translocating ATPase (EC 3.6.1.3) activity assay
Proton-translocating ATPase activity was localized with a Wachstein and Meisel-type procedure [13]. Briefly, sections of 30 μm were incubated for 15–30 min at 37°C in a solution containing 48 mM tris-maleate buffer, 3 mM lead nitrate, 2.3 mM ATP, 5 mM $MgSO_4$ and sucrose [14].

NADH-oxidase assay (EC 1.6.99.3)
The procedure of NADH-oxidase localization has previously been described by Briggs *et al.* [15]. Briefly, 30 μm frozen sections were cut and preincubated for 10 min in a solution containing 0. 1 M tris-maleate buffer (pH 7.5), 7% sucrose and 1 mM aminotriazole. The sections were then incubated for 2 h in an incubation medium containing 0.1 M tris-maleate buffer (pH 7.5), 7% sucrose, 1 mM cerium chloride, 0.71 mM NADH and 10 mM aminotriazole [14].

After all cytochemical incubations, the sections were postfixed with osmium tetroxide, dehydrated in graded series of ethanol and routinely embedded in Epon. Ultrathin sections, either unstained or briefly counterstained with uranyl acetate and lead citrate, were examined by electron microscopy.

For proton-translocating ATPase and NADH oxidase activity stunned myocardium from dogs was used as internal control. The left anterior descending artery was occluded for 15 min followed by a reperfusion period of 150 min. After this period of myocardial stunning, the hearts were fixed by retrograde perfusion for 5 min with 2% glutaraldehyde in 0. 1 M cacodylate buffer followed by 5 min perfusion with cacodylate buffer supplemented with 7.5% sucrose as described by Vandeplassche [14].

Calcium distribution studies

The localization of exchangeable calcium was assessed with the oxalate-pyroantimonate (OPA) procedure, as previously described by Van Reempst *et al.*, [16]. In brief, biopsies were fixed by immersion in potassium oxalate-buffered 3% glutaraldehyde (pH 7.4), postfixed in 2% osmium tetroxide containing 1% potassium pyroantimonate, dehydrated in graded series of ethanol and embedded in Epon.

Phospholipid-bound calcium was localized with the phosphate pyroantimonate (PPA) technique [17]. In brief, the biopsy samples were fixed by immersion in potassium phosphate-buffered 3% glutaraldehyde, postfixed in 2% osmium tetroxide containing 1% potassium pyroantimonate, dehydrated in graded series of ethanol and embedded in Epon.

Ultrathin sections, either unstained or briefly counterstained with uranyl acetate and lead citrate, were examined by electron microscopy.

Results

Cardiomyocytes from chronic hibernating myocardium (CHM) underwent numerous structural changes of which depletion of myofibrils and accumulation of glycogen were

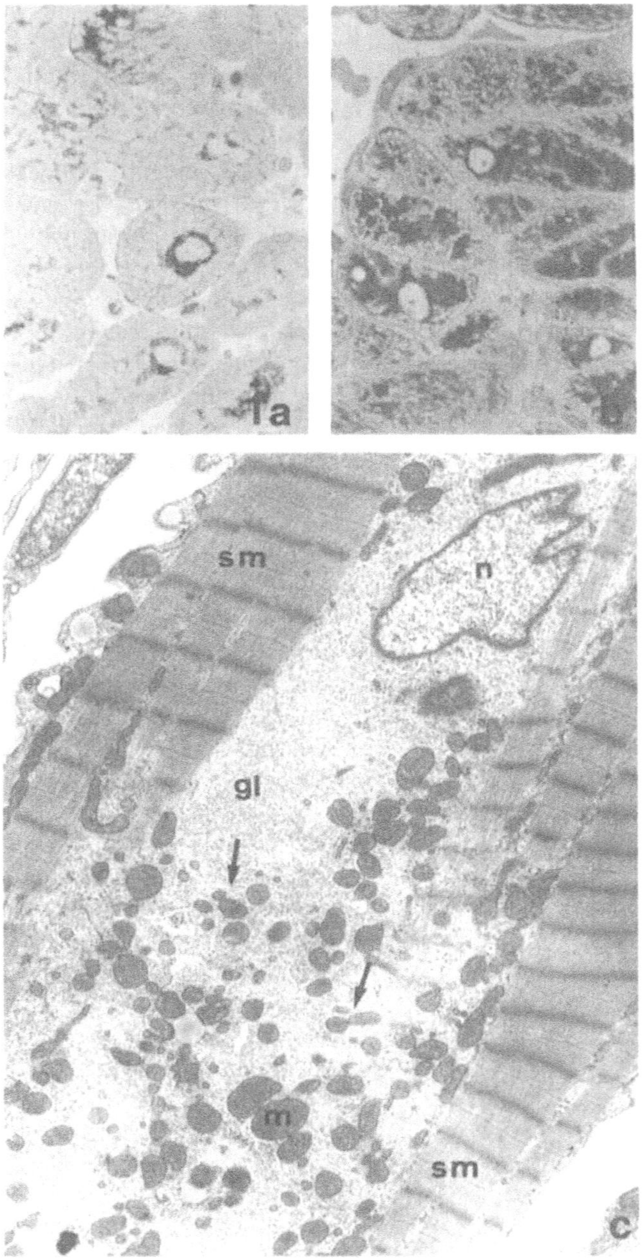

Fig. 1. Light and electron microscopy. (a) Light microscopy of human myocardium with normally structured cardiomyocytes. PAS staining (gray areas) is almost absent within these cardiomyocytes. The sarcomeres, stained with toluidine blue, are present throughout the cytoplasm (×: 360). (b) Light microscopy of a section from chronic hibernating myocardium showing severe myolysis. Glycogen accumulation (dark areas; PAS positive) in the sarcomere-free regions of the myocytes (×: 360). (c) Electron micrograph of a structurally altered cardiomyocyte from CHM, showing residual sarcomeres (sm) at the periphery of the cell, glycogen accumulation (gl), depletion of sarcoplasmic reticulum and numerous small mitochondria (arrows) next to normal mitochondria n (m), n=nucleus (×: 6600).

the most remarkable (Fig. 1). In the group of 21 patients with CHM used for this study a mean of 27% of the cardiomyocytes was affected by myolysis, i.e., showed loss of myofibrils for more then 20% of the cell surface area (excluding the nucleus).

At the ultrastructural level, the sarcolemma-glycocalyx complex appeared to be intact. The T-tubular invaginations were reduced in number and became virtually absent in the more severely affected cardiomyocytes. Remarkably, typical degenerative changes such as cytoplasmic vacuolization, cytosolic edema, mitochondrial swelling, membrane disruption, accumulation of secondary lysosomes, membrane whorls, and lipid droplets were seldomly seen in these structurally altered cardiomyocytes. Numerous small donut-shaped mitochondria were observed in cardiomyocytes with a decreased amount of myofibrils (Fig. 1). Damage to the mitochondrial structure indicative of ischemia, such as swelling of the mitochondria with loss of intramatrix granules, blebbing and disruption of the outer membrane, and formation of flocculent densities (Jennings granules) were not observed. No changes characteristic of apoptosis such as shrinking of cytoplasm, the formation of cytoplasmic blebs, condensation of nuclear chromatin or the formation of apoptotic bodies were documented.

Nuclear alterations

In myolytic cardiomyocytes from CHM 3 types of nuclei could be classified based on differences in chromatin distribution. Type 1 nuclei were the predominant type seen in all

Table 1. Nuclear alterations in cardiomyocytes from CHM

Patient nr.	%type 1 nuclei	% type 2 nuclei	% type 3 nuclei
1	76.6	0.0	23.4
2	86.4	12.3	1.2
3	91.5	8.5	0.0
4	70.4	20.9	8.7
5	73.6	9.4 1	7.0
6	69.1	24.3	6.6
7	60.8	33.3	5.9
8	62.5	34.1	3.4
9	61.9	3.6	34.5
10	60.9	5.5	13.6
11	33.3	4.0	62.7
12	78.3	20.3	1.4
13	73.1	25.4	1.5
14	73.0	26.0	1.0
15	84.0	16.0	0.0
16	77.3	12.5	10.2
17	78.6	14.3	7.1
18	88.1	0.0	11.9
19	82.1	9.0	9.0
20	93.8	3.1	3.1
21	87.1	9.7	3.2

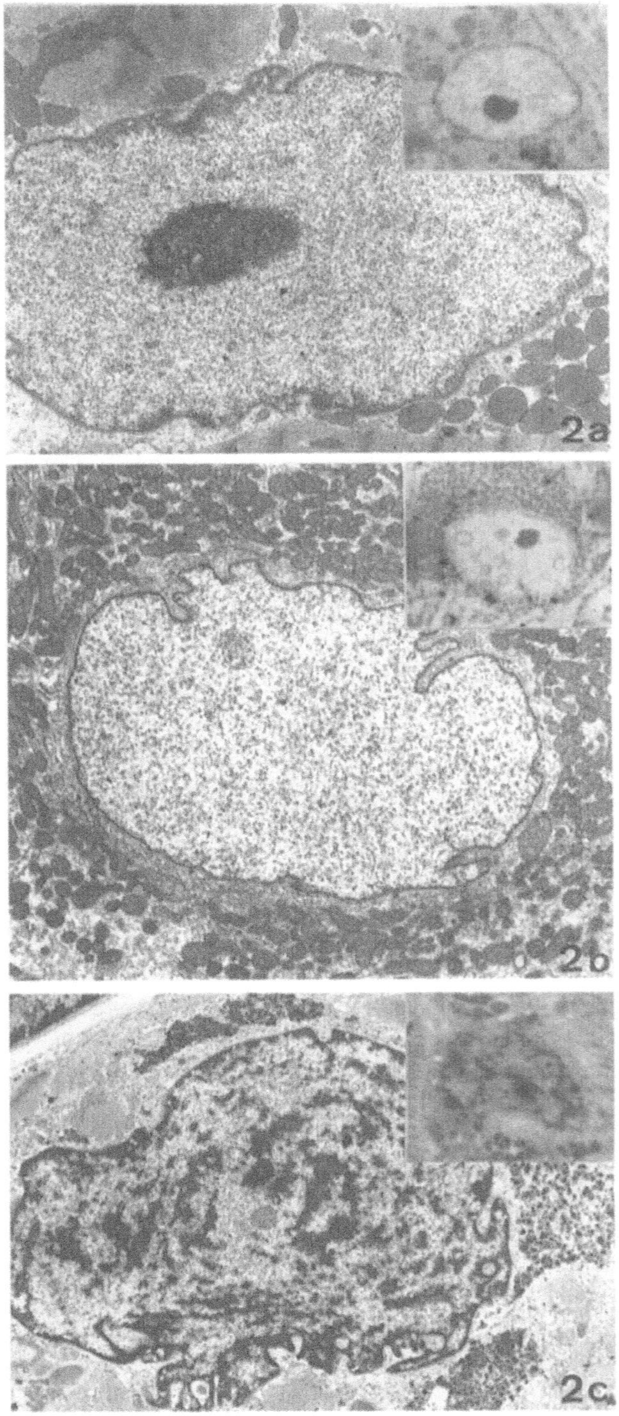

Fig. 2. Electron microscopic details of CHM cardiomyocyte nuclei with accompanying light microscopic inserts. (a) Nucleus showing only a tiny rim of heterochromatin near the nuclear membrane (type l) (×: 6750). (Insert ×: 1500). (b) Nucleus of a cardiomyocyte with homogeneously dispersed heterochromatin (type 2) (×: 6500). (Insert ×: 1000). (b) Nucleus with clumping of heterochromatin throughout the nucleoplasm (type 3) (×: 5700). (Insert ×: 1420).

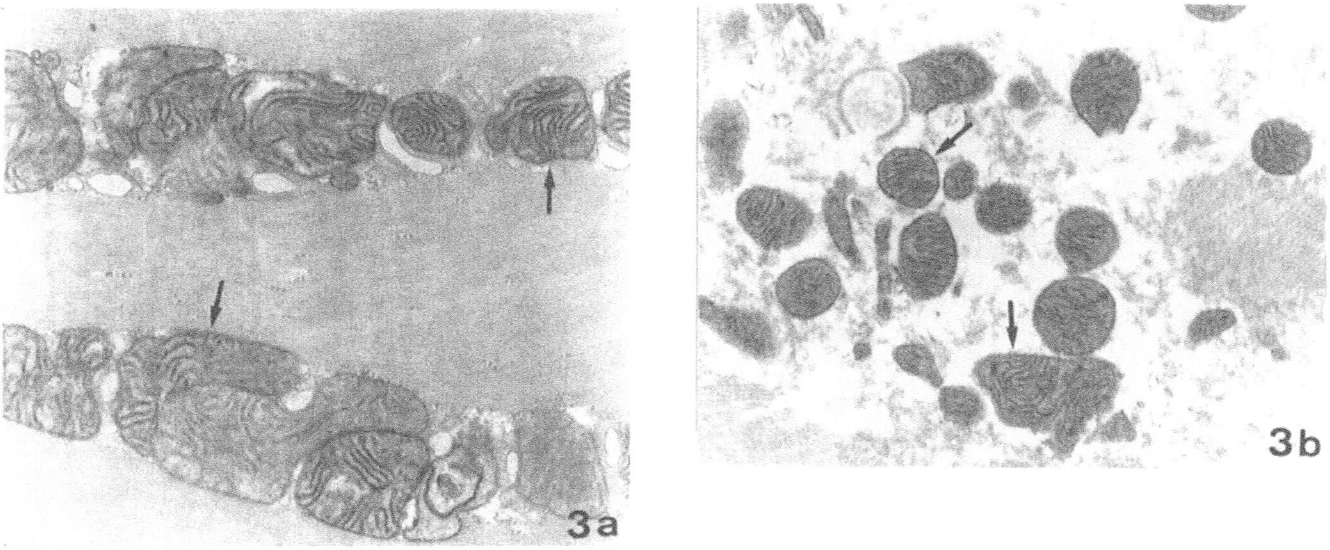

Fig. 3. Cytochrome c oxidase. Activity in mitochondria (arrows) of a normally structured human cardiomyocyte (×: 22700). (b) Activity in morphologically altered mitochondria (arrows) of a myolytic cardiomyocyte from CHM. The cytochrome oxidase activity remains preserved (×: 16720).

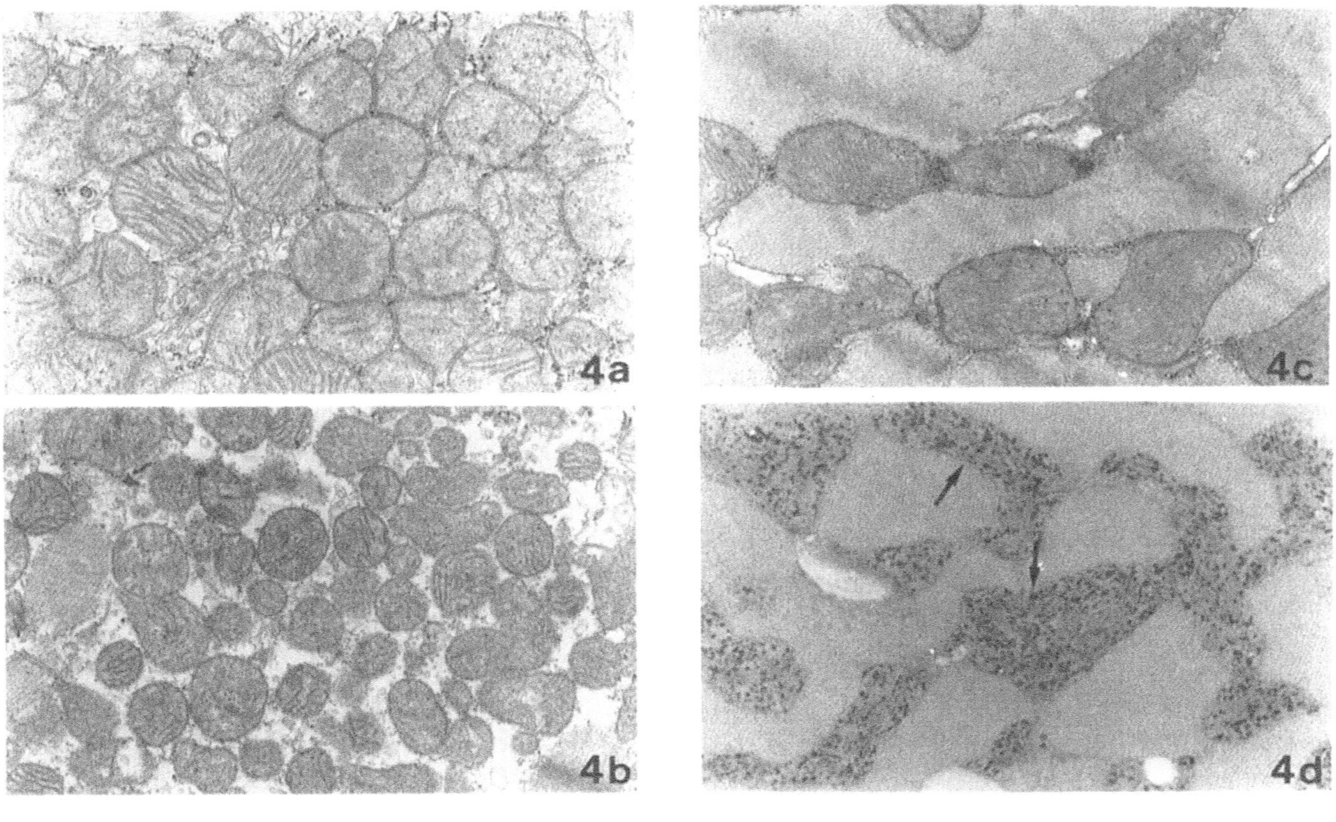

Fig. 4. Proton-translocation ATPase. No activity is detected in mitochondria of normal human myocardium (a) (×: 21600), in structurally altered mitochondria of hibernating cells (b) (×: 17300), in mitochondria of normal dog myocardium (c) (×: 19450). In contrast, in stunned dog myocardium the activity (black precipitate) is detected in the mitochondrial matrix (arrows) (d) (×: 17300) (c,d). With courtesy of Vandeplassche *et al.* [14].

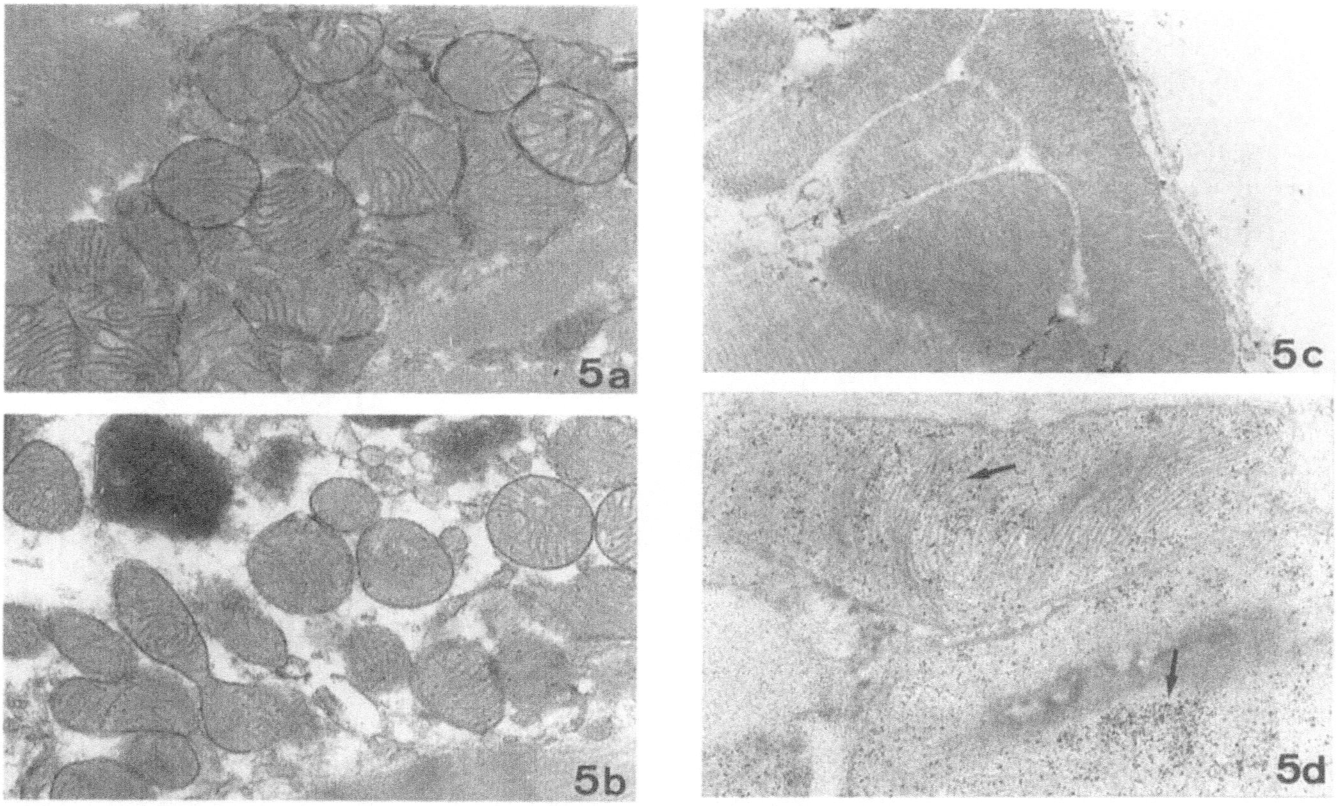

Fig. 5. NADH-oxidase activity. No activity is detected in mitochondria of normal human cardiomyocytes (a) (×: 17250), in structurally altered mitochondria of hibernating cardiomyocytes (b) (×: 16900), in mitochondria of normal dog cardiomyocytes (c) (×: 27650). Mitochondria from cardiomyocytes of stunned dog myocardium show fine granular NADH-oxidase reaction product (arrows) (d) (×: 62200). (c,d). With courtesy of Vandeplassche *et al.* [14].

myolytic cardiomyocytes from all patients (Table 1). The majority of these nuclei present heterochromatin in a thin rim near the nuclear membrane (Fig. 2a). In only a minority of cells was the heterochromatin distribution nearly indistinguishable from that of unaffected (nonmyolytic) cells. In type 2 nuclei from myolytic cardiomyocytes, the heterochromatin was found to be evenly dispersed, resulting in a homogeneous light appearance of the whole nucleoplasm (Fig. 2b). Type 3 nuclei were found in a limited number of patients and were characterized by abundant clumping of chromatin near the nuclear membrane and throughout the whole nucleoplasm (Fig. 2c). These type 3 nuclei, although abnormal, showed no clear resemblance with nuclear changes during apoptosis or necrosis (oncosis) [18].

Other morphologic changes representative of apoptosis such as the formation of apoptotic bodies and cytoplasmic blebs were not seen in these well delineated hibernating areas.

The number of nuclei of each type as counted in cardiomyocytes from CHM with more than 20% myolysis, are presented in Table 1. Except for 1 patient all had more than 60% of type 1 nuclei. Seven patients showed more than 20% of type 2 nuclei. In only 3 of the 21 patients more than 20% of the nuclei were type 3.

Enzyme cytochemistry

Cytochrome c oxidase activity

In normal cardiomyocytes mitochondria are located between the rows of sarcomeres (Fig. 3a), and cytochrome c oxidase activity is uniformly displayed. In the small donut-shaped mitochondria of structurally altered cardiomyocytes from CHM cytochrome c oxidase activity appeared to be present in a normal distribution (Fig. 3b).

Proton translocating ATPase activity

Proton-translocating ATPase activity was not detected in the mitochondrial matrix of cardiomyocytes from normal myocardium (Fig. 4a). In mitochondria from CHM the enzyme activity was similarly absent (Fig. 4b). In ischemic myocardium proton-translocating ATPase activity can be detected as a result of uncoupling of the mitochondria. In stunned myocardium from dogs, the enzyme reaction product was

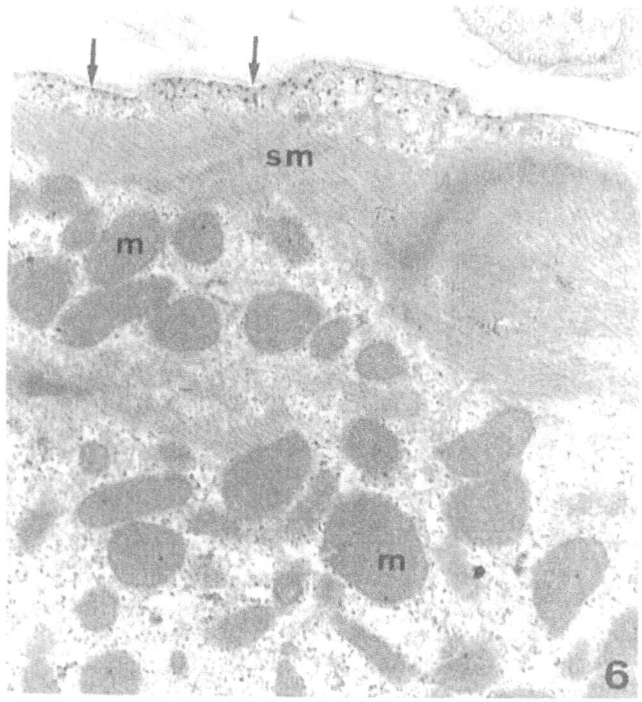

Fig. 6. Calcium localization. Assessment of calcium distribution with the PPA precipitation technique in a hibernating cell showing a peripheral strand of sarcomeres (sm) and many mitochondria (m). Note the association of calcium precipitate with the sarcolemma and sarcolemma-derived vesicles (arrows). Mitochondria are almost devoid of precipitate (×: 20500).

intense in the mitochondria (Fig. 4d). In contrast, mitochondria of cells derived from normal dog myocardium did not show any reaction product (Fig. 4c).

NADH-Oxidase
In normal cardiomyocytes the NADH-oxidase reaction was virtually absent (Fig. 5a). In the abnormally shaped small mitochondria from CHM no NADH-oxidase activity was demonstrated (Fig. 5b). In contrast cardiomyocytes derived from stunned myocardium in the dog showed intense NADH-oxidase reaction product (Fig. 5d). In normal cells of control dogs, no activity was detected in mitochondria (Fig. 5c).

Calcium distribution
When subcellular calcium distribution was assessed with the PPA technique calcium precipitate was localized along the inner leaflet of the sarcolemma, the sarcolemma-derived vesicles, T-tubules and intercalated disks of normally structured cardiomyocytes. Mitochondria were free of calcium precipitate or presented only a few deposits. In myolytic cardiomyocytes from CHM a similar distribution of calcium was found along the sarcolemma. Mitochondria and other intracellular organelles were almost free of calcium precipitate (Fig. 6).

The OPA technique detected only small calcium deposits in mitochondria of either normally structured or myolytic cardiomyocytes.

Discussion

It has been suggested that the structural changes observed in cardiomyocytes from chronic hibernating myocardium (CHM) are not degenerative in nature, but involve a phenotypic switch form a functional (contractile) to a non-functional (hibernating) phenotype akin to the embryonic or neonatal state of development [3–6]. In the current study, degenerative changes as observed in acute ischemic myocardium and different cardiomyopathies were not observed in areas which could be unambiguously identified as hibernating. However some doubts remain about the integrity and long-term viability of the dedifferentiated cardiomyocytes.

Are the changes in nuclear morphology related to apoptosis?

In hibernating cardiomyocytes from CHM 3 types of nuclei are present. The majority of the cardiomyocytes contained nuclei in which the heterochromatin redistributed and became less clumped. In only a minority of patients was chromatin clumping seen. Some authors [7, 8] correlate condensation of nuclear chromatin in CHM with apoptosis and degeneration. Extensive clumping of nuclear chromatin and other characteristics of apoptosis such as the formation of apoptotic bodies and cytoplasmic blebs were not observed in cardiomyocytes from CHM by us. Studies on the expression of apoptosis-related molecules were not indicative for apoptosis in these structurally altered cells. Nicking of DNA typical of apoptotic cells was absent from the above specified areas of chronic hibernating myocardium, and apoptosis related molecules like BCL-2, P53, PCNA were not expressed in CHM [19]. A patchy nuclear A-type lamin expression as observed in a few of the nuclei from CHM may possibly correspond to the nuclei showing clumping of their nuclear chromatin [6], since nuclear lamins are known to regulate the chromatin topology.

The majority of nuclei adopted a morphology with an embryonic-like homogenous distribution of the heterochromatin. The presence of an embryonic-like nuclear morphology was previously suggested and supported by the expression patterns of nuclear lamins in myolytic cardiomyocytes from chronic hibernating myocardium [6]. Eighty percent of the nuclei of cardiomyocytes from CHM were negative for A-type lamin and positive for B-type lamin expression. Nuclear A-type lamins are known to be expressed in a differentiation-related fashion and contribute to the nuclear integrity and chromatin organization. A-type lamin expression is absent in undifferentiated cells and cells at early stages of development [20–24]. The absence of A-type lamin expression forms an argument for the dedifferentiation state of cardiomyocytes from CHM. The presence of B-type lamin expression together with the absence of the characteristics for cell death [25, 26] suggests that the dedifferentiated cardiomyocytes from CHM are unlikely to represent degenerative stages, either apoptotic or necrotic.

Chronic hibernating myocardium and acute ischemia

Signs of acute ischemia, including damage to mitochondria (loss of intramatrix granules, clarification of mitochondrial matrices, disruption of cristae), depletion of glycogen stores, cellular edema, pyknosis of the nuclei, swelling of sarcoplasmic reticulum, damage to the sarcolemmal structures and contraction band necrosis, were not observed in this study of CHM. Mitochondria from both cells with normal morphology and structurally altered cells demonstrated similar preserved cytochrome c activity and absence of proton translocating ATPase and NADH-oxidase activities, suggesting absence of overt damage or uncoupling [14, 27, 28]. Exposure of mitochondria to hydrogen peroxide released after an ischemic insult results in a direct inhibition of mitochondrial proton-translocating ATPase complex [29]. The high activities of ATPase in stunned myocardium may represent a very early sign of uncoupling of mitochondrial oxidative phosphorylation, whereby the ATP synthetase complex reverses its activity and hydrolyzes ATP [14].

Cardiomyocytes from CHM are not calcium overloaded

It has been proposed that calcium bound to the inner side of the lipid bilayer plays an important role in the integrity of the sarcolemma [17]. The loss of the ability of the sarcolemma to bind calcium is considered to be the first sign of loss of intracellular calcium homeostasis because it occurs already during the early ischemic phase and precedes other signs of degeneration [17]. The persistent binding of calcium to the inner layer of the sarcolemma of the hibernating cells indicated that the asymmetry of the phospholipid bilayer is still intact and that the hibernating cells do not suffer from oxygen shortage. The sarcolemma-associated calcium, most likely representing the fraction of calcium bound to annexin VI, was unaltered. Annexin VI, a calcium dependent phospholipid-binding protein, has been considered to play a role in sodium/calcium exchange [30].

Under ischemic conditions mitochondria scavenge calcium when the cytosol becomes overloaded [17]. Mitochondria from CHM are virtually devoid of a calcium precipitate, which suggests the absence of calcium overload in the hibernating cardiomyocytes. The absence of changes in mitochondrial calcium deposition together with a preserved mitochondrial structure (vide supra) strongly supports the idea that hibernating cells are not ischemic, a concept suggested previously [31–33].

Hibernating cardiomyocytes are ischemia tolerant

Biopsies from 21 patients with CHM showed structural alterations such as a reduced amount of myofibrils, glycogen accumulation, depletion of sarcoplasmic reticulum, reduction of T-tubules and the presence of numerous small mitochondria. In a previous study [34] it was shown that the structurally altered cardiomyocytes are protected against acute oxygen shortage. The latter study with biopsies taken before and at the end of coronary artery bypass surgery indicated that dedifferentiated cardiomyocytes are more resistant to acute ischemic injury. In particular, mitochondria from dedifferentiated cardiomyocytes are not swollen and are devoid of calcium overload, the sarcolemma-glycocalyx complex remains intact, and nuclei do not show chromatin condensation. In contrast, neighbouring normal differentiated cardiomyocytes were structurally injured by ischemia, indicating that dedifferentiated cardiomyocytes have a greater tolerance to ischemia [34].

Interestingly, embryonic and neonatal hearts have a better functional recovery from an ischemic insult than adult ones. Embryonic myocardium is known to maintain higher adenosine triphosphate and adenosine monophosphate levels after short periods (15–60 min) of ischemia [35–37]. Embryonic myocardium works at lower PO_2 and uses glucose (partially from glycogen stores) as energy source. Embryonic cardiomyocytes tolerate hypoxic conditions better because they have a higher glycolytic capacity, and an enhanced glycogen storage together with a lower contractile function to match the lowered blood supply [37].

The hypoxic environment in CHM might induce an embryonic program to resist the ischemic insult in a way similar to embryonic/neonatal myocardium. Hence, the adaptation of the cardiomyocytes to the lowered oxygen supply – by adopting an embryonic program – may result in a 'survival' state that prevents myocardial degeneration.

The present enzyme-histochemical studies indicated almost no changes suggesting no acute ischemia, mitochondrial uncoupling, apoptosis or altered intracellular calcium distribution. The morphologic changes found seem most consistent with a dedifferentiated state of the structurally altered cells, rather than ongoing oncosis or apoptosis. It is unknown whether or not the delay in restoration of cardiac function is caused by the structural changes in the cardiomyocytes. However, this might be speculated. Further studies will be needed to determine mechanisms of chronic myocardial hibernation.

Conclusion

Dedifferentiated cardiomyocytes from CHM show no signs of obvious degeneration. Although it can not be concluded that mitochondria are fully functional the tested mitochondrial enzyme activities remain present and the overall calcium homeostasis seems preserved. It is proposed that the dedifferentiated character of the hibernating cells allows them not only to remain viable for a prolonged period of time but also to better tolerate acute ischemic insults. In other words, hibernating cells can be considered cells optimally 'preconditioned' to acute ischemia.

Acknowledgment

Part of this work was supported by the Dutch Heart Foundation grant NHS 96-155.

References

1. Rahimtoola SH: A perspective on the three large multicenter randomized clinical trials of coronary bypass surgery for chronic stable angina. Circulation 72 (Suppl V): 123–135, 1985
2. Rahimtoola SH: From coronary artery disease to heart failure: Role of the hibernating myocardium. Am J Cardiol 75: 16E–22E, 1995
3. Borgers M, Thoné F, Wouters L, Ausma J, Shivalkar B, Flameng W: Structural correlates of regional myocardial dysfunction in patients with critically coronary artery stenosis: Chronic hibernation? Cardiovasc Pathol 2: 237–245, 1993
4. Ausma J, Schaart G, Thoné F, Shivalkar B, Flameng W, Depré, Vanoverschelde J, Ramaekers F, Borgers M: Chronic ischemic viable myocardium in man: Aspects of dedifferentiation. Cardiovasc Pathol 4: 29–37, 1995
5. Ausma J, Fürst D, Thoné F, Flameng W, Weber K, Ramaekers F, Borgers M: Molecular changes of titin in chronic hibernating myocardium. J Mol Cell Cardiol 27: 1203–1212, 1995
6. Ausma J, van Eys GJJM, Broers JLV, Thoné F, Flameng W, Ramaekers FCS, Borgers M: Nuclear lamin expression in chronic hibernating myocardium in man. J Mol Cell Cardiol 28: 1297–1305, 1996
7. Elsasser A, Klovekorn W-P, Zimmerman R, Munkel B, Muller K-D, Strasser R, Schaper J: 'Hibernating myocardium': Changes of transcriptional activity. Circulation 94: 1–186, 1996
8. Schwarz ER, Schaper J, vom Dahl J, Altehoefer C, Grohmann B, Schoendube F, Sheehan FH, Uebis R, Buell U, Messmer BJ, Schaper W, Hanrath P: Myocyte degeneration and cell death in hibernating human myocardium. J Am Coll Cardiol 27: 1577–1585, 1996
9. Vanoverschelde JLJ, Wijns W, Depré C, Essamri B, Heyndricks S, Borgers M, Bol A, Melin J: Mechanisms of chronic regional post-ischemic dysfunction in humans: New insights form the study on non-infarcted collateral dependent myocardium. Circulation 87: 1513–1523, 1993
10. Maes A, Flameng W, Nuyts J, Borgers M, Shivalkar B, Ausma J, Bormans G, Schiepers C, De Roo M, Mortelmans L: Histological alterations in chronically hypoperfused myocardium. Correlation with PET findings. Circulation 90: 735–745, 1994
11. Flameng W, Wouters L, Sergeant P, Lewi P, Borgers M, Thoné F, Suy R: Multivariate analysis of angiographic histologic and electrocardiographic data in patients with coronary heart disease. Circulation 70: 7–17, 1984
12. Seligman A, Karnovsky M, Wasserkrug H, Hanker 1: Nondroplet ultrastructural demonstration of cytochrome oxidase activity with a polymerizing osmiophilic reagent diaminobenzidine (DAB). J Cell Biol 38: 1–14, 1968
13. Wachstein M, Meisel E: On the histochemical demonstration of glucose-6-phosphatase. J Histochem Cytochem 10: 592, 1956
14. Vandeplassche G, Hermans C, Thoné F, Borgers M: Stunned myocardium has increased mitochondrial NADH oxidase and ATPase activities. Cardioscience 2: 47–53, 1991
15. Briggs R, Drath D, Karnovsky ML, Karnovsky MJ: Localization of NADH oxidase on the surface of human polymorphonuclear leukocytes by a new cytochemical method. J Cell Biol 67: 566–586, 1975
16. Van Reempts J, Borgers M, Offner F: Ultrastructural localization of calcium in the rat retina with a combined oxalate-pyroantimonate technique. Histochem J 14: 517–522, 1982
17. Borgers M, Liu GS, Xhonneux R, Thoné F, Van Overloop P: Changes in ultrastructure and Ca^{++} distribution in the isolated working rabbit heart after ischemia. Am J Pathol 126: 92–102, 1987
18. Gottlieb RA, Buriesin KO, Kioner, RA, Babior BM, Engler RL: Reperfusion injury induces apoptosis in rabbit myocytes. J Clin Invest 94: 1621–1628, 1994
19. Dispersyn GD, Ausma J, Thoné F, Flameng W, Vanoverschelde J-U, Ramaekers FCS, Borgers M: Chronic hibernating myocardium and chronic atrial fibrillation; a prelude to apoptosis? Submitted
20. Stick R, Hansen P: Changes in the nuclear lamina composition during early development of Xenopus laevis. Cell 41: 191–200, 1985
21. Lehner CF, Stick R, Eppenberger M, Nigg EA: Differential expression of nuclear matrix proteins during chicken development. J Cell Biol 105: 571–687, 1987
22. Stewart C, Burke B: Teratocarcinoma stem cells and early mouse embryos contain only a single major lamin polypeptide closely resembling lamin B. Cell 51: 383–392, 1987
23. Guilly M-N, Kolb J-P, Gosti F, Godeau F, Courvalin J-C: Lamins A and C are not expressed at early stages of human lymphocyte differentiation. Exp Cell Res 189: 145–148, 1990
24. Röber R-A, Weber K, Osbom M: Differential timing of nuclear lamin A/C expression in the various organs of the mouse embryo and the young animal: A developmental study. Development 105: 365–378, 1989
25. Oberhammer FA, Hochegger K, Froschi G: Chromatin condensation during apoptosis is accompanied by degradation of lamin A+B, without enhanced activation of cdc2 kinase. J Cell Biol 126: 827–837, 1994
26. Tinnemans MMJF, Lenders M-HJH, ten Velde GPM, Ramaekers FCS, Schufte B: Alterations in cytoskeletal and nuclear matrix associated proteins during apoptosis. Eur J Cell Biol 68: 35–46,1995
27. Vandeplassche G, Hermans C, Thoné F, Borgers M: Mitochondrial hydrogen peroxide generation by NADH-oxidase activity following regional myocardial ischemia in the dog. J Mol Cell Cardiol 21: 383–392, 1989
28. Vandeplassche G, Thoné F, Borgers M: Cytochemical evidence of NADH-oxidase activity in the isolated working rabbit heart subjected to normothermic global ischemia. Histochemical J 22: 11–17, 1990
29. Hyskop PA, Hinshaw DB, Hasley WA, Schraufstafter IU, Sauerheber Spragg RG, Jackson JH, Cochrane CC: Mechanisms of oxidant-mediated cell injury. J Biol Chem 263: 1665–1675, 1988

30. Gunteski-Hamblin A-M, Song G, Walshe RA, Frenzke M, Boivin, GP, Dorn GW, Kaetzel MA, Horseman ND, Dedman JR: Annexin VI overexpression targeted to heart alters cardiomyocyte function in transgenic mice. Am J Physiol 270: H1091–H1100, 1996
31. Borgers M, De Nollin S, Thoné F, Wouters L, Van Vaeck L, Flameng W: Distribution of calcium in a subset of chronic hibernating myocardium in man. Histochem J 25: 312–318,1993
32. Schelbert HR: Positron emission tomography for the assessment of myocardial viability. Circulation (Suppl l) 84: 1-122–131, 1991
33. Kloner RA, Przykienk K, Patel B: Altered myocardial states. Am J Med (suppl A) 86: 14–22, 1989
34. Borgers M, Flameng W: Morphology of acute and ischemic myocardium in man. In: HM Piper, CJ Preusse (eds). Ischemia-reperfusion in cardiac surgery. Kluwer Academic Publishers, Boston, Dordrecht, 1993, pp 353–375
35. Mask WK, Abt-Elfaftah AS, Jessen M, Brunsting LA, Lekven J, Wechsler AS: Embryonic versus adult myocardium: Adenine nucleotide degradation during ischemia. Ann Thorac Surg 48: 109–112, 1989
36. Abt-Elfattah ASA, Murphy CE, Salter DR, Goldstein JP, Godwin CK, Wechsler AS: Age and species related differences in adenine nucleotide degradation during myocardial global ischemia. Fed Proc 45: 1039, 1986
37. Grice WN, Konishi T, Apstein CS: Resistance of neonatal myocardium to injury during normothermic and hypothermic ischemic arrest and reperfusion. Circulation 76 (Suppl V): 150–155, 1987

Molecular and Cellular Biochemistry **186:** 169–175, 1998.
© 1998 *Kluwer Academic Publishers.*

Role of cyclic nucleotide phosphodiesterases in ischemic preconditioning

Amanda Lochner,[1] Sonia Genade,[1] Erna Tromp,[1] Lionel Opie,[2] Johan Moolman,[1] Sylvia Thomas[3] and Thomas Podzuweit[3]

[1]*Department of Medical Physiology and Biochemistry, University of Stellenbosch, Faculty of Medicine, Tygerberg; MRC Programme for Experimental Biology, Tygerberg, Republic of South Africa;* [2]*MRC Ischaemic Heart Laboratory, University of Cape Town;* [3]*Max-Planck Institute, Bad Nauheim, Germany*

Abstract

Several signal transduction pathways have been implicated in the mechanism of protection induced by ischemic preconditioning (PC). For example, stimulation of a variety of G-protein coupled receptors results in stimulation of protein kinase C (PKC) which has been suggested to act as common denominator in eliciting protection. PC also significantly attenuated cAMP accumulation during sustained ischemia, suggesting involvement of an anti-adrenergic mechanism. The aim of this study was to evaluate the β-adrenergic signal transduction pathway (as evidenced by changes in tissue cAMP and cAMP- and cGMP-phosphodiesterase) during the PC protocol as well as during sustained ischemia. Isolated perfused rat hearts were preconditioned by 3 × 5 min global ischemia (PC1,2,3) interspersed by 5 min reperfusion, followed by 25 min global ischemia. Tissue cAMP- and cGMP-PDE activity as well as cAMP and cGMP levels were determined at different time intervals during the PC protocol and sustained ischemia. Tissue cAMP increased with each PC ischemic event and normalized upon reperfusion, while PDE activity showed the opposite, viz a reduction during ischemia and an increase during reperfusion. Except for PC1, tissue cGMP showed similar fluctuations. Throughout 25 min sustained ischemia, cAMP- and cGMP-PDE activities were higher in PC than in nonpreconditioned hearts, associated with a significantly lesser accumulation in cAMP and higher cGMP levels in the former. Fluctuations in cyclic nucleotides during preconditioning were associated with concomitant changes in PDE activity, while the attenuated β-adrenergic response of preconditioned hearts during sustained ischemia may partially be due to increased PDE activity. (Mol Cell Biochem **186**: 169–175, 1998)

Key words: preconditioning, cAMP and cGMP, cAMP- and cGMP-phosphodiesterases

Introduction

Despite intensive efforts, elucidation of the exact mechanism(s) of ischemic preconditioning (PC) has still not been accomplished. Involvement of several signal transduction pathways has been implicated. For example, stimulation of a variety of G protein-coupled receptors (e.g. adenosine A_1, α_1-adrenergic, muscarinic, bradykinin receptors) results in activation of protein kinase C which has been suggested to act as the common denominator in eliciting protection [1, 2]. However, the significance of PKC activation in the phenomenon of PC has recently been questioned and other signal transduction pathways proposed, such as MAPkinase (for a review see reference [3]).

The β-adrenergic response to myocardial ischemia may also be involved: Activation of the β-adrenergic signal transduction pathway occurs within minutes after the onset of myocardial ischemia [4]. Furthermore, in isolated rat and intact rabbit hearts prior preconditioning significantly attenuated cAMP accumulation during sustained ischemia [5, 6]. This observation may be due to upstream effects (e.g. reduced noradrenaline release or desensitization of the β-receptor [5], changes in G_s, or G_i[7] or adenylyl cyclase activity [8]) and/or downstream mechanisms such as less

Address for offprints: A. Lochner, Department of Medical Physiology and Biochemistry, P.O. Box 19063, Tygerberg 7505, Republic of South Africa

inhibited phosphodiesterase activity.

The changes in downstream factors involved in cAMP metabolism/breakdown in preconditioning are unknown. The aim of this study was therefore to evaluate the cAMP and cyclic nucleotide phosphodiesterases (PDES) during the PC protocol as well as during sustained ischemia in a well characterized rat heart model.

As the activity of the PDEs is also influenced by cGMP and other reputed mediators of preconditioning (such as NO or bradykinin) [9], which in turn increase tissue cGMP levels, we also studied changes in the myocardial content of this second messenger.

Materials and methods

Animals

Wistar rats weighing 150–250 g were used. Before anesthesia (30 mg pentobarbital i.p.), rats were allowed free access to food and water. The investigation conforms with the 'Guide for the care and use of laboratory animals' published by the US National Institutes of Health (NIH publication no 85-23, revised 1985).

Perfusion technique

The perfusion technique was as described previously [6]. Briefly, after removal hearts were mounted onto the aortic cannula and perfused retrogradely in a non-recirculating manner at 100 cm H_2O (temperature 37°C). The left atrium was cannulated to allow atrial perfusion according to the working heart model (preload 15 cm H_2O; afterload 100 cm H_2O). Krebs Henseleit bicarbonate solution was used as buffer, containing in mM: NaCl 119; $NaHCO_3$ 24.9; KCL 4.74; KH_2PO_4 1.19; $MgSO_4$ 0.6; Na_2SO_4 0.59; $CaCl_2$ 1.25; glucose 10. Oxygenation was done with 95% O_2/5% CO_2. Normothermic zero-flow global ischemia was utilized for both preconditioning and sustained ischemia. Hearts were surrounded by a small tightly stoppered water-jacketed chamber (temperature 37°C) to prevent cooling.

Perfusion protocol

In all hearts, 15 min retrograde perfusion (Langendorff technique) was followed by 15 min as working hearts (Fig. 1). This study is a follow-up of a previous study [6] in which mechanical function during reperfusion was studied; therefore the 15 min working heart perfusion was maintained in the protocol. Non-preconditioned hearts were subsequently perfused retrogradely for a further 30 min, whereas the hearts to be preconditioned were subjected to three episodes of 5 min global ischemia (PC1–, 2–, 3–), interspersed by 5 min retrograde perfusion (PC1+, 2+, 3+) (total duration 30 min). All hearts (at least 6 at each time point) were then subjected to different periods of total global ischemia (5–35 min) and freeze clamped at different time intervals as indicated in 'Results'.

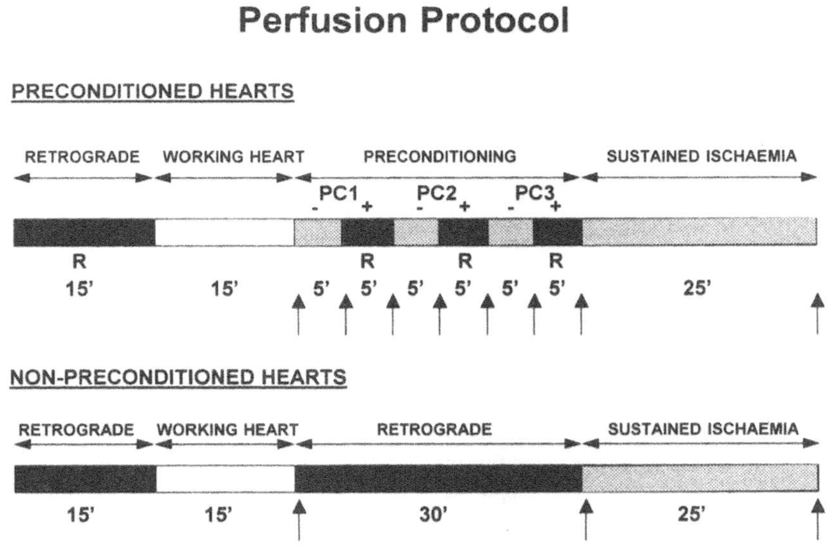

Fig. 1. Perfusion protocol. PC1–, PC2–, PC3–: Preconditioning for 5 min without reperfusion; PC1+, PC2+, PC3+: Preconditioning for 5 min with reperfusion. Arrows indicate time intervals during preconditioning protocol when hearts were freeze-clamped. Hearts were also freeze-clamped at different time intervals during sustained ischemia.

Biochemical analyses

Freeze-clamped hearts were stored in liquid nitrogen until extraction. Tissue cAMP (100–200 mg) was extracted with 1.2 ml 6% perchloric acid. Extracts were neutralized and the cAMP determined using Amersham's cAMP (^3H) assay system. This assay is based on the competition between unlabelled cAMP and a fixed quantity of (^3H) cAMP for binding to a protein with high affinity and specificity for cAMP. A standard curve ranging from 0.125–16 picomol cAMP was included in each assay. For cGMP analyses, 100 –200 mg of tissue was extracted with 1.2 ml 5% trichloro-acetic acid. The extracts were washed 4 times with 3 ml ether before analysis with Amersham's cGMP [^{125}I] assay system. This assay is based on the competition between unlabelled cGMP and a fixed quantity of [^{125}I] cGMP for a limited number of binding sites on a cGMP-specific anti-body. The concentration of unlabelled cGMP in the samples was determined by interpolation from a standard curve (2–128 fmol cGMP).

PDE assays were performed on freeze-dried material which was homogenized in water (50 μl/mg dry weight). Since the homogenate was concentrated, it contained endogenous cyclic nucleotides as well as Mg^{2+} in a concentration close to that found *in vivo*.

The PDE activity of the homogenate was determined using a radiotracer dilution method. The reaction was initiated by adding 100 μl of homogenate to 20 μl incubation mixture containing c[8–^{3}H]AMP (SA 962 GBq/mmol) and c[8–^{3}H]GMP (SA 363 GBq/mmol) (ca 200 000 cpm). After 10 min incubation (temperature 25°C) the reaction was terminated by addition of 20 μl 60% perchloric acid. After centrifugation (2 min, 12 000 g) the supernatant was separated by reversed phase HPLC (Purospher RP 18, 5 microspher, Merck) with on-line radiochemical detection of labelled nucleotides. The mobile phase consisted of (A) 100 mM ammoniumphosphate pH 5.5 and (B) Buffer A plus 6% acetonitrile. A linear gradient was used from 100% A to 100% B in 20 min (flow rate 0.5 ml/min). PDE activity was calculated as the percentage of cleavage products to total radioactivity (which was equivalent to the radioactive cyclic nucleotide initially present in the reaction mixture).

Statistics

All data are given as the mean and standard error of the mean (S.E.M.). Changes during the preconditioning protocol were analyzed using one way analysis of variance (ANOVA) and the Bonferroni correction applied. The rest of the data was analyzed by two way analysis or variance. Rank transformation of data was used for evaluation of cGMP and cAMP changes during sustained ischemia[10].

Results

Changes during preconditioning protocol

Hearts exposed to three episodes of 5 min global ischemia, showed repeated significant increases in myocardial content of cAMP and almost complete reversal by each reperfusion period.

Tissue cGMP was significantly increased at PC2– and PC3– but not at PC1–. With the exception of PC1, reperfusion caused a significant reduction in tissue cGMP, although not to baseline values (Fig. 2a).

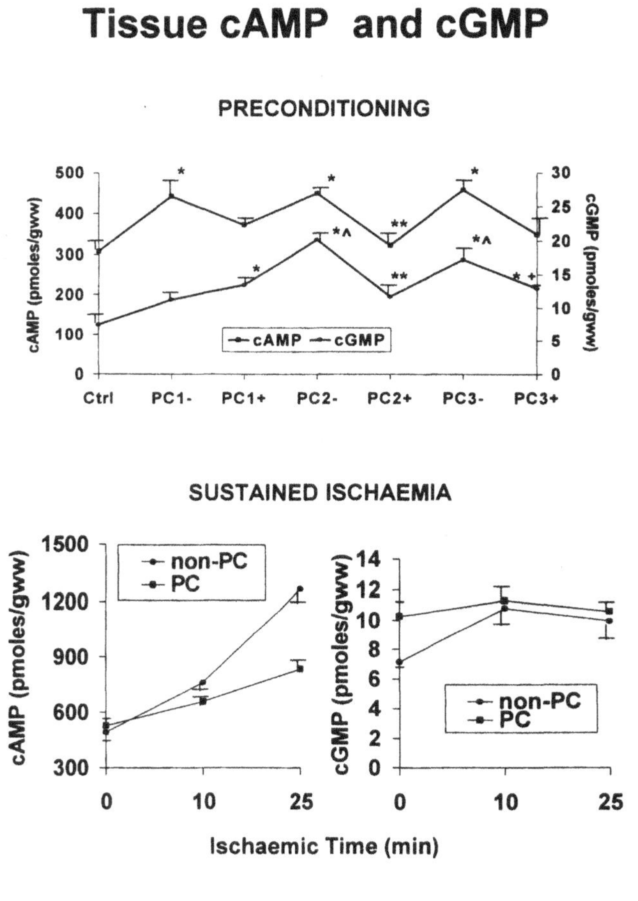

Fig. 2. (a) Tissue cAMP and cGMP during preconditioning. Controls refer to hearts freeze-clamped after 30 min control perfusion (i.e. immediately before onset of PC1–). Six hearts were studied at each time interval. *p < 0.05 vs. control; **p < 0.05 vs PC2–; ^p < 0.05 vs PC1–; +p < 0.05 vs. PC3–. (b) Tissue cAMP and cGMP during sustained ischemia. Zero-time hearts refer to hearts freeze-clamped immediately before onset of sustained ischemia. Six hearts were studied at each time interval. Two-way analysis of variance of cAMP showed significant interaction between time and groups (p = 0.002). Tissue cAMP was significantly lower in PC hearts at both 10 and 25 min ischemia (p = 0.0024 and 0.0001 respectively). Tissue cGMP of PC hearts was significantly higher than that of non-preconditioned hearts throughout sustained ischaemia (2-way ANOVA p = 0.0356).

The marked fluctuations in myocardial cAMP and cGMP levels during ischemia-reperfusion cycles were reflected by opposite changes in both cAMP- and cGMP-hydrolyzing PDE activities. At PC1–, both phosphodiesterases showed a significant reduction in their activities, followed by a return to baseline values during PC1+. Similar observations were made at the second and third preconditioning episodes, however, inhibition of PDEs was less than during PC 1– (Figs 3 and 4).

Changes during sustained ischemia

Tissue cAMP content at the onset of sustained ischemia was not significantly different in preconditioned and non-preconditioned hearts, while cGMP levels of preconditioned hearts were significantly higher than those of non-preconditioned hearts (10.3 ± 0. 9 vs. 7.2 ± 0.5 pmoles/g wet weight, $p = 0.0167$).

Two way analysis of variance of cAMP data showed a significant interaction between time and groups ($p = 0.002$). The increase in cAMP in both groups was significant compared to the values at the onset of sustained ischemia. However, tissue cAMP was significantly lower in PC hearts at both 10 and 25 min ischemia ($p = 0.0024$ and 0.0001 respectively).

The tissue cGMP content of PC hearts was significantly higher than that of nonpreconditioned hearts ($p = 0.0356$) throughout sustained ischemia, while the changes with time were marginally significant ($p = 0.0536$) (Fig. 2b).

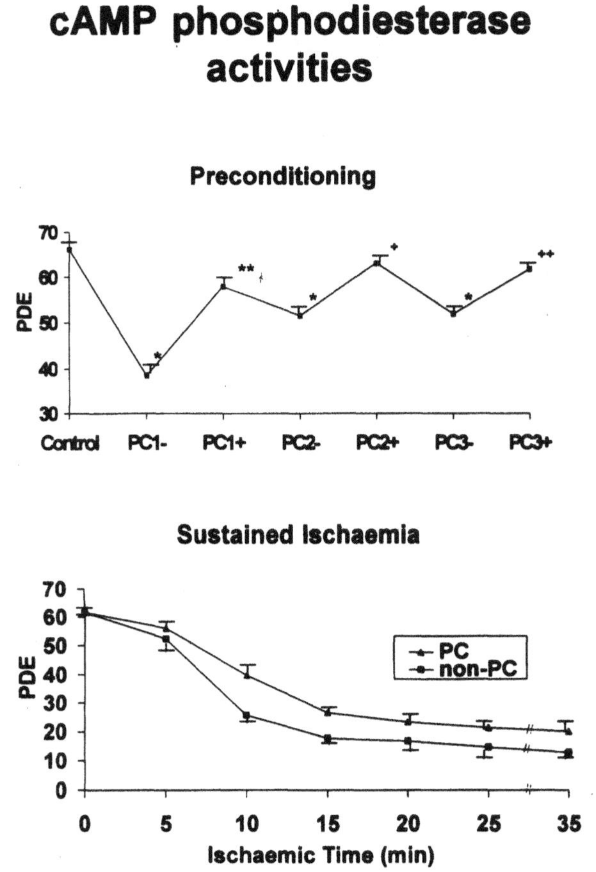

Fig. 3. cAMP-PDE activities during preconditioning and sustained ischemia (6 hearts/time interval). PDE activity is expressed as the percentage of cleavage products to total radioactivity (which is equivalent to the radioactive nucleotide initially present in the reaction mixture). Statistical analysis of data (ANOVA) during preconditioning. $^*p < 0.05$ vs. control; $^{**}p < 0.05$ vs PC1; $^+p < 0.05$ vs. PC2–; $^{++}p < 0.05$ vs. PC3–. Two way analysis of variance of PDE activities during sustained ischaemia, showed that they were significantly less inhibited in preconditioned than in non-preconditioned hearts ($p = 0.0001$).

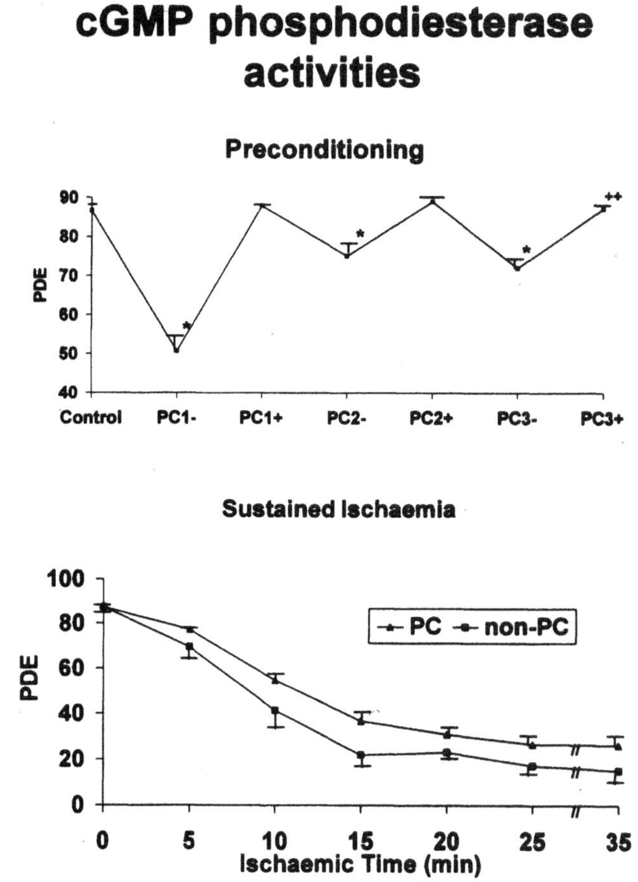

Fig. 4. cGMP-PDE activities during preconditioning and sustained ischemia (6 hearts/time interval). PDE activity is expressed as the percentage of cleavage products to total radioactivity (which is equivalent to the radioactive nucleotide initially present in the reaction mixture). Statistical analysis of data (ANOVA) during preconditioning. $^*p < 0.05$ vs. control; $^{++}p < 0.05$ vs PC3–. Two way analysis of variance of PDE activities during sustained ischaemia, showed that they were significantly less inhibited in preconditioned than in non-preconditioned hearts ($p = 0.0001$).

In both preconditioned and non-preconditioned hearts, cAMP- and cGMP-PDE activities declined with increasing periods of global ischemia. This inhibition was maximal after 15 min. Over the time period studied (5–35 min), both PDE activities were significantly less inhibited in preconditioned than in non-preconditioned hearts ($p = 0.0001$) (Figs 3 and 4).

Discussion

Changes during preconditioning

Exposure of the heart to three successive episodes of preconditioning is expected to elicit some of the many changes associated with ischemia and reperfusion. Although increases in tissue cAMP and cGMP during long periods of myocardial ischemia have been described previously [11, 13], this is the first description of the profiles of both these second messengers during repeated short episodes of ischemia and reperfusion as occurs during the preconditioning protocol (Figs 2–4).

The rapid rise in tissue cAMP during 5 min global ischemia may be due to several factors, for example, dual sensitization of the β-adrenergic system (activation of adenylyl cyclase and upregulation of the β-receptor population)[4], release of endogenous catecholamines [14] or inhibition of PDEs [15]. The latter may play an important role in the cyclic changes observed: The rise and fall of cAMP was associated with similar, but opposite fluctuations in cAMP- and cGMP-PDE activities.

cGMP levels showed similar cyclic changes during PC2 and PC3 as the cAMP levels. Activation of nitric oxide synthase (NOS), together with the significant inhibition of cGMP-PDE during PC2– and PC3– (Fig. 4) are possibly responsible for the rise in cGMP, whereas the rise in PDE activity and normalization of NOS activity during reperfusion may account for the reduction in cGMP during the reperfusion episodes. DePré *et al.* [16] have recently shown that activation of NOS (suggested to be the soluble fraction of the ENOS isoenzyme) occurs within 5 min of ischemia and returns to normal levels during reperfusion. The mechanism of the ischemia-induced activation of NOS is not known, but could occur secondary to increased $[Ca^{2+}]_2$[17]. Why a relatively small change in tissue cGMP levels was observed at PC1–, in the presence of marked inhibition of cGMP-PDE activity remains to be established. This rather modest change in cGMP may explain why one episode of preconditioning was less effective than three episodes in eliciting protection in the isolated perfused rat heart [18].

As far as we know, this is the first report on involvement of phosphodiesterases in preconditioning of the rat heart. It should be pointed out that PDE as measured in this study indicates the total cAMP and cGMP hydrolyzing capacity of the isoenzymes present in heart muscle. It can be seen that the PDEs of extracts prepared from non-ischemic tissue were active under conditions of the assay, decreasing during ischemia and recovering upon reperfusion. The rat myocardium is reported to contain at least 4 PDE isoenzymes [19] of which PDE 1 (Ca^{2+}–CaM stimulated) and PDE 2 (cGMP stimulated) hydrolyze cGMP, while PDEs 3 and 4 hydrolyze cAMP only. Selective inhibition of PDE 5 had no effect on cardiac function [20], suggesting that this isoform is not important in the myocardium. By offering cAMP or cGMP as substrate, one can estimate which isoforms may be involved.

Our findings suggest that inhibition of PDEs during ischemia is caused by small molecular inhibitors and is not due to phosphorylation or covalent modification of the enzyme moiety. Previous studies from the laboratory of Podzuweit *et al.* suggested that acidosis is a major determinant of PDE inhibition in ischemia [15]. This would also explain why PDE inhibition is readily reversed by reperfusion as hydrogen ions are washed out. Thus, it is likely that all PDE isoenzymes are inhibited in ischemia due to the developing acidosis.

The significance of these recurring changes in cAMP and cGMP, alone or in combination, in eliciting the protection conferred by preconditioning remains to be established. Simultaneous generation of these two cyclic nucleotides could lead to considerable crosstalk. For example, it has been shown that NO (and thus cGMP) attenuates β-receptor mediated responses [21], while inhibition of NOS activity could enhance the β-response to isoproterenol in myocytes by 30% [22].

The observation that transient β-adrenergic stimulation can precondition the isolated rat heart [23, 24] suggests that these changes in cAMP content, at least, may be important: It was suggested that the increased workload produced by β-stimulation may cause a degree of demand ischemia which, in turn, produced preconditioning, probably via attenuation of Ca^{2+} overload during reperfusion [23].

It is possible that cAMP or cGMP (or both) may act as a messenger or signal to elicit protection to subsequent ischemic damage. The signal however is readily abolished during preconditioning reperfusion, which makes it difficult to explain a role for them in continued protection during sustained ischemia. Elevation in these cyclic nucleotides may also directly or indirectly (via cross-talk) stimulate phosphorylation of intracellular proteins/structures or activate other pathways that may change the response of myocytes to ischemia.

Finally, although coincidence does not prove causality, the repeated elevation in both cAMP and cGMP during preconditioning, strongly suggests a causal role in eliciting protection. Exactly *how* this occurs, remains to be established.

Changes during sustained ischemia

As observed before [5, 6] preconditioning leads to a significant reduction in tissue cAMP during sustained ischemia. Tissue cGMP, on the other hand, was significantly higher in preconditioned hearts (Fig. 2b). It is not yet known whether these changes are merely the consequence or reflection of the preconditioning-induced protection or whether they are protective by themselves.

The attenuated β-adrenergic response during sustained ischemia may be due to diminished release of catecholamines [14], 'increase' in Gi[7], desensitization of the β-adrenergic receptor (unlikely according to Sandhu *et al.*) or less inhibited breakdown by cAMP- and cGMP-PDE, as shown in Figs 3 and 4. Since cGMP levels remained somewhat elevated despite the increased cGMP-PDE activity, its generation appears to override its breakdown. Depré *et al.* [12] showed that activation of NOS is maintained throughout sustained ischemia.

As discussed above, inhibition of PDE during ischemia may be caused, in part, by acidosis [15]. Therefore the higher pH observed in preconditioned hearts [25] probably accounts for the attenuation of inhibition of PDEs during sustained ischemia. The increased cGMP levels in preconditioned hearts may also contribute to activation of the cGMP-stimulated PDE [26]. However, the rise in cGMP also causes inhibition of PDE3 with little effect on total cAMP degrading capacity [27].

How important are these changes in conferring protection? The significance of the attenuated inhibition of PDE activity during sustained ischemia is unknown and should be determined by use of appropriate blockers. If attenuation of cAMP accumulation during sustained ischemia induces protection, then β-adrenergic blockade should elicit or β-stimulation abolish protection. We have previously shown that attenuation of β-adrenergic effects during ischemia by prior treatment with propranolol or reserpination elicited functional protection similar to that of preconditioning [6]. In contrast, increasing cAMP levels during sustained ischemia by administration of either isoproterenol [15] or forskolin [6] after preconditioning but before the onset of sustained ischemia, causing high cAMP levels throughout this period, failed to abolish preconditioning. Should cAMP be involved in preconditioning it probably acts during the preconditioning protocol, prior to sustained ischemia.

However, it is possible that cAMP acts in synergism with cGMP during sustained ischemia: these two cyclic nucleotides have opposing effects on the cardiac slow Ca^{2+} channel (cAMP stimulating and cGMP inhibiting [28]) . The lowering of tissue cAMP and elevation in cGMP may reduce Ca^{2+} influx during ischemia as well as during reperfusion. Elevation of endogenous cGMP has long been known to lower intracellular calcium [29]. As discussed above, NO (and thus cGMP) is an important endogenous inhibitory regulator of the actions of adrenergic agonists and may thus play a role in the attenuated β-adrenergic response during sustained ischemia.

In summary, the results obtained in the present study show that the preconditioning protocol was characterized by marked fluctuations in tissue cAMP and cGMP (partly caused by concomitant changes in cAMP- and cGMP-PDE activities). Sustained ischemia in preconditioned hearts was characterized by an attenuated β-adrenergic response and elevated tissue cGMP, associated with less PDE inhibition. The significance of these events remains to be established, but we suggest that both of these cyclic nucleotides are involved in the protection conferred by preconditioning.

Acknowledgements

This study was supported financially by the Harry Crossley Foundation, the University of Stellenbosch and the South African Medical Research Council.

References

1. Downey JM, Cohen MV: Mechanisms of preconditioning: Correlates and epiphenomena. In: M.M. Marber and D.M. Yellon (eds). Ischaemia: Preconditioning and adaptation. UCL Molecular Pathology Series, BIOS Scientific Publishers Limited, Oxford, UK, 1996, pp 21–34
2. Lawson CS, Downey JM: Preconditioning: state of the art myocardial protection. Cardiovasc Res 27: 542–550, 1993
3. Brooks G, Hearse DJ: Role of protein kinase C in ischemic preconditioning: Player or spectator. Circ Res 79: 627–630, 1996
4. Strasser RH, Krimmer J, Dullaeus-Braun R, Marquetant R, Kübler W: Dual sensitization of the adrenergic system in early myocardial ischaemia: Independent regulation of the β-adrenergic receptors and the adenylyl cyclase. J Mol Cell Cardiol 22: 1405–1423, 1990
5. Sandhu R, Thomas U, Diaz R, Wilson GJ: Effect of ischemic preconditioning of the myocardium on cAMP. Circ Res 78: 137–147, 1996
6. Moolman JA, Genade S, Tromp E, Lochner A: A comparison between ischaemic preconditioning and anti-adrenergic interventions; cAMP, energy metabolism and functional recovery. Bas Res Cardiol 91: 219–233, 1996
7. Niroomand R, Weinbrenner C, Weis A, Bangert M, Schwenke C, Marquetant R, Beyer T, Strasser RH, Kübler W, Rauch B: Impaired function of inhibitory G-proteins during acute myocardial ischaemia of canine hearts and its reversal during reperfusion and a second period of ischemia. Circ Res 76: 861–870, 1995
8. Iwase T, Murakami T, Tomita T, Miki S, Nagai K, Sasayamas: Ischemic preconditioning is associated with a delay in ischemia-induced reduction of β-adrenergic signal transduction in rabbit hearts. Circulation 88: 2827–2837, 1993
9. Parrat JR: Possibilities for the pharmacological exploitation of ischaemic preconditioning. J Mol Cell Cardiol 27: 991–1000, 1995
10. Convover WJ, Iman RL: Rank transformations as a bridge between parametric and nonparametric statistics. American Statistician 36: 124–133, 1981

11. Krause EG, Ziegelhöffer A, Fedelsova M, Styk J, Kostolanski S, Gabauer I, Blasig I, Wollenberger A: Myocardial cyclic nucleotide levels following coronary artery ligation. Adv Cardiol 25: 119–129, 1978
12. Dupré C, Hue L: Cyclic GMP in the perfused rat heart. Effect of ischaemia, anoxia and nitric oxide synthase inhibitor. FEBS Lett 345: 241–245, 1994
13. Podzuweit T, Muller A, Thomas S, Bader R: Ischaemia induced inhibition of cyclic nucleotide phosphodiesterases in the pig heart. Cause of cAMP accumulation and arrhythmias? Circulation 92: 0641, (Abstract), 1995
14. Seyfarth M, Richard G, Mizsnyak A, Kurz T, Schömig A: Transient ischaemia reduces norepinephrine release during sustained ischaemia. Circ Res 78: 573–580, 1996
15. Podzuweit T, Thomas S, Binz K, Moller A: Ischaemia and acidosis cause inhibition of cyclic nucleotide phosphodiesterases in pig heart. J Mol Cell Cardiol 28: 115, (Abstract), 1996
16. Depré C, Fiérain L, Hue L: Activation of nitric oxide synthase by ischaemia in the perfused heart. Cardiovasc Res 33: 82–87, 1997
17. Kelly RA, Balligand J-L, Smith TW: Nitric oxide and cardiac function. Circ Res 79: 363–380, 1996
18. Liu YL, Downey JM: Ischaemic preconditioning protects against infarction in rat heart. Am J Physiol 263: H1107–H1112, 1992
19. Shahid M, Nicholson CD: Comparison of cyclic nucleotide phosphodiesterase isoenzymes in rat and rabbit ventricular myocardium: Positive inotropic and phosphodiesterase inhibitory effects of Org 30029, milrinone and rolipram. Naunyn Schmiedebergs Arch Pharmacol 342: 698–705, 1990
20. Cohen AH, Hanson K, Morris K, Fouty B, McMurty IF, Clarke W, Rodman DM: Inhibition of cyclic 3′5′-guanosine monophosphate-specific phosphodiesterase selectively vasodilates the pulmonary circulation in chronically hypoxic rats. J Clin Invest 97: 172–179, 1996
21. Ebihara Y, Karmazyn M: Inhibition of β- but not α-mediated adrenergic responses in isolated hearts and cardiomyocytes by nitric oxide and 8-bromo-cyclic GMP. Cardiovasc Res 32: 622–629, 1996
22. Balligand I-L, Kelly RA, Marsden PA, Smith TW, Michel T: Control of cardiac muscle cell function by an endogenous nitric oxide signalling system. Proc Natl Acad Sci 90: 347–351, 1993
23. Asimakis GK, Inners-McBride K, Conti VR, Yang C: Transient β-adrenergic stimulation can precondition the rat heart against post-ischaemic contractile dysfunction. Cardiovasc Res 28: 1726–1734, 1994
24. Miyawaki H, Ashraf M: Isoproterenol mimics calcium preconditioning-induced protection against ischemia. Am J Physiol 272: H927–H936, 1997
25. Steenbergen C, Perlman ME, London RE, Murphy E: Mechanism of preconditioning. Ionic alterations. Circ Res 72: 112–125, 1993
26. Hardman JG, Beavo JA, Gray JP, Chrisman TD, Patterson WD, Sutherland EW: The formation and metabolism of cyclic GMP. Ann NY Acad Sci 185: 27–35, 1971
27. Beavo JA, Reifsnyder DH: Primary sequence of cyclic nucleotide phosphodiesterase isozymes and the design of selective inhibitors. Trends Pharmacol Sci 11: 150–155, 1990
28. Sperelakis N. Regulation of calcium slow channels of heart by cyclic nucleotides and effects of ischaemia. Adv Pharmacol 31: 1–24, 1994
29. Lincoln TM: Cyclic GMP and mechanism of vasodilation. Pharmacol Ther 41: 479–502, 1989

Molecular and Cellular Biochemistry **186:** 177–184, 1998.
© 1998 *Kluwer Academic Publishers.*

Myocardial protection during ischemia and reperfusion

David J. Hearse

Director of Cardiovascular Research, The Rayne Institute, St Thomas' Hospital, London, UK

Abstract

The objectives of this article are to: (i) discuss the origins and the nature of ischemic injury, (ii) identify factors influencing the evolution of injury, (iii) consider various cellular targets for ischemic injury, (iv) assess the overall importance of 'reperfusion injury', (v) discuss conceptual approaches to cardioprotection and (vi) to identify new ideas and approaches in the realm of myocardial protection. In the human heart, myocardial ischemia may take many forms, it may exist for periods as short as a few seconds or minutes, it may last for hours or it may persist for years. In terms of discussing interventions to combat myocardial ischemia, this article will focus on: (i) *regional* ischemia as occurs during evolving myocardial infarction and (ii) whole heart or *global* ischemia as occurs during cardiac surgery and transplantation. (Mol Cell Biochem **186:** 177–184, 1998)

Key words: myocardial infarction, ischemia, reperfusion injury, cardioplegic solutions, ischemic preconditioning

The complexity of ischemic injury

Understanding and successfully manipulating myocardial ischemic injury requires a consideration of the temporal, spatial, cellular and molecular characteristics of ischemia.

The temporal characteristics of myocardial ischemia and the need for early reperfusion

Myocardial ischemia initiates a continuum of progressively more severe cellular changes that, unless interrupted by early reperfusion, inevitably culminate in cell death and tissue necrosis (Fig. 1). Following the total or partial occlusion of a coronary artery, metabolic and functional changes are initiated within seconds. These changes, which become progressively more severe with time, are initially of a reversible nature. However, with increasing durations of ischemia, they become more severe and eventually irreversible injury occurs. From this point on, the cells are destined to tissue necrosis. From this it follows that, the longer the duration of elapsed ischemia, the fewer will be the number of reversibly injured cells and the smaller will be the amount of tissue potentially available for salvage by reperfusion and return to normal contractile function.

Time is of the essence – successful and complete salvage of ischemic tissue *can only be achieved by early reperfusion* at a time when tissue injury is restricted to the reversible phase. Under such conditions (depending upon the duration and severity of preceding ischemia), cardiac function norm-

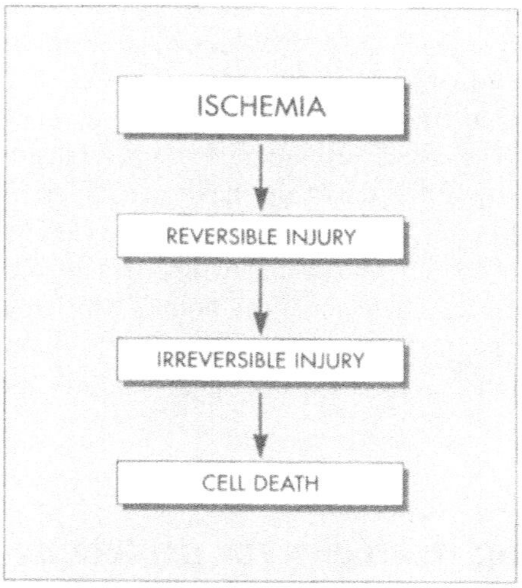

Fig. 1. Reversible and irreversible injury: myocardial ischemia initiates a sequence of events that, unless interrupted by early reperfusion, lead to cell death and tissue necrosis.

Address for offprints: Professor David J. Hearse DSc, Director of Cardiovascular Research, The Rayne Institute, St Thomas' Hospital, London SE1 7EH, UK

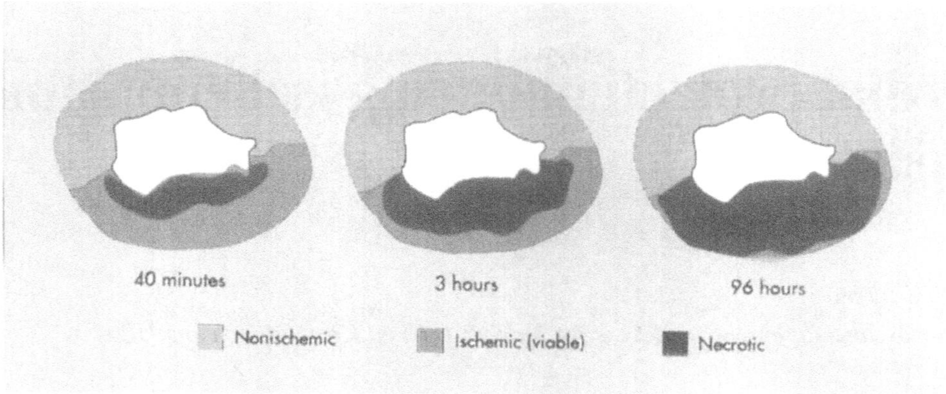

Fig. 2. The wavefront of cell death: irreversible ischemic injury usually originates in the endocardium and migrates with time towards the epicardial surface. Redrawn from [1].

ally returns within a relatively short time. With longer durations of ischemia, recovery may take some time (myocardial 'stunning') but, so long as the injury is still reversible, a full recovery will eventually occur. Reperfusion of irreversibly injured tissue, by definition, cannot evoke a recovery of cardiac function and may even be detrimental (see 'reperfusion injury' below).

The spatial characteristics of ischemic injury and the anatomy of cell death

Within the heart, ischemic injury does not evolve in a uniform manner. Regional differences in metabolism and energy requirements render the endocardium most vulnerable to injury. For this reason, myocardial injury and tissue necrosis usually originate in the endocardium and, with time, migrate as a 'wave front of cell death' towards to the epicardial surface. Thus, the classical dog heart study of Reimer and Jennings [1] charted the temporal evolution of necrotic injury transmurally across the canine heart over a period of 40 min to 96 h of regional ischemia (Fig. 2). This study illustrated beautifully not only the spatial nature of ischemic injury but also its temporal characteristics – both of which must be taken fully into account in the application and assessment of any intervention designed to beneficially modify the ischemic process.

Defining strategies for protecting the ischemic heart

It is apparent from the preceding sections that there are two key objectives that must be implemented in any attempt to prevent the deterioration of ischemic injury to tissue necrosis. Firstly, since *early reperfusion is an absolute prerequisite for the survival of a ischemic tissue,* every attempt must be made to reverse ischemic injury by reperfusing the affected tissue at the earliest possible opportunity. Secondly, if early reperfusion is not possible, attempts should be made to *slow the rate of the evolution of ischemic injury* so as to delay, for as long as possible, the onset of irreversible injury. In this way, more cells will remain in a reversible state of injury for a longer period of time and more tissue will be able to recover when reperfusion is eventually achieved.

In the case of regional ischemia and evolving myocardial infarction, reperfusion may occasionally occur spontaneously, however, most reperfusion requires intervention using thrombolytic or surgical procedures. Such procedures may take considerable time to implement and this underlines the potential importance of slowing the rate of development of ischemic injury until reperfusion can be affected. With global ischemia in cardiac surgery, reperfusion can be instituted at will, however, the extended ischemic times required for complex surgical procedures requires that injury-slowing procedures are applied to prevent infarction during the course of surgery. As will be discussed later, the use of interventions to slow the rate of development of ischemic injury has been highly successful in open heart surgery and cardiac transplantation. Developing such interventions to successfully manipulate the rate of ischemic injury requires a detailed knowledge of the many factors that influence the evolution of injury.

Factors influencing the rate of evolution of ischemic injury

There are a host of complex, interacting factors which are known to influence the rate at which tissue injury occurs in the ischemic heart. These factors conspire together to result

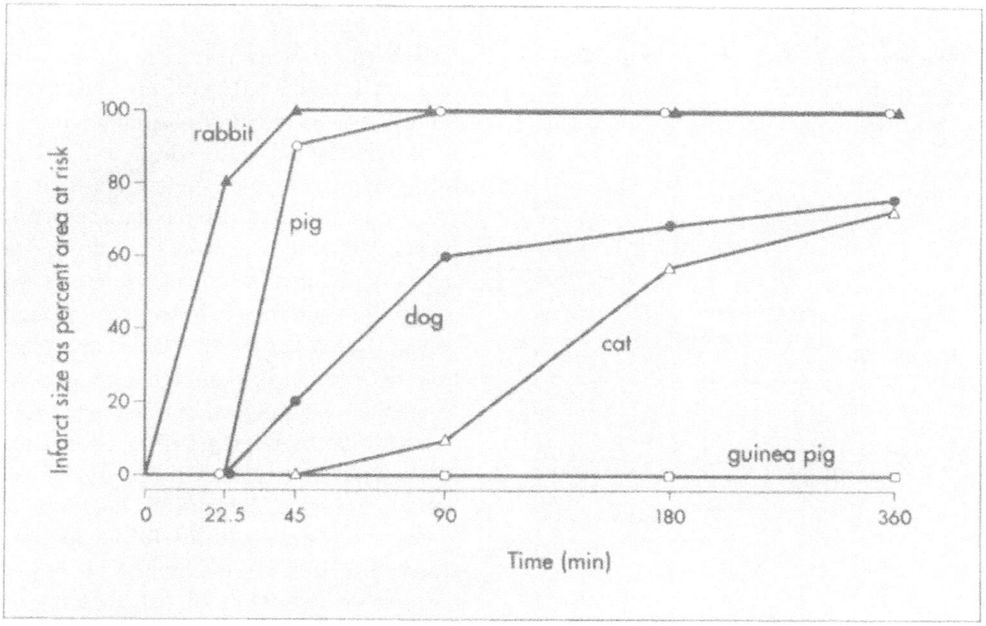

Fig. 3. Species differences in the evolution of ischemic injury: the effect of differing degrees of collateral flow on the rate of development of necrosis. Redrawn from [2].

in highly variable rates of injury both within and between hearts – even in the face of an identical ischemic insult. Some of the major factors influencing the rate of evolution of injury are: (i) the extent (if any) of residual coronary or non-coronary collateral flow, (ii) various diseases that may coexist with the ischemia such as hypertrophy, diabetes and hypertension, (iii) heart rate, metabolic rate and tissue temperature, (iv) variations in the metabolic response to ischemia (particularly in relation to patterns of substrate metabolism), (v) nutritional and hormonal status, (vi) the presence of a variety of cardioactive drugs, (vii) previous ischemia and various adaptive processes and (viii) the age, sex and species of the tissue. Variable combinations of all of the above factors means that *there is no universal time after which ischemic tissue becomes irreversibly injured.* This important generalisation, together with the vital role played by collateral flow in determining the rate of evolution of ischemic injury, is beautifully illustrated in the classical study by Schaper *et al.* [2] which shows (Fig. 3) how the evolution of necrosis within an area of regional ischemia occurs extremely rapidly in animals that have little or no collateral flow (such as the rat and rabbit) whilst it occurs slowly in animals that have collaterals (such as the dog). In animals such as the guinea pig, necrosis does not occur because the heart is totally collateralised. Thus, 90 min following the occlusion of a coronary artery the extent of necrosis within the zone of ischemia may range between 0 and 100%. However, despite this, the curves in Fig. 3 still bear testimony to the need for very early reperfusion and/or intervention to prevent the progression of ischemic injury to necrosis.

Developing interventions to manipulate the rate of ischemic injury prior to reperfusion

When early reperfusion is not possible, attempts should be made to slow the rate of ischemic injury using interventions that are not reliant upon the restitution of flow. Successful development of such interventions requires a detailed knowledge of the molecular characteristics of the cascade of metabolic and cellular changes that are induced by ischemia. Armed with this knowledge, it should be possible to identify key detrimental changes that might act as targets for selective intervention. Figure 4 shows just a few of the multitude of changes that occur following the onset of myocardial ischemia [3]. The nature and sequence of these changes may vary greatly from heart to heart and the figure is only intended to convey the complexity, temporal nature and increasing severity of the ischemic process. In order to successfully slow such a process it is necessary to target metabolic changes that are occurring during the early (reversible) phase of ischemia and such an approach has been subject of many studies aimed ameliorating the potentially devastating consequences of ischemia.

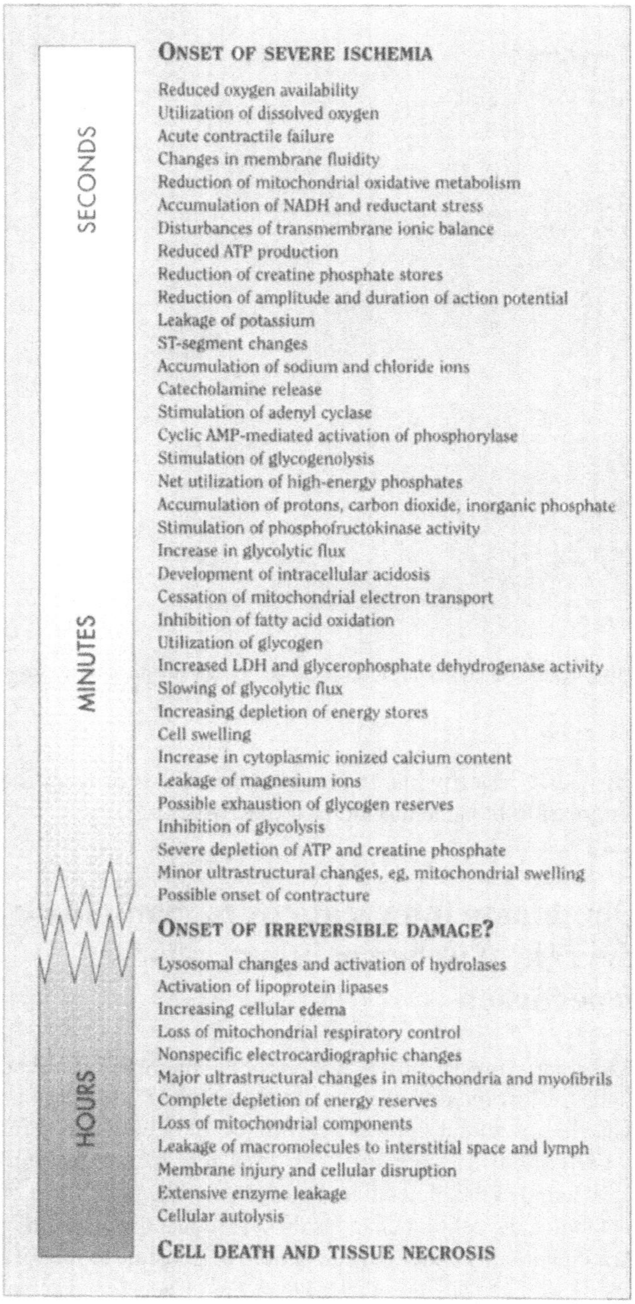

Fig. 4. An illustration of some of the metabolic changes induced by ischemia: a vast number of cellular and metabolic changes occur, these are initiated within seconds, minutes or hours of the onset of ischemia. The nature and sequence of the changes are highly variable. Redrawn from [3].

Molecular targets for anti-ischemic intervention

There have been many studies with hundreds of different interventions, all of which have been suggested to possess the ability to slow the progression of myocardial ischemic injury. In general one or more of the following approaches have been adopted: (i) limiting the accumulation of protons and toxic metabolites during the ischemic process, (ii) modifying the extent of nutrient and energy depletion, (iii) manipulating cellular sodium and calcium overload, (iv) modifying the regulation of various enzymes and proteins, (v) combating loss of potassium, co-factors and trace elements, (vi) limiting damage to nuclear material, (vii) combating the activation of lytic enzymes and attack by leukocytes and, (viii) countering cell swelling and membrane disruption. Thus, using a variety of species and experimental models, both *in vivo* and *in vitro*, the administration of agents such as calcium antagonists, buffers, osmotic agents, substrates, extracellular high energy phosphates and magnesium have all been shown to reduce the degree of ischemic injury and, upon reperfusion, enhance the recovery of function. Investigators who have made the effort to undertake dose-response studies for such interventions have reaped rewards in that optimal concentrations of a single intervention can often be shown to confer a remarkable degree of protection. Figure 5 shows the dose-response relationship for the ability of various concentrations of magnesium to enhance the post-ischemic recovery of contractile function in the rat heart after a fixed period of ischemia [4]. From this figure it is clear that, simply by increasing the extracellular magnesium concentration immediately before ischemia from its normal value of approximately 1.2 to 15.0, results in a tripling of post-ischemic functional recovery. It is also noteworthy that, like so many anti-ischemic interventions, excessive concentrations can lead to a loss of protection. As will be discussed

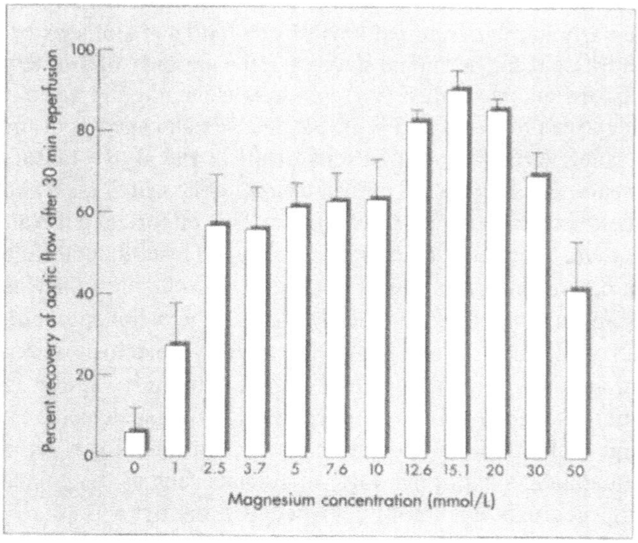

Fig. 5. Dose-response relationship for the ability of extracellular magnesium to reduce ischemic injury and enhance post-ischemic recovery. Redrawn from [4].

later, the concept of injury slowing by anti-ischemic interventions has been very successfully applied to cardiac preservation during open heart surgery and transplantation.

Cellular targets for injury

Cardiologists might be criticised for being preoccupied with protecting the myocyte (possibly as consequences of the ease of measuring the biological and physiological activity of this cell type) whilst ignoring the consequence of ischemia on other cellular constituents of the heart. However, for understanding and effectively controlling ischemic injury it is vital to remember that the heart is composed of a multitude of interacting cell types in addition to the myocyte. These include cells of the endothelium, vascular smooth muscle, conducting tissue, connective tissue, mast cells and even the blood-borne elements that are carried through the heart. Each of these would be expected to be vulnerable to ischemic injury and therefore, the rational development of any anti-ischemic intervention should ideally attempt to target and protect as many of these cell types as possible. Unfortunately, this obvious requirement has rarely been considered, furthermore, little attention is given to the real possibility that interventions that confer cellular protection to one cell type (for example the ability of potassium to protect the myocyte) might, in high concentrations, prove detrimental to other cell types (such as the endothelium).

Recently, there has been increasing attention focused on the possibility that the microcirculation (and in particular the endothelium) might be particularly vulnerable to injury during ischemia and reperfusion [5]. Indeed, the possibility is now emerging that the microcirculation might be the primary determinant of the survival of the heart as a whole. Realisation of such a possibility should open many new avenues in the protection of the heart against ischemic injury.

Is it insufficient to consider only ischemia? Cellular events during reperfusion

Despite many thousand of experiments and a multitude of putative anti-ischemic interventions, few (if any) interventions facilitate a complete recovery from ischemic injury. This has lead investigators to question whether the failure of heart to recover during reperfusion necessarily only reflects the severity of the ischemic injury. This concern has been supported by an increasing number of experimental studies that have indicated that, interventions given *only at the time of reperfusion*, may improve ischemic recovery. A survey of the literature reveals that there are now many interventions that appear to be able to improve the rate or extent of post-ischemic function recovery (or limit the size of an infarct measured at the end of reperfusion) when administered only at the time of reperfusion.

Successful reperfusion interventions include: (i) the modification of extracellular potassium, calcium, sodium or magnesium, (ii) the manipulation of substrate availability and utilisation with agents such as glutamate, aspartate or pyruvate, (iii) the control of free radicals and oxidant stress with agents such as superoxide dismutase, catalase, vitamin E, vitamin C and allopurinol, (iv) the manipulation of physical factors such as oxygen tension, temperature, osmolarity and pH, (v) the control of blood-borne elements such as leukocytes and platelets and (iv) the administration of vasoactive agents such as adenosine, nitrates and dipyridamol as a means of enhancing coronary flow and its distribution. Success in improving the outcome of reperfusion with interventions administered only at the time of reperfusion has lead investigators to suggest that some component of reperfusion injury may be damaging and, as such, subtract from the obvious benefit of the restoration of flow.

Reperfusion-induced injury

Although an absolute pre-requisite for tissue survival, the possibility exists that reperfusion may increase injury *over and above that sustained during ischemia*. Certainly, there is convincing evidence that reperfusion is able to precipitate a number of unfavourable cellular changes and the term 'reperfusion injury' has fallen into common usage [6, 7] However, the existence and importance of reperfusion injury (and hence the need for intervention) has become highly controversial with some investigators denying its existence whilst other claim that it as an important determinant of tissue survival. Some of the confusion and controversy relating to the to the existence and importance of reperfusion injury can be resolved if the phenomenon is divided into *reversible* reperfusion injury versus *irreversible* (or lethal) reperfusion injury.

Reversible reperfusion injury is defined as transient (but nonetheless potentially detrimental) injury, induced by the act of reperfusion, from which the heart can eventually recover. A classic example of such injury is myocardial stunning [8, 9] This is the prolonged contractile dysfunction that persists after reperfusion despite the absence of necrosis. It may take several hours or days for the heart to recover from this contractile deficit but, so long as sufficient time is allowed, a full recovery will eventually occur. Figure 6 shows a typical example of the extent and duration of stunning in a dog heart following a brief period of regional ischemia. It is evident that, despite the short duration of ischemia (15 min), several days are required for a full recovery of function.

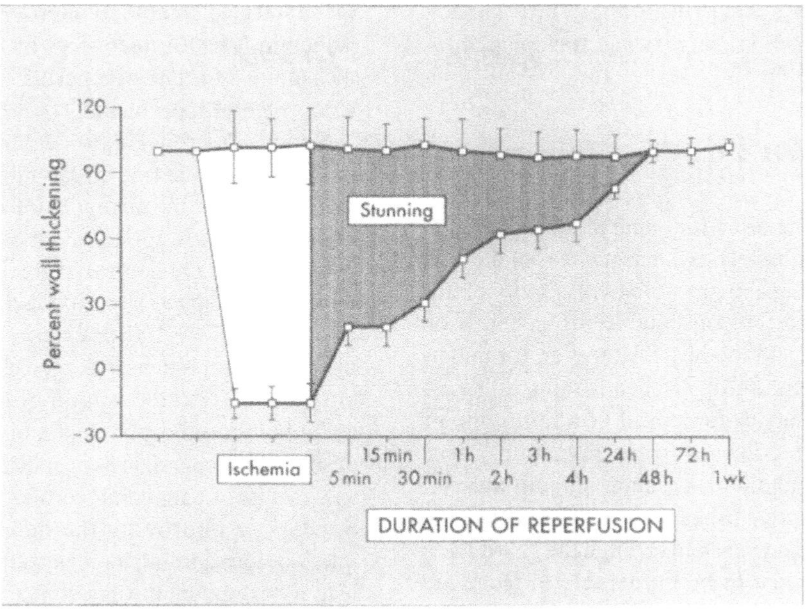

Fig. 6. Myocardial stunning: a brief period (15 min) of regional ischemia in the dog heart can cause reversible contractile dysfunction from which the heart may take several days to recover. Redrawn from [9].

Although reversible, myocardial stunning represents a significant clinical problem which may be a factor in precipitating left ventricular failure with its attendant morbidity and mortality. It is thought to be particularly important during reperfusion in cardiac surgery and transplantation, it may also occur following angina, successful thrombolysis and exercise-induced ischemia. As a consequence, there has been much interest in the mechanism(s) underlying stunning and in interventions that might attenuate the phenomenon.

A variety of possible mechanisms have been suggested to be responsible for the myocardial stunning these include: (i) an impaired resynthesis of adenosine triphosphate, (ii) functional sympathetic denervation, (iii) heterogeneous reperfusion, (iv) loss of creatine kinase activity, (v) damage to myocardial collagen, (vi) free radicals and oxidant stress, (vii) perturbations of calcium homeostasis and (viii) modification of contractile protein calcium sensitivity. It is now becoming clear the last two of these mechanisms (possibly working in concert) are most likely to be the most important contributors to myocardial stunning [10]. The contribution of free radicals and oxidant stress to myocardial stunning and the fact that injury is occurring during reperfusion are clearly illustrated in the important study by Bolli and colleagues [11] in which an antioxidant intervention was given to the dog heart *only at the time of reperfusion* and this resulted in a dramatic attenuation of stunning.

A second example of reversible reperfusion injury is the transient burst of various arrhythmias which frequently occur during the early moments of reperfusion. These arrhythmias may include ventricular fibrillation (a complication which is frequently encountered during reperfusion in cardiac surgery). Although potentially lethal, these arrhythmias are ‘reversible’ in the sense that they often terminate spontaneously, can be electrically converted or can be prevented with drugs. Again, as with myocardial stunning, there have been many studies of the mechanism(s) underlying the genesis of arrhythmias during ischemia and reperfusion. In the case of reperfusion arrhythmias, it would appear that the most likely mechanisms, as with stunning, involve free radical-induced oxidant stress and (possibly as a consequence) disturbances of ionic homeostasis [10]. As with stunning, reperfusion arrhythmias can be effectively prevented through the administration of a variety of antioxidant interventions or through the manipulation of ion composition or pH during the early minutes of reperfusion.

Lethal reperfusion injury: Does it exist?

The existence of this irreversible form of reperfusion injury (which some believe to be capable of killing cells that were viable in the moments before reperfusion) is particularly controversial. This author is of the opinion that, at the present time, there is little or no conclusive evidence for the existence of this particular form of reperfusion injury [6]. This is not to deny its existence but simply to highlight the fact that definitive experiments to demonstrate its existence and quantitative importance have yet to be reported.

Defining the goals and concepts for cardiac protection during ischemia and reperfusion

From the preceding sections it should be apparent that the primary goals for achieving cardiac protection during ischemia must be: (i) restoration of flow to the tissue as soon as possible, (ii) using interventions to slow the rate of development of ischemic injury until such a time as reperfusion can be affected and (iii) minimising the severity of reperfusion injury so as to facilitate as rapid a recovery of function as possible.

A number of conceptual approaches can be identified to help facilitate the achievement of these goals. These include: (i) the application of anti-ischemic agents to slow tissue injury, (ii) the use of reperfusion interventions to accelerate recovery from reperfusion injury and (iii) the exploitation of endogenous (adaptive) responses.

The use of interventions to slow the evolution of ischemic injury

A large portfolio of potential anti-ischemic interventions (designed to target various deleterious ischemia-induced cellular changes) has now been established. Although, when used on their own, many interventions have been reported to significantly reduce ischemic injury, it is perhaps naive to expect a single intervention to achieve extensive or long lasting protection during severe ischemia. For this reason, a poly-pharmacological approach to anti-ischemic therapy would seem appropriate. Whilst, for a variety of reasons, this has rarely been attempted in the context of regional ischemia and evolving myocardial infarction, poly-pharmacology has formed the cornerstone of the very successful development of anti-ischemic cardioplegic interventions for use during cardiac surgery and transplantation. As reviewed elsewhere [12], three basic approaches have been combined and exploited by surgeons so to greatly enhance the ability of human heart to withstand extended periods of severe global ischemia. This has involved: (i) the use of cardioplegic agents such as the potassium to rapidly arrest the heart during ischemia and thereby conserve energy for tissue protection and subsequent recovery, (ii) the inclusion in cardioplegic solutions of agents such as buffers, substrates and drugs to combat as many deleterious ischemia-induced changes as possible and (iii) the combination of chemical cardioplegia with hypothermia so as to further slow the rate of cellular reactions that contribute to ischemia-induced injury. As mentioned above, this has been particularly successful both in the context of cardiac surgery and cardiac transplantation where human hearts, which would fail to recover after less than one hour of unmodified global ischemia, now routinely withstand up to 4 h of a ischemic arrest with good post-ischemic functional recovery.

Using interventions to minimise reperfusion injury and enhance the rate of post-ischemic recovery

As mentioned previously (particularly in the realm of cardiac surgery) it would appear to be appropriate to advocate the use of 'reperfusion solutions' to ameliorate any unfavourable components of the reperfusion process. In this way, it should be possible to use agents such as: (i) anti-oxidants to attenuate stunning – an approach which would seem preferable to overriding stunning through the administration of inotropic agents such as calcium or catecholamines with their attendant risk of increasing heart rate and arrhythmogenesis, (ii) vasoactive drugs to improve the extent and quality of reperfusion by manipulating vascular responses during reperfusion and (iii) ion and pH regulation to promote the rapid reestablishment of ionic homeostasis.

Exploiting endogenous responses: an additional weapon for cardioprotection?

The final, and perhaps most challenging, conceptual approach to protecting the myocardium during ischemia and reperfusion, arises from the recognition that the heart has an inherent ability to protect itself against the ravages of ischemia. This protection can be achieved by exploiting a variety of endogenous adaptive responses, two of which are currently attracting considerable attention [13, 14]. These are: (i) the ability of the heart to express 'stress proteins' and (ii) the phenomenon of ischemic preconditioning. Although not the subject of this article, it is now known that the heart responds to a variety of stresses (for example transient heat shock) by the expression of a spectrum of stress response proteins. These proteins, which are maximally expressed some 24 h after the initial stress, have been shown to be responsible for a substantial enhancement of the ability of the heart to resist ischemic injury. The nature of the mechanisms underlying this powerful adaptive response are currently under investigation and might lead to the development of mimetics capable of affording additional protection to the ischemic heart. Likewise, the phenomenon of ischemic preconditioning, is an adaptive response whereby the resistance of the heart to injury during ischemia and reperfusion can be *temporarily* enhanced by prior (less than 2 h) exposure to one or more *brief* (less than 5 min) episodes of ischemia and reperfusion. Myocardial preconditioning has been shown to be able to protect against tissue necrosis (measured as a limitation of infarct size following reperfusion), it has been show to be capable of improving post-ischemic recovery of contractile function (probably as a secondary consequence to the reduction of infarct size) and, in some species, it has also been shown to exert a profound protective effect against both ischemia- and reperfusion-induced arrhythmias. The brief periods of ischemia that precondition the heart can be substituted by a number of pharmacological and biochemical mimetics. Thus, brief exposure to adenosine, norepinephrine, bradykinin and a variety of other receptor agonists can produce striking protection and such studies are greatly aiding in the resolution of the mechanisms underlying this important adaptive phenomenon. Once the mechanism of preconditioning is definitively established, it may be possible to develop pharmacological agents that exploit the adaptive response for sustained periods of time. However, the extent to which such interventions (or indeed ischemic preconditioning itself) can be additive to other forms of cardioprotection remains to be resolved.

References

1. Reimer KA, Jennings RB: The 'wavefront phenomenon' of myocardial ischemic death: II Transmural progression of necrosis within the framework of ischemic bed size (myocardium at risk) and collateral flow. Lab Invest 40: 633–644, 1979
2. Schaper W: Experimental infarcts and the microcirculation. In: DJ Hearse, DM Yellon (eds.). Therapeutic approaches to myocardial infarct size limitation. Raven Press, New York, 1984, pp 79–90
3. Hearse DJ: Cellular damage during myocardial ischemia: metabolic changes leading to enzyme leakage. In: DJ Hearse, J de Leiris (eds.). Enzymes in cardiology: diagnosis and research. John Wiley, London, 1979, pp 1–19
4. Hearse DJ, Stewart DA, Braimbridge MV: Myocardial protection during ischemic cardiac arrest: the importance of magnesium in cardioplegic infusates. J Thoracic Cardiovas Surg 75: 877–885, 1978
5. Hearse DJ, Maxwell L, Saldanha C, Gavin JB: The myocardial vasculature during ischemia and reperfusion: a target for injury and protection. J Mol Cell Cardiol 25: 759–800, 1993
6. Hearse DJ: Ischemia, reperfusion and the determinants of tissue injury. Cardiovas Drugs Therapy 4: 767–776, 1991
7. Hearse DJ: Myocardial injury during ischemia and reperfusion: concepts and controversies. In: DM Yellon, RB Jennings (eds.). Myocardial Protection: the Pathophysiology of Reperfusion and Reperfusion Injury. Raven Press, New York, 1992, pp 13–32
8. Bolli R: Mechanism of myocardial stunning. Circulation 82: 723–738, 1990
9. Hearse DJ, Bolli R: Reperfusion-induced injury: manifestations, mechanisms and clinical relevance. Cardiovas Res 26: 101–108, 1992
10. Hearse DJ: Stunning: a radical re-view. Cardiovas Drugs Therapy 5: 853–876, 1991
11. Bolli R, Patel BS, Jeroudi MO, Lai EK, McCay PB: Demonstration of free radical generation in stunned myocardium of intact dogs with the use of the spin trap α-phenyl N-tert-butyl nitrone. J Clin Invest 82: 476–485, 1988
12. Hearse DJ, Braimbridge MV, Jynge P: Protection of the ischemic myocardium: cardioplegia. Raven Press, New York, 1981
13. Yellon DM, Latchman DS: Stress proteins and myocardial protection during ischemia and reperfusion. In: DM Yellon, RB Jennings (eds.). Myocardial Protection: the Pathophysiology of Reperfusion and Reperfusion Injury. Raven Press, New York, 1992, pp 185–195
14. Przyklenk K, Kloner RA, Yellon DM: Ischemic Preconditioning: the concept of endogenous cardioprotection. Kluwer Academic Publishers, Boston, 1994

Molecular and Cellular Biochemistry **186:** 185–193, 1998.
© 1998 *Kluwer Academic Publishers.*

Features of short-term myocardial hibernation

Gerd Heusch and Rainer Schulz
Department of Pathophysiology, Centre of Internal Medicine, University of Essen School of Medicine, Essen, Germany

Abstract

When severe ischemia, such as that resulting from a sudden and complete coronary artery occlusion, is prolonged for more than 20–40 min, myocardial infarction develops, and there is irreversible loss of contractile function. When myocardial ischemia is less severe but nevertheless prolonged, the myocardium is dysfunctional but can remain viable. In such ischemic and dysfunctional myocardium, contractile function is reduced in proportion to the reduction in regional myocardial blood flow; i.e. a state of 'perfusion-contraction matching' exists. The metabolic status of such myocardium improves over the first few hours, as myocardial lactate production is attenuated and creatine phosphate, after an initial reduction, returns towards control values. Ischemic myocardium, characterized by perfusion-contraction matching, metabolic recovery and lack of necrosis, has been termed 'short-term hibernating myocardium'. Short-term hibernating myocardium can respond to inotropic stimulation with increased contractile function, although at the expense of renewed worsening of the metabolic status. This occurrence of increased regional contractile function at the expense of metabolic recovery during inotropic stimulation can be used to identify short-term hibernating myocardium. When inotropic stimulation is prolonged, short-term hibernation is impaired and myocardial infarction develops. The mechanisms responsible for the development of short-term myocardial hibernation remain unclear at present. Significant involvement of adenosine and activation of ATP-dependent potassium channels have been excluded. The role of triggering events and acidosis is controversial. Short-term hibernating myocardium is, however, characterized by reduced calcium responsiveness. (Mol Cell Biochem **186**: 185–193, 1998)

Key words: inotropic reserve, ischemia, myocardial metabolism, myocardial hibernation

The concept of myocardial hibernation and its evolution

The term 'hibernation' is borrowed from zoology and implies an adaptive reduction of energy expenditure through reduced activity in conditions of reduced energy supply. In the context of coronary artery disease, myocardial hibernation refers to an adaptive reduction of myocardial contractile function in response to a reduction of myocardial blood flow. Thus, in myocardial hibernation the observed impairment of myocardial contractile function is not regarded as the consequence of an ongoing energetic deficit, but instead as a regulatory event which serves to maintain myocardial integrity and viability.

The concept of myocardial hibernation was entirely founded on clinical grounds when Shahbudin Rahimtoola in the early eighties reviewed the results of coronary bypass surgery trials and identified patients with coronary artery disease and chronic left ventricular dysfunction that improved upon revascularization [1, 2]. However, the clinical concept of hibernation quickly merged with a number of experimental observations. Also in the early eighties, several laboratories found that the long-held concept of myocardial ischemia as an imbalance between energy supply - as determined by myocardial blood flow - and energy demand - as determined by contractile function - was not applicable on the regional myocardial level. In fact, the reduction in regional contractile function was proportional to the reduction in regional myocardial blood flow [3–6] (for review see reference [7]). In that context Gallagher coined the term 'absolute ischemia' [5] and Ross later used the term 'perfusion-contraction matching' [8]. Such perfusion-contraction matching was observed for a period of 5 h of moderate ischemia in conscious, chronically instrumented dogs with eventual full recovery of contractile function upon reperfusion and without the occurrence of infarction in the dysfunctional

Address for offprints: G. Heusch, Department of Pathophysiology, Centre of Internal Medicine, University of Essen School of Medicine, Hufelandstraße 55, 45122 Essen, Germany

myocardium [9]. Further experimental evidence for the concept of myocardial hibernation was gained from studies demonstrating the recovery of metabolic markers during ongoing persistent ischemia [10, 11].

The concept of myocardial hibernation was popularized by the 1986 editorial by Braunwald and Rutherford [12], and since then has received ever increasing attention and interest, both from clinicians and basic scientists.

Transition from an imbalance between supply and demand towards short-term myocardial hibernation

Within the first few seconds following an acute reduction of myocardial blood flow, energy demand of the hypoperfused myocardium clearly exceeds the reduced energy supply. However, this imbalance between energy supply and demand is an inherently unstable condition since ischemia induces mechanisms which reduce contractile function and thus energy demand [13]. In the subsequent steady state condition, the reduction of contractile function is proportional to the reduction of myocardial blood flow [3–6], i.e., a state of perfusion-contraction-matching exists (Table 1) [8, 14].

In chronically instrumented, conscious dogs, a reduction in myocardial blood flow associated with a decrease in regional contractile function of 40% can be maintained for 5 h without the development of necrosis within the dysfunctional myocardium. Regional contractile function recovers during reperfusion, but full recovery requires 7 days [9]. In a very recent study from our laboratory with 24 h of controlled coronary hypoperfusion in an anesthetized pig model, the close matching between blood flow and function seen during early ischemia was lost after 12 h. In contrast, a close correlation of function and myocardial oxygen consumption was maintained, and infarction was absent. These recent data suggest that mechanisms other than downregulation of function preserve myocardial viability during prolonged ischemia [15].

Table 1. Characterization of short-term hibernating myocardium

Balance between the reduced regional myocardial blood flow and the reduced contractile function (perfusion-contraction-matching)
Recovery of contractile function during reperfusion
Recovery of metabolic parameters (creatine phosphate, lactate) during persistent ischemia
Recruitable inotropic reserve at the expense of metabolic recovery

Metabolism of short-term hibernating myocardium

Within the first 5 min of an acute coronary inflow reduction, coronary venous pH and lactate extraction are reduced and coronary venous pCO_2 increases, but these parameters gradually return towards control values during continued moderate ischemia [10]. Likewise, during constant moderate ischemia the myocardial creatine phosphate concentration is significantly decreased immediately after the onset of ischemia, but gradually recovers over time towards preischemic values, whereas regional contractile function is persistently reduced [11, 16–19].

These studies are consistent with the idea that there is only a transient phase of an energy supply-demand imbalance during early myocardial ischemia. The ensuing reduction in regional myocardial function may induce a downregulation of energy demand and allow stabilization at a new metabolic level characterized again by an energy balance, as indicated by the recovery of metabolic markers. Indeed, in a recent study from our laboratory with controlled coronary hypoperfusion for 90 min in an anesthetized pig model, changes in the free energy of ATP hydrolysis were determined from biochemical measurements of ATP, creatine phosphate, creatine and inorganic phosphate in myocardial biopsies. After an initial reduction it returned towards preischemic values during continued ischemia [20], confirming prior model calculations [17].

Recruitment of an inotropic reserve at the expense of metabolic recovery in short-term hibernating myocardium

Although baseline contractile function is depressed, the hypoperfused myocardium retains its responsiveness to an inotropic challenge (Fig. 1.) [19]. When, after 85–90 min of sustained moderate ischemia in anesthetized pigs, dobutamine is infused selectively into the ischemic region, contractile function increases, although regional blood flow remains reduced. Thus, energy is available in the ischemic myocardium which is not used to maintain baseline function, but permits the increase in contractile function upon an inotropic challenge. These results suggest that the decrease in contractile function secondary to a reduction in myocardial blood flow is not simply the consequence of an ongoing energy deficit, but rather reflects an active adaptive process of the myocardium. Imposition of an inotropic stimulus on the short-term hibernating myocardium disrupts this adaptive process, as indicated by the once more decreased myocardial creatine phosphate concentration and increased

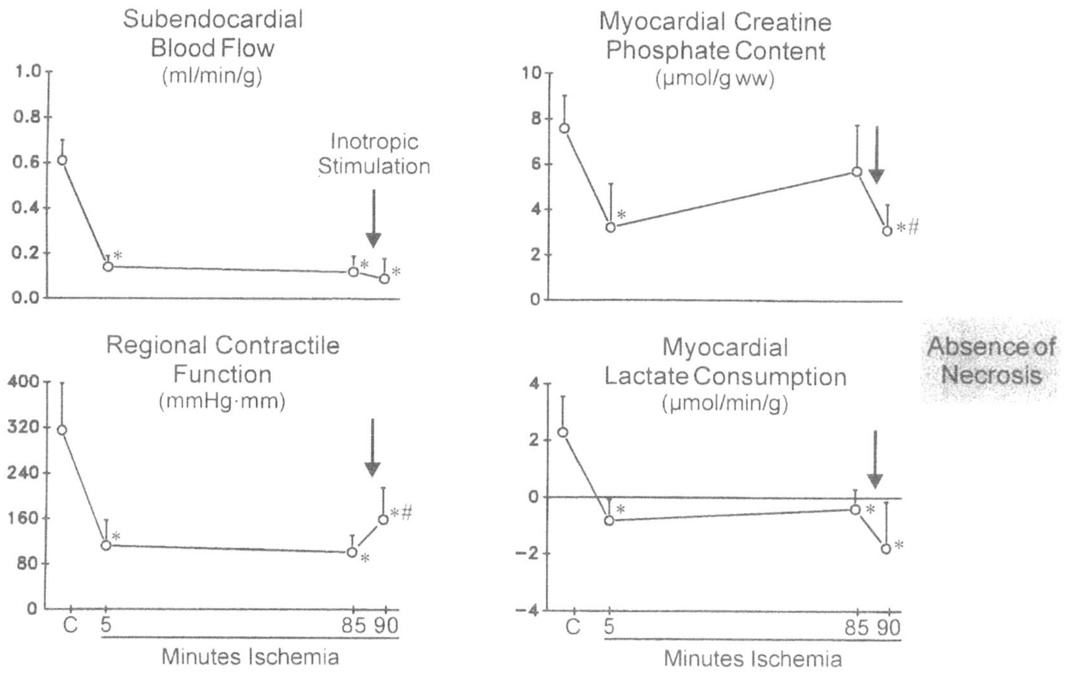

Fig. 1. The reduction in subendocardial blood flow of the anterior myocardium at 5 min of ischemia is associated with decreased regional contractile function. The creatine phosphate concentration falls and lactate consumption is reversed to net lactate production. With the extension of ischemia to 85 min at a constant subendocardial blood flow, there is a tendency for a further decrease in regional contractile function. In contrast, lactate production is attenuated and the creatine phosphate concentration recovers to a value no longer significantly different from control. Infusion of dobutamine after 90 min of ischemia increases regional contractile function at an unchanged subendocardial blood flow. This increase in contractile function, however, is associated once more with increased lactate production and a decreased creatine phosphate concentration. (Adapted from reference [45])

lactate production (Fig. 1). A similar response to that with dobutamine was observed with intracoronary calcium infusion as regional contractile function was increased, but creatine phosphate once more decreased and lactate production increased [21].

A disruption of the developing metabolic balance after prolonged moderate ischemia is also observed with a chronotropic challenge [17]. Increasing heart rate by atrial pacing in this particular study, however, increased the severity of ischemia and reduced regional myocardial blood flow, function and creatine phosphate concentration and increased lactate production. Since the severity of ischemia was increased, the observed metabolic deterioration during pacing does not conclusively demonstrate active adaptation of energy demand during the preceding more moderate ischemia.

Limitations of short-term hibernation

As already stated, the development of such delicate balance between regional myocardial blood flow and function during early ischemia is disturbed by unfavourable alterations in energy supply and demand. When energy supply in anesthetized pigs is further reduced by an additional reduction of myocardial blood flow after 5 minutes of ischemia caused by a blood flow reduction known to produce myocardial hibernation if maintained for 90 min, necrosis develops. The lower limit of myocardial blood flow compatible with the development of short-term myocardial hibernation amounts to approximately 25% of baseline in this experimental setting [22].

Also, a further increase in energy demand by continuous inotropic stimulation with dobutamine induces necrosis (Fig. 2.) [22]. Thus, both an additional reduction in energy supply by an increasing severity of ischemia and an enhanced energy expenditure by continuous inotropic stimulation impair the development of short-term myocardial hibernation and precipitate myocardial infarction.

Experimental identification of short-term hibernating myocardium

In contrast to dysfunctional, irreversibly damaged myocardium, the short-term hibernating myocardium retains an inotropic

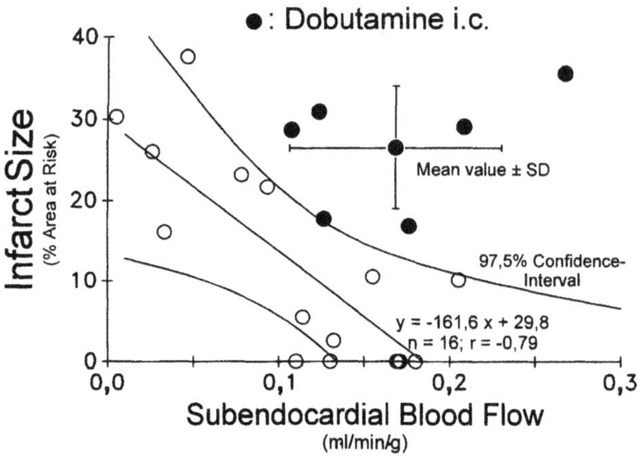

Fig. 2. Relationship between subendocardial blood flow (ml/min/g) in the left ventricular myocardium at risk (x-axis) and infarct size (expressed as a percentage of the area at risk, y-axis). Open circles represent data from control animals undergoing 90 min of ischemia of varying severity. With increasing severity of ischemia infarct size is increased. In animals subjected to 90 min of ischemia with 85 min of dobutamine infusion (closed circles), infarct size is increased above the level expected from the blood flow reduction alone. (Adapted from reference [22])

reserve. Recruitment of such inotropic reserve is at the expense of metabolic recovery (Fig. 1) [19, 22]. The extent of postsystolic wall thickening in ischemic myocardium also permits one to distinguish irreversibly damaged from viable, hibernating myocardium [23]. The extent of postsystolic wall thickening at the end of a 90 min ischemic period correlates inversely to the extent of necrosis, and positively to the concentration of myocardial creatine phosphate, the dobutamine-recruitable inotropic reserve and the functional recovery upon reperfusion. Thus a greater postsystolic wall thickening correlates with successful hibernation of more myocardium (Fig. 3).

Adenosine and short-term myocardial hibernation

Adenosine is a key trigger and mediator of ischemic preconditioning [24]. Endogenous adenosine is released during ischemia and might act through a number of secondary mechanisms, such as inhibition of norepinephrine release from sympathetic nerve terminals, inhibition of adenylate cyclase, or inhibition of L-type calcium channels, to attenuate the decrease in high energy phosphates and the increase in intracellular calcium, and thereby preserve myocardial viability. However, a role of endogenous adenosine in the development of short-term hibernation has been excluded. In an anesthetized pig preparation with 90 min of controlled coronary hypo-

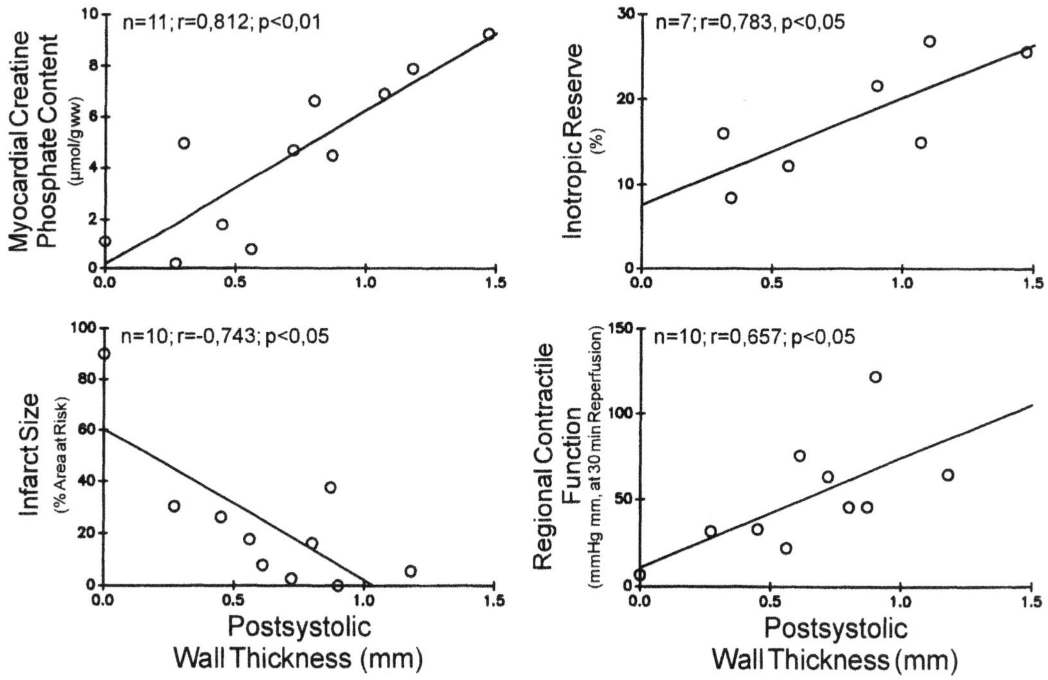

Fig. 3. Correlation between the extent of postsystolic wall thickening at the end of a 90 min ischemic period and the extent of necrosis, myocardial creatine phosphate concentration, dobutamine-recruitable inotropic reserve and functional recovery upon reperfusion. The extent of postsystolic wall thickening at the end of a 90 min ischemic period correlates inversely to the extent of necrosis, and positively to the myocardial creatine phosphate concentration, the dobutamine-recruitable inotropic reserve and the functional recovery upon reperfusion. Thus, a greater postsystolic wall thickening correlates with successful hibernation of more myocardium. (Adapted from reference [23])

perfusion, increased catabolism of endogenous adenosine by intracoronary infusion of adenosine deaminase did not alter perfusion-contraction matching and the recovery of metabolic parameters (creatine phosphate, lactate), did not impair inotropic reserve, and did not induce necrosis (Fig. 4) [25].

K_{ATP} - channel activation and short-term myocardial hibernation

Activation of K_{ATP} channels is a key event in ischemic preconditioning; conversely, blockade of K_{ATP} channels prevents ischemic preconditioning [26]. Activation of K_{ATP} channels reduces action potential duration and calcium influx into cardiomyocytes. The attenuation of calcium overload during ischemia may act to preserve myocardial integrity and viability. In an anesthetized pig preparation with 90 min of controlled coronary hypoperfusion, K_{ATP} channels were indeed activated, as indicated by reduced action potential duration. However, blockade of K_{ATP} channels with glibenclamide abolished the reduction in action potential duration, but failed to alter perfusion-contraction matching, affect the recovery of metabolic parameters (creatine phosphate, lactate), impair inotropic reserve, or induce necrosis (Fig. 5) [25]. Thus, activation of K_{ATP} channels is not the underlying mechanism of short-term hibernation.

β-Adrenoceptors and short-term myocardial hibernation

β-Adrenoceptor density and affinity are not altered in a porcine model of short-term hibernation [22]. The maintained inotropic response to dobutamine also excludes an uncoupling of β-adrenoceptor activation from the subsequent signal cascade. However, detailed analyses of the adrenergic signal cascade in hibernation are not available.

Acidosis and triggering events in short-term myocardial hibernation

A metabolic adaptation to severe, sustained (4 h) low-flow ischemia was evident in isolated buffer-perfused rabbit hearts when there was a preceding short episode (10 min) of no-flow

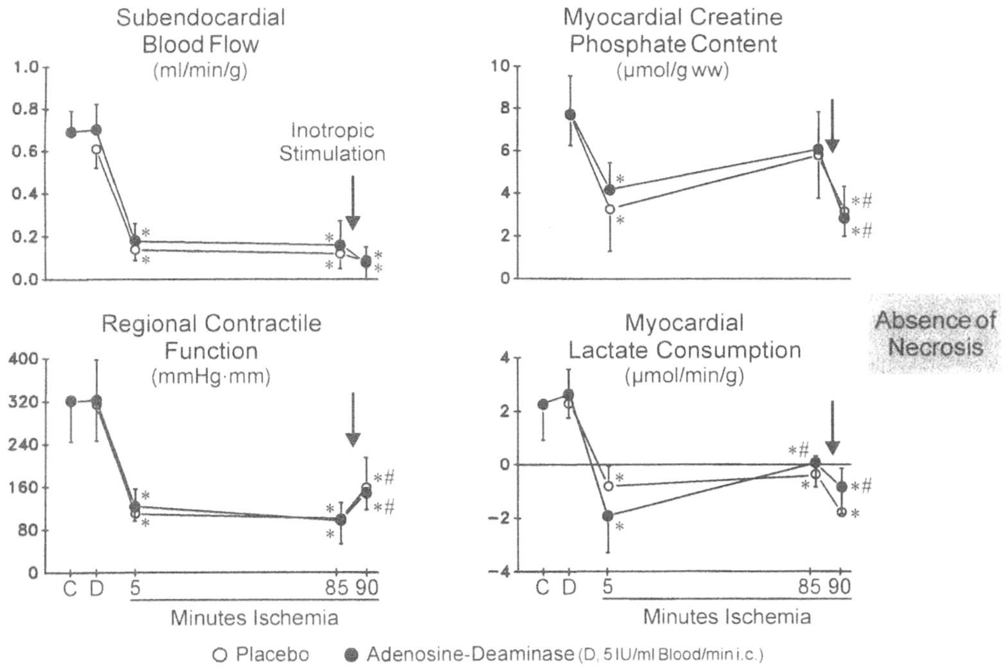

Fig. 4. The reduction in subendocardial blood flow of the anterior myocardium after 5 min of ischemia is associated with decreased regional contractile function in placebo swine and swine receiving adenosine deaminase. Within the first 5 min of ischemia, the creatine phosphate concentration decreases and lactate consumption is reversed to net lactate production in both groups. With the extension of ischemia to 85 min at a constant subendocardial blood flow, there is a tendency for a further decrease in regional contractile function in both groups. In contrast, lactate production is attenuated and the creatine phosphate concentration recovers to a value no longer significantly different from the respective control value. Infusion of dobutamine after 90 min of ischemia increases regional contractile function at an unchanged subendocardial blood flow to a comparable extent in both groups of swine. This increase in contractile function, however, is associated once more with increased lactate production and a decreased creatine phosphate concentration. (Adapted from reference [25]).

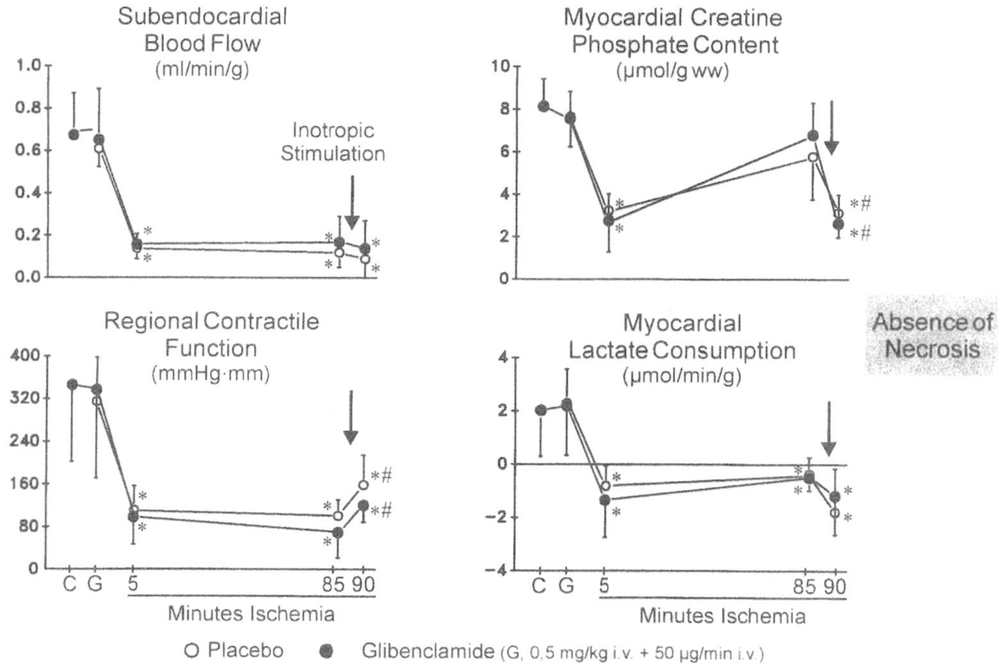

Fig. 5. The reduction in subendocardial blood flow of the anterior myocardium at 5 min of ischemia is associated with decreased regional contractile function in placebo swine and swine receiving the ATP-dependent potassium channel antagonist glibenclamide. Within the first 5 min of ischemia, the creatine phosphate concentration decreases and lactate consumption is reversed to net lactate production in both groups of swine. With the extension of ischemia to 85 min at a constant subendocardial blood flow, there is a tendency for a further decrease in regional contractile function in both groups. In contrast, lactate production is attenuated and the creatine phosphate concentration recovers to a value no longer significantly different from the respective control value. Infusion of dobutamine after 90 min of ischemia increases regional contractile function at an unchanged subendocardial blood flow to a comparable extent in both groups of swine. This increase in contractile function, however, is associated once more with increased lactate production and a decreased creatine phosphate concentration. (Adapted from reference [25])

ischemia. In these hearts, the early decline in contractile function was more pronounced and significantly faster than in control hearts that did not have the brief episode of no-flow ischemia. The rapid decline in contractile function during the brief episode of no-flow ischemia was accompanied by a greater decrease in interstitial and intracellular pH, and the contractile quiescence was attributed to a faster development of myocardial acidosis. The faster achievement of contractile quiescence, in turn, was thought to facilitate the balance between energy supply and demand. Indeed, hearts with a preceding episode of low-flow ischemia had better recovery of contractile function and less creatine kinase release upon reperfusion [27, 28].

Protection from infarction was also found in anesthetized pigs when sustained (80 min) low-flow ischemia was preceded by a short (10 min) episode of no-flow ischemia. When compared to hearts undergoing 90 min of low-flow ischemia without a preceding episode of no-flow ischemia, infarct size was reduced, but this reduction was not related to acidosis and could be prevented by the K_{ATP} channel blocker glibenclamide, indicating that the protection by the initial stimulus/trigger of very severe ischemia was related to ischemic preconditioning without an intervening reperfusion [29].

A few other studies have suggested that the development of hibernation does not require a critical trigger. In anesthetized pigs, a gradual slow reduction of coronary arterial inflow over 35 min with a constant reduced blood flow for an additional 25 min resulted in decreased regional contractile function, but with minimal lactate production and no decrease in the ratio of subendocardial creatine phosphate to ATP concentrations [30]. Also in anesthetized pigs, coronary venous pH and pCO_2 were better preserved and lactate production and infarct size were reduced when sustained severe ischemia for 60 min was preceded by a slowly and continuously increasing severity of flow reduction over 70 min [31].

Thus, the role of an initial trigger as well as the importance of acidosis in the development of short-term hibernation remain to be better defined. An initial trigger of severe ischemia, however, may represent a link between hibernation and ischemic preconditioning.

Excitation-contraction coupling and short-term myocardial hibernation

In the search for possible mechanisms of hibernation it is not surprising that attention has been directed to calcium. In isolated buffer-perfused ferret hearts moderately decreased coronary perfusion resulted in decreased calcium transients and decreased contractile force without any change in high energy phosphates [32]. In anesthetized pigs with controlled coronary hypoperfusion for 90 min, regional contractile responses to intracoronary calcium were substantially reduced, indicating impaired calcium responsiveness. This reduced calcium responsiveness was largely attributable to a decrease in maximal developed force and not to a decrease in calcium sensitivity, as the relations between fractional increase in contractile function and intracoronary calcium dose were not different in normoperfused and short-term hibernating myocardium (Fig. 6) [21]. Thus, short-term hibernating myocardium has reduced calcium responsiveness and possibly also reduced calcium transients.

Short-term hibernation vs. stunning

Both the short-term hibernating and the stunned myocardium are characterized by reversible contractile dysfunction. In short-term hibernating myocardium ischemia is still ongoing, whereas in stunned myocardium blood flow is fully or almost fully restored. In contrast to dysfunctional, irreversibly damaged myocardium, both short-term hibernating and stunned myocardium retain an inotropic reserve. To further distinguish short-term hibernating from stunned myocardium, either regional myocardial blood flow must be measured [33] or the metabolic changes associated with an inotropic challenge must be analyzed. The recruitment of an inotropic reserve in short-term hibernating myocardium is at the expense of metabolic recovery (Fig. 1) [19, 22], while in stunned myocardium no metabolic deterioration occurs during inotropic stimulation [34–36]. Also, whereas prolonged inotropic stimulation causes infarction in short-term hibernating myocardium [22], it does not cause necrosis in stunned myocardium [37].

Transition from short-term hibernation to long-term hibernation

There are a few experimental studies which have tried to investigate the transition process from short-term hibernation to long-term hibernation by subjecting swine to either a prolonged partial coronary artery stenosis for a duration ranging from 1 day to 32 weeks [38–42] or a progressive narrowing of the coronary artery with ultimate complete occlusion by use of an ameroid constrictor [43, 44]. These studies, however, failed to continuously monitor myocardial blood flow and contractile function over the entire time period of coronary hypoperfusion. When resting regional myocardial blood flow was measured [40–44], it was either reduced at the beginning [40, 41] and the end [40, 41, 44] of the stenosis protocol or was normal [43] or almost normal at the beginning and end [42] of the protocol. Despite normal or almost normal resting blood flow, however, coronary reserve was impaired [40, 42–44]. With impaired coronary reserve there could have been repetitive episodes

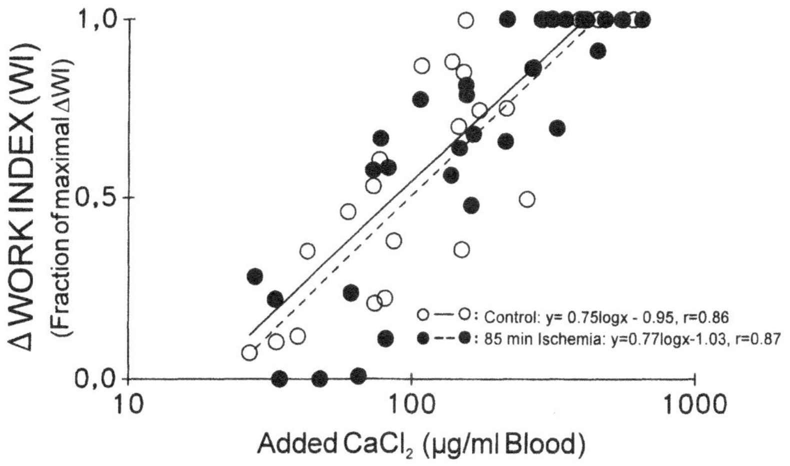

Fig. 6. Relation between the added intracoronary calcium chloride (on a logarithmic scale) and the increase in regional myocardial work (as a fraction of the maximal increase). The relations obtained under control conditions and at 85 min of ischemia are superimposable. At 85 min of ischemia, perfusion-contraction matching and viability were maintained, and creatine phosphate had recovered towards control values, indicating short-term hibernation.

of exercise-induced ischemia and reperfusion-induced stunning, thus providing an alternative explanation for the observed chronic contractile dysfunction during persistent coronary stenosis. Indeed, in one study episodes of apparent stunning were documented [43]. However, since myocardial blood flow and contractile function were not measured continously, these studies do not prove or disprove the existence of true 'myocardial hibernation'.

Conclusion and perspective

Triggers and mediators of myocardial hibernation are still largely unclear. We propose a conceptual three-stage model of hibernation: (a) Very rapidly after a reduction in coronary blood flow, a biochemical mechanism is turned on which reduces contractile function and restores the energy balance between energy supply/blood flow and energy demand/ contractile function. Adenosine and activation of K_{ATP} channels are not important here, but acidosis and reduced calcium responsiveness deserve to be studied in more detail. (b) With prolongation of ischemia, the expression of a number of genes and subsequently proteins is increased. Alterations in the amount and activity of calcium handling proteins may decrease contractile function and thus replace or add to the initial biochemical mechanism in reducing contractile function. No data are available. (c) Increased expression of genes and proteins involved in apoptosis may induce controlled cardiomyocyte death and thus improve blood supply to the surviving cardiomyocytes. No data are available. The semantic distinction between hibernation, stunning and ischemic preconditioning may be clearer than the reality. Triggering events may link hibernation and ischemic preconditioning, and with some residual blood flow a component of reperfusion / stunning may be superimposed on hibernation.

References

1. Rahimtoola SH: Coronary bypass surgery for chronic angina. Circulation 65: 225–241, 1982
2. Rahimtoola SH: A perspective on the three large multicenter randomized clinical trials of coronary bypass surgery for chronic stable angina. Circulation 72 (Suppl V): V-123–V-135, 1985
3. Vatner SF: Correlation between acute reductions in myocardial blood flow and function in conscious dogs. Circ Res 47: 201–207, 1980
4. Weintraub WS, Hattori S, Agarwal JB, Bodenheimer MM, Banka VS, Helfant RH: The relationship between myocardial blood flow and contraction by myocardial layer in the canine left ventricle during ischemia. Circ Res 48: 430–438, 1981
5. Gallagher KP, Matsuzaki M, Osakada G, Kemper WS, Ross Jr. J: Effect of exercise on the relationship between myocardial blood flow and systolic wall thickening in dogs with acute coronary stenosis. Circ Res 52: 716–729, 1983
6. Canty JM: Coronary pressure-function and steady-state pressure-flow relations during autoregulation in the unanesthetized dog. Circ Res 63: 821–836, 1988
7. Heusch G: The relationship between regional blood flow and contractile function in normal, ischemic, and reperfused myocardium. Basic Res Cardiol 86: 197–218, 1991
8. Ross Jr. J: Myocardial perfusion-contraction matching. Implications for coronary heart disease and hibernation. Circulation 83: 1076–1083, 1991
9. Matsuzaki M, Gallagher KP, Kemper WS, White F, Ross Jr. J: Sustained regional dysfunction produced by prolonged coronary stenosis: Gradual recovery after reperfusion. Circulation 68: 170–182, 1983
10. Fedele FA, Gewirtz H, Capone RJ, Sharaf B, Most AS: Metabolic response to prolonged reduction of myocardial blood flow distal to a severe coronary artery stenosis. Circulation 78: 729–735, 1988
11. Pantely GA, Malone SA, Rhen WS, Anselone CG, Arai A, Bristow J, Bristow JD: Regeneration of myocardial phosphocreatine in pigs despite continued moderate ischemia. Circ Res 67: 1481–1493, 1990
12. Braunwald E, Rutherford JD: Reversible ischemic left ventricular dysfunction: Evidence for the 'hibernating myocardium'. J Am Coll Cardiol 8: 1467–1470, 1986
13. Guth BD, Schulz R, Heusch G: Time course and mechanisms of contractile dysfunction during acute myocardial ischemia. Circulation 87 (Suppl IV): IV35–IV42, 1993
14. Ross J Jr.: Mechanisms of regional ischemia and antianginal drug action during exercise. Prog Cardiovasc Dis 31: 455–466, 1989
15. Schulz R, Rose J, Vahlhaus C, Post H, Backenköhler U, Heusch G: No maintenance of perfusion-contraction matching during 24 h sustained moderate myocardial ischemia in pigs. Faseb J, 1998
16. Downing SE, Chen V: Myocardial hibernation in the ischemic neonatal heart. Circ Res 66: 763–772, 1990
17. Arai AE, Pantely GA, Anselone CG, Bristow J, Bristow JD: Active downregulation of myocardial energy requirements during prolonged moderate ischemia in swine. Circ Res 69: 1458–1469, 1991
18. Downing SE, Chen V: Acute hibernation and reperfusion of the ischemic heart. Circulation 85: 699–707, 1992
19. Schulz R, Guth BD, Pieper K, Martin C, Heusch G: Recruitment of an inotropic reserve in moderately ischemic myocardium at the expense of metabolic recovery: A model of short-term hibernation. Circ Res 70: 1282–1295, 1992
20. Martin C, Schulz R, Rose J, Heusch G: Inorganic phosphate content and free energy change of ATP hydrolysis in regional short-term hibernating myocardium. Cardiovasc Res 1998, in press
21. Heusch G, Rose J, Skyschally A, Post H, Schulz R: Calcium responsiveness in regional myocardial short-term hibernation and stunning in the in situ porcine heart - inotropic responses to post-extrasystolic potentiation and intracoronary calcium. Circulation 93: 1556–1566, 1996
22. Schulz R, Rose J, Martin C, Brodde OE, Heusch G: Development of short-term myocardial hibernation: Its limitation by the severity of ischemia and inotropic stimulation. Circulation 88: 684–695, 1993
23. Rose J, Schulz R, Martin C, Heusch G: Post-ejection wall thickening as a marker of successful short term hibernation. Cardiovasc Res 27: 1306–1311, 1993
24. Schulz R, Rose J, Post H, Heusch G: Involvement of endogenous adenosine in ischemic preconditioning in swine. Pflügers Arch 430: 273–282, 1995
25. Schulz R, Rose J, Post H, Heusch G: Regional short-term hibernation in swine does not involve endogenous adenosine or KATP channels. Am J Physiol 268: H2294–H3201, 1995
26. Schulz R, Rose J, Heusch G: Involvement of activation of ATP-

dependent potassium channels in ischemic preconditioning in swine. Am J Physiol 267: H1341–H1352, 1994
27. Ferrari R, Cargnoni A, Bernocchi P, Pasini E, Curello S, Ceconi C, Ruigrok TJC: Metabolic adaptation during a sequence of no-flow and low-flow ischaemia: a possible trigger for hibernation. Circulation 94: 2587–2596, 1996
28. van Binsbergen XA, van Emous JG, Ferrari R, van Echteld CJA, Ruigrok TJC: Metabolic and functional consequences of successive no-flow and sustained low-flow ischaemia; a 31P MRS study in rat hearts. J Mol Cell Cardiol 28: 2373–2381, 1996
29. Schulz R, Post H, Sakka S, Wallbridge DR, Heusch G: Intraischemic preconditioning. Increased tolerance to sustained low-flow ischemia by a brief episode of no-flow ischemia without intermittent reperfusion. Circ Res 76: 942–950, 1995
30. Arai AE, Grauer SE, Anselone CG, Pantely GA, Bristow JD: Metabolic adaptation to grandual reduction in myocardial blood flow. Circulation 92: 244–252, 1995
31. Ito BR: Gradual onset of myocardial ischemia results in reduced myocardial infarction. Association with reduced contractile function and metabolic downregulation. Circulation 91: 2058–2070, 1995
32. Kitakaze M, Marban E: Cellular mechanism of the modulation of contractile function by coronary perfusion pressure in ferret hearts. J Physiol 414: 455–472, 1989
33. Bolli R: Myocardial 'Stunning' in man. Circulation 86: 1671–1691, 1992
34. Ambrosio G, Jacobus WE, Bergmann CA, Weisman HF, Becker LC: Preserved high energy phosphate metabolic reserve in globally stunned hearts despite reduction of basal ATP content and contractility. J Mol Cell Cardiol 19: 953–964, 1987
35. Arnold JMO, Braunwald E, Sandor T, Kloner RA: Inotropic stimulation of reperfused myocardium with dopamine: Effects on infarct size and myocardial function. J Am Coll Cardiol 6: 1036–1044, 1985
36. Görge G, Papageorgiou I, Lerch R: Epinephrine-stimulated contractile and metabolic reserve in postischemic rat myocardium. Basic Res Cardiol 85: 595–605, 1990
37. Bolli R, Zhu W-X, Myers ML, Hartley CJ, Roberts R: Beta-adrenergic stimulation reverses postischemic myocardial dysfunction without producing subsequent deterioration. Am J Cardiol 56: 964–968, 1985
38. Bolukoglu H, Liedtke AJ, Nellis SH, Eggleston AM, Subramanian R, Renstrom B: An animal model of chronic coronary stenosis resulting in hibernating myocardium. Am J Physiol 263: H20–H29, 1992
39. Liedtke AJ, Renstrom B, Nellis SH, Subramanian R: Myocardial function and metabolism in pig hearts after relief from chronic partial coronary stenosis. Am J Physiol 267: H1312–H1319, 1994
40. Mills I, Fallon JT, Wrenn D, Sasken H, Gray W, Bier J, Levine D, Berman S, Gilson M, Gewirtz H: Adaptive responses of coronary circulation and myocardium to chronic reduction in perfusion pressure and flow. Am J Physiol 266: H447–H457, 1994
41. Chen C, Li L, Chen LL, Preda JV, Chen MH, Fallon ST, Weyman AE, Waters D, Gillam L: Incremental doses of dobutamine induce a biphasic response in dysfunctional left ventricular regions subtending coronary stenoses. Circulation 92: 756–766, 1995
42. Liedtke AJ, Renstrom B, Nellis SH, Hall JL, Stanley WC: Mechanical and metabolic functions in pig hearts after 4 days of chronic coronary stenosis. J Am Coll Cardiol 26: 815–828, 1995
43. Shen Y-T, Vatner SF: Mechanism of impaired myocardial function during progressive coronary stenosis in conscious pigs. Hibernation versus stunning. Circ Res 76: 479–488, 1995
44. Fallafollita JA, Perry BJ, Canty JM: 18F-2-deoxyglucose deposition and regional flow in pigs with chronically dysfunctional myocardium. Evidence for transmural variations in chronic hibernating myocardium. Circulation 95: 1900–1909, 1997
45. Schulz R, Heusch G: Acute adaptation to ischemia: Short-term myocardial hibernation. Basic Res Cardiol 90: 29–31, 1995

Molecular and Cellular Biochemistry **186:** 195–199, 1998.
© 1998 *Kluwer Academic Publishers.*

Hibernating myocardium: Its pathophysiology and clinical role

Roberto Ferrari,[1, 2] Federica Ferrari,[1] Massimo Benigno,[1] Patrizia Pepi[1] and Odoardo Visioli[1]

[1]*Cattedra di Cardiologia, Universita' degli Studi di Brescia, Brescia;* [2]*Fondazione Clinica del Lavoro, Centro di Fisiopatologia Cardiovascolare 'S. Maugeri', Gussago, Brescia, Italy*

Abstract

Myocardial hibernation, as first defined by Rahimtoola, is a state of chronic contractile dysfunction in patients with coronary artery disease which is fully reversible upon reperfusion. Clinical conditions consistent with the existence of myocardial hibernation include unstable and stable angina, myocardial infarction heart failure, and anomalous origin of coronary arteries. The mechanisms of hibernation are not known. Morphological alterations have been described in the hibernating area of patients, but these information are strongly affected by the diagnostic criteria utilized to screen patients. It has been postulated that hibernation is an adaptive phenomenon occurring during ischemia. In this context, downregulation of contraction is not regarded as a consequence of energetic deficit, but as a regulatory event aimed at reducing energy expenditure, thereby maintaining integrity and viability. Thus, hibernation might bear a relationship to the phenomenon of low-flow perfusion-contraction matching, or repetitive stunning or preconditioning. Clear-cut evidence for the mechanism of hibernation in the clinical setting seems likely to remain elusive, because of the nature of the studies needed to document it. Current experimental evidence supports the view that hibernation, stunning, preconditioning, or their coexistence can be responsible for regional myocardial contractile dysfunction which is reversible upon reperfusion. These are all adaptive and protective phenomena independent of their terminology and strict definitions and do not always apply to the extremely complex situation of myocardial ischemia in man. (Mol Cell Biochem **186:** 195–199, 1998)

Key words: hibernating myocardium, myocardial ischemia, heart failure, myocardial viability, ischemic cardiomyopathy

Introduction

Differentiation of viable myocardium from non-viable myocardium in patients with coronary artery disease and left ventricular dysfunction is an issue of clinical relevance in this age of myocardial revascularization. Impaired left ventricular function may occur transiently during an attack of angina and is usually reversible. It may persist over hours or even days in patients after an episode of ischemia followed by reperfusion, leading to the so-called condition of 'stunning'. In patients with persistent limitation of coronary flow, left ventricular dysfunction may be present over months or years, or indefinitely in subjects with fibrosis, scar formation and remodeling after myocardial infarction, leading to the syndrome of heart failure. Chronic left ventricular dysfunction does not necessarily mean permanent or irreversible cell damage. Hypoperfused myocytes can remain viable but akinetic. This type of dysfunction has been called 'hibernating myocardium' [1].

The diagnosis of hibernation is clinically relevant because it has therapeutic implications, as regional and global left ventricular function due to hibernation will improve after revascularization and it is associated with improved survival [2]. The true clinical gold standard for hibernation is the improvement in systolic function of dysfunctional myocardial segments after revascularization. Such a retrospective standard is, however, insufficient. An accurate prospective diagnosis of patients with potentially reversible left ventricular dysfunction is essential for the identification of ideal candidates for revascularization procedures. Thus, the

Address for offprints: R. Ferrari, Cattedra di Cardiologia, Universita' degli Studi di Brescia, c/o Spedali Civili, P.le Spedali Civili, 1, 25123 Brescia, Italy

hibernating myocardium poses a series of clinical issue. The three fundamental questions are: 1) how and with what accuracy this entity can be recognized?; 2) how should it be best treated? and 3) can reperfusion of hibernating myocardium be considered an alternative to transplantation?

How hibernating can be explained

The molecular factors responsible for hibernation and for the chronically-depressed contractile function have not yet been defined. The hypothesized mechanism of a down-regulated contractile performance matching the reduced energy supply is, in part, supported by experimental and clinical studies using positron emission tomography (PET). These studies show that during prolonged underperfusion metabolism is shifted from FFA to glucose with the recruitment of glycolysis. In humans, a mismatch of flow and glucose metabolism predicts recovery of mechanical function after revascularization. There is little data, however, to support or disprove this theory that a perfect balance between reduced oxygen supply and contractile function prevents myocardial injury [3]. This is because there are neither adequate animal models for chronic persistent ischemia nor any clinical studies showing whether hibernating myocardium is truly a chronic condition.

If the above mechanism is the only one responsible for hibernation, this condition should occur in every patient with reduction of coronary flow. Obviously this is not the case. In addition, in clinical practice hibernation is often present in patients with a history of acute ischemia, such as infarction or prolonged angina pain. This complicates not only the distinction between viable and non-viable myocardium, but also the understanding of the pathophysiological mechanism underlying hibernation.

Different hypotheses have been suggested for the factor responsible for down-regulation of contraction during hibernation. An early theory suggests that decrement in coronary perfusion pressure reduces sarcomere length because of distention in the adjacent coronary microvasculature. Therefore, the extent of the contraction would be reduced by the Frank Starling mechanism. Several lines of evidence argue against this suggestion. Another theory considers that a reduction of energy stores might cause down-regulation of contraction. 31P-NMR studies in live animals, however, suggest that net ATP and CP stores are not depleted unless coronary flow is reduced to very low levels. Furthermore, it has been shown that in hibernating myocardium, ATP levels are normal [4]. Our own studies suggest that changes in intracellular pH, phosphate (Pi) and myocardial NAD/NADH ratio are responsible for contractile and possibly metabolic down-regulation in hibernation [5–7].

Doubts exist whether this new metabolic state (hibernation) should be considered a true ischemic condition. In strict terms, ischemia is a condition that exists when fractional uptake of oxygen is not sufficient to meet the rate of mitochondria oxidation, which, in the heart, is largely determined by the mechanical or physical activity of the myocytes [5]. It is likely that in hibernating myocardium, the residual flow is able to deliver enough oxygen to meet the reduced rate of mitochondrial oxidation. This concept will explain why, the hibernating myocardium does not produce lactate (a typical marker of ischemia) and shows indirect signs of metabolic activity. It also explains the full recovery after reperfusion and the retention of a contractile reserve. Thus, hibernation is not a state of chronic ischemia but it represents a new metabolic state that is consequent to an ischemic condition but it is not actually ischemic. In strictly molecular terms, hibernation represents a chronic hypoperfusion of akinetic but aerobic myocytes.

A difficult concept to conceive, in clinical terms, is how the mechanism discussed above or some other unknown mechanism could provide an exact balance between energy demand and supply for months to years. This would probably require subcellular adaptive changes to occur. As a result, this fascinating problem is presently under investigation using a molecular biology approach. Unfortunately, the electrophysiological changes occurring in the hibernated zone are not known at the present. An interesting hypothesis is that the ischemia induced activation of ATP/ADP modulated K^+ channels which could down-regulate contractility during chronic left ventricular dysfunction.

Today, a number of methods are available to assess viability and contractile reserve in regions with hypokinetic or akinetic wall motion. These include cardiac imaging techniques that evaluate myocardial viability on the basis of myocardial perfusion, cell membrane integrity and metabolic activity and the infusion of a low-dose positive inotropic agent during echocardiography to evaluate contractile reserve. These methods provide greater precision in the assessment of hibernation than can be achieved by analysis of coronary anatomy, regional function or specific electrocardiographic changes. The single limits and advantages, and the comparison of the results obtained by each technique, have recently been discussed in detail [8].

PET provides the capacity to quantitate non-invasively regional blood flow and to assess regional metabolic activity independent of flow. In addition, it provides enhanced image resolution and routine correction for body attenuation. In this way it overcomes the two major limitations of thallium imaging: poor resolution and photon attenuation. Viable myocardium is identified by PET on the basis of enhanced or preserved metabolic activity in underperfused and dysfunctional myocardial regions. Usually, the metabolic shift from FFA to glucose is a marker of myocardial ischemia. These metabolic changes, however, may not apply to hibernating myocardium where a new state of perfusion-contraction coupling is achieved in the absence of ischemia. 18-F-flurodeoxyglucose (FDG) has emerged as a marker of regional exogenous glucose utilization in hypoperfused regions. FDG is a glucose analogue that undergoes the same transport and phosphorylation of glucose but does not enter glycolysis or glycogen synthesis. An increase of FDG uptake in regions with reduced perfusion, the so-called FDG-blood flow mismatch, provides a metabolic signal for identifying viable myocardium [8].

Recently, absolute regional myocardial blood flow data became available from PET studies using either 13NH3 or H215O as flow tracers. The majority of these studies clearly shows a significant reduction in resting regional myocardial blood flow in those dysfunctional areas classified as hibernating myocardium as compared to intraindividual normal remote areas, with an average reduction in blood flow by between 20–30% of baseline. Others, however, failed to show a significant reduction in resting blood flow in the dysfunctional area, both as compared to that in a remote reference region and as compared to normal flow values obtained from healthy volunteers [9–12]. This challenges the idea that hibernation is the result of chronic hypoperfusion and suggest that repeated episodes of ischemia and subsequent commutative stunning may also be involved. While these arguments are important to explain the physiopathology, they seem to be irrelevant from the clinical point of view as hibernation is always treated by revascularization.

Thallium myocardial imaging is an established and clinically important method to assess perfusion and sarcolemmal integrity and hence to detect myocardial viability. Regional thallium activity on redistribution images acquired either early (3–4 h) or late (8–72 h) after stress is commonly used to demonstrate the distribution of viable myocardial cells and the extent of myocardial fibrosis. Different opinions exist on which protocol should be used for the detection of hibernating myocardium. This, obviously depends on the underlying clinical question and on the meaning of the term hibernation. In our experience, the clinical question in a patient with left ventricular dysfunction, possibly due to hibernation, is one of viability and not of inducible ischemia. Therefore, we use rest-redistribution thallium imaging to discriminate viable from non-viable myocardium [13].

Low-dose dobutamine infusion to enhance regional systolic contraction during two-dimensional echocadiography has been successfully used to identify viable versus necrotic myocardium during the first week following successful thrombolysis. Thus, this technique is able to unmask contractile reserve in stunned myocardium in which coronary flow has been restored. In patients with hibernating myocardium, with reduction of coronary blood flow severe enough to produce sustained regional contractile dysfunction, the administration of a positive inotropic agent has been hampered by the risk of increasing myocardial demand in the setting of exhausted coronary flow reserve, thereby producing myocardial ischemia and persistent regional dysfunction. Despite this potential limitation, recent data suggest that low-dose dobutamine echocardiography is a safe and accurate method for assessing hibernating myocardium [14]. We have evaluated its role in identifying hibernation in 33 CAD patients with positive thallium rest-redistribution test and persistent left ventricular dysfunction [15]. Reversible dysfunction was identified by functional improvement of akinetic areas immediately after surgical revascularization, detected by intraoperative epicardial echocardiography and both 2 weeks and 3 months later, by means of transthoracic echocardiography. Infusion of dobutamine predicted reversible dysfunction in 178 of the 205 akinetic segments that recovered after surgery with a sensitivity of 86.8% and identified 89 of the 109 segments that failed to recover with a specificity of 81.6%.

In our study echo-dobutamine predicted the immediate functional recovery after revascularization with higher accuracy than thallium rest redistribution. On the contrary, scintigraphic imaging was better correlated than echo-dobutamine with long-term recovery of function [13]. This discrepancy stresses that the two tests have different clinical significance as they explore two different aspects of the same issue: viability and contractile reserve of the akinetic myocytes. We believe that demonstration of viability provides the ultimate rationale for revascularization. Presence of contractile reserve provides information on the overall operative risk, thus overriding the additional concept that global systolic ejection fraction is a crucial determinant of surgical risk. Global ejection fraction, in fact, can be related to different pathophysiogical mechanisms.

How should hibernating myocardium best be treated?

It is clear that hibernating myocardium should be reperfused. It is our conviction, however, that, in a decision-making process, several steps should be fulfilled before recommending myocardial revascularization. 1) the symptomatic status of the patient is important. This should not be limited to the presence of angina pectoris or of inducible ischemia, but also to the presence of signs of heart failure in absence of other symptoms. 2) coronary anatomy has to be suitable for revascularization either by open-heart surgery or angioplasty. 3) the extension and severity of the ventricular dysfunction has to justify an interventional procedure, the risk of which is not negligible. 4) the demonstration of viable myocardium by one or more of the possible imaging modalities is essential, particularly to predict long-term recovery after revascularization. 5– the demonstration of a contractile reserve by low-dose dobutamine echocardiography is also important as it confirms viability and provides further information on the surgical risk.

No data exist regarding a possible pharmacological treatment. Theoretically drugs which improve coronary perfusion and protect the myocyte against ischemia might be useful.

Can reperfusion of hibernating myocardium be considered as an alternative to transplantation?

Revascularization of patients with end-stage ischemic cardiomyopathy and clinical manifestation of congestive heart failure rather than angina may be considered an alternative to transplantation. Data regarding this possibility are few. In our study [15], one-third of patients had dyspnea as the only symptom, a severe reduction of ejection fraction and some of these patients were potential candidates for transplantation. Examining the existing literature on revascularization of severe ischemic cardiomyopathies and pooling together the data, it seems that perioperative mortality rate is between 15 and 20%, with 72–75% survival at 3 years. There is a clear improvement of symptoms and of global ejection fraction. Demonstration of viability, however, is essential for a positive outcome [16]. However, we believe that to answer this fundamental question we need a proper clinical study of CAD patients awaiting transplantation in which large areas of hibernating myocardium have been demonstrated. They should then be randomized to revascularization or standard medical therapy.

Conclusion

The discovery by Rahimtoola [1] that chronic left ventricular regional dysfunction due to prolonged hypoperfusion can recover after full reperfusion has radically changed the current pathophysiological concept of myocardial ischemia and the actual treatment of CAD patients. We need, however, proper studies aimed to evaluate the role of the different diagnostic methodologies for detection of myocardial hibernation and to assess the clinical efficacy of revascularization of hibernating myocardium in terms of reduction of morbidity and mortality. From the clinical point of view, attention and resources should be concentrated on those cases with large zones of hibernating myocardium in which the differentiation of viable from non-viable myocardium and the stratification of operative risk is the real overriding clinical concern. It is this subset of patients which will justify further investigation in this fascinating field and which hopefully, will improve our therapeutic possibilities.

Acknowledgements

This work was supported by the National Research Council (C.N.R) target project 'Prevention and control of disease factors' n. 91.00156 PF 41. The authors thank Miss Roberta Bonetti for secretarial assistance in preparing the manuscript.

References

1. Rahimtoola S: The hibernating myocardium. Am Heart J 117: 211–221, 1989
2. Eitzman D, Al-Aouar Z, Kanter HL: Clinical outcome of patients with advanced coronary artery disease after viability studies with positron emission tomography. J Am Coll Cardiol 20: 559–565, 1992
3. Ross J Jr: Myocardial perfusion-contraction matching: implications for coronary heart disease and hibernation. Circulation 83: 1076–1082, 1991
4. Schulz R, Guth BD, Pieper K, Martin C, Heusch G: Recruitment of an inotropic reserve in moderately ischemic myocardium at the expense of metabolic recovery: a model of short-term hibernation. Circ Res 70: 1282–1295, 1992
5. Ferrari R, Visioli O: Stunning: damaging or protective to the myocardium? Cardiovasc Drugs Ther 5: 939–946, 1991
6. Ferrari R, Cargnoni A, Curello S, Ceconi C, Volpini M, Visioli O: Metabolic adaptation of underperfused isolated rabbit heart: an insight into molecular mechanisms underlying hibernation. Circulation 88: 1004 (abstr.),1993
7. Ferrari R, La Canna G, Giubbini R, Ceconi C, Pardini A, Coletti G, Berra P, Alfieri O: In: M Yacoub, J Papper (eds.). Hibernating myocardium. Annual of Cardiac Surgery, 1994, pp 28–32
8. Dilsizian V, Bonow R. Current diagnostic techniques of assessing myocardial viability in patient with hibernating and stunned myocardium. Circulation 87: 1–20, 1993
9. Marinhoi NVS, Keogh BE, Costa DC, Lammertsma AA, Ell PJ, Camici PG: Pathophysiology of chronic left ventricular dysfunction. New insights from the measurement of absolute myocardial blood flow and glucose utilization. Circulation 93: 737–744, 1996
10. Sambuceti G, Marzullo P, Giorgetti A, *et al.*: Global alteration in perfusion response to incrousing oxygen consumption in patients with single-vessel coronary artery disease. Circulation 90: 1696–1705, 1994
11. Vanoverschelde JL, Wijns WC, Depre C, *et al.*: Mechanisms of chronic regional postiischemic dysfunction in humans. New insights from the study of noninfarcted collateral-dependent myocardium. Circulation 87: 1513–1523, 1993
12. Gerver BL, Vanoverschelde JL, Bol A, *et al.*: Myocardial blood flow, glucose uptake, and recruitment of inotropic reserve in chronic left ventricular ischemic dysfunction. Implications for the pathophysiology of chronic myocardial hibernation. Circulation 94: 651–659, 1996
13. Giubbini R, Alfieri O, Ferrari R, Milan E, Rossini P, Lorusso R, Berra P, La Canna G: Thallium rest-redistribution and echo-dobutamine tests to predict functional improvement after bypass surgery in patients with left ventricular dysfunction. Circulation 88: 2883 (abstr.), 1993
14. Cigarroa CG, deFilippi CR, Brickner ME, Alvarez LG, Wait MA,

Grayburn PA: Dobutamine stress echocardiography identifies hibernating myocardium and predicts recovery of left ventricular function after coronary revascularization. Circulation 88: 430–436, 1993
15. La Canna G, Alfieri O, Giubbini R, Gargano M, Ferrari R, Visioli O: Echocardiography during infusion of dobutamine for identification of reversible dysfunction in patients with chronic coronary artery disease. JACC 23: 617–626, 1994
16. Ferrari R, La Canna G, Giubbini R, Ceconi C, Pardini A, Coletti G, Berrra P, Alfieri O: In: M Yacoub, J Pepper (eds.). Hibernating myocardium. Annual of Cardiac Surgery, 1994, pp 28–32

Molecular and Cellular Biochemistry **186**: 201–203, 1998.
© 1998 *Kluwer Academic Publishers.*

Non-ischemic myocardial preconditioning

Raúl Domenech and Pilar Macho

Instituto de Ciencias Biomédicas, Facultad de Medicina, Universidad de Chile, Santiago, Chile

Abstract

The reduction of infarct size produced by brief ischemic episodes prior to a sustained occlusion of a coronary artery, called ischemic preconditioning, is a well known phenomenon that occurs in several species, but its mechanism is still under investigation. Recent reports support the idea that this protection can also be obtained by non-ischemic maneuvers like distention of the left ventricle and metabolic stimulation of myocardial cells. The features of non-ischemic preconditioning (temporal limitation, second window, tolerance development, remote preconditioning and efficiency of the protection), as opposed to those of ischemic preconditioning, are still to be determined. Neither is it known if non-ischemic preconditioning occurs in humans. From a physiological point of view the protective effect of an increase in metabolic rate of the heart means a constant feed-back mechanism in the myocardial cell that counteracts the presumptive damage consequent to the increase in metabolism. Therefore, in the presence of a sudden coronary occlusion the metabolic rate of the heart immediately before the occlusion would have a dual role of increasing the degree of ischemia and of protecting against it. (Mol Cell Biochem **186**: 201–203, 1998)

Key words: adenosine, myocardial preconditioning, myocardial infarction, protein kinase C

Ischemic myocardial preconditioning was described in 1986 by Murry *et al.* [1] as the decrease in myocardial infarct size produced by a prolonged coronary occlusion by effect of brief periods of ischemia prior to the occlusion. This phenomenon has been observed in several animal species [2] and there is strong evidence that it also occurs in man [3, 4]. Although the mechanisms of ischemic preconditioning are not yet known, there are several hypotheses regarding the probable mediators [2] which participation varies according to species. However, the triggering event for this mechanism is the myocardial metabolic alteration during the brief periods of ischemia. Recently, some reports have provided experimental evidence for myocardial preconditioning in the absence of ischemia and pharmacological interventions. Thus, Ovize *et al.* [5] demonstrated a preconditioning effect of ventricular wall distention obtained by acute vascular volume overload in dogs. The effect seems to be mediated by stretch-activated ion channels since it was completely prevented by the administration of Gd^{3+}. The authors hypothesized that the activation of these channels could limit myocyte edema associated with ischemia through loss of KCl and water, thereby delaying ischemic cell death. The absence of ischemia during the volume overload was demonstrated by the increase in transmural blood flow without a decrease in the inner/outer wall flow ratio. An important observation in these experiments was that Gd^{3+} partly attenuated the reduction in infarct size obtained by conventional ischemic preconditioning, a result suggesting that these stretch-activated channels also participate in the protection obtained with ischemic preconditioning. This hypothesis sounds reasonable because of the segmental ventricular distention that occurs during ischemia.

It seems reasonable to think that the metabolic trigger for preconditioning during ischemic preconditioning can be mimicked by increasing the metabolic demands of the myocardium in the absence of ischemia. Some studies have shown a decrease in the degree of ischemia, the extent of hemodynamic alterations, and the incidence of arrhythmias by increasing the heart rate in different species [6, 7]. Szekeres *et al.* [6] reported that repetitive overpacing of the heart in conscious rabbits reduces the pacing-induced ST-segment elevation and the increase in left ventricular end-diastolic pressure observed during the first period of pacing. Furthermore, Vegh *et al.* [7] reported a protective effect of overpacing on the development of ST-segment changes and arrhythmias produced by coronary occlusion in dogs.

Address for offprints: R. Domenech, Instituto de Ciencias Biomédicas, Facultad de Medicina, Universidad de Chile, Casilla 16038, Santiago 9, Chile

However, in the above experiments [6, 7] overpacing was used to induce ischemia, and, therefore, a presumptive participation of the increase in myocardial oxygen consumption, consequent to tachycardia, in the preconditioning effect can not be deduced from them.

Iwamoto *et al.* [8] reported a preconditioning effect of stimulation of the left stellate cardiac nerve in dogs. However, this effect occurred only when the increase in coronary flow caused by the adrenergic stimulation was prevented, i.e., when a supply-demand imbalance was allowed to occur. These experiments demonstrate that an absolute reduction in resting energy supply is not required for preconditioning, but the possibility exists that relative ischemia (demand ischemia) was induced during the adrenergic activation and this may have been the actual cause of preconditioning.

Recently Koning *et al.* [9] induced myocardial protection with 30 min of rapid ventricular pacing immediately before a prolonged coronary occlusion in open-chest pigs. The absence of ischemia in these experiments was demonstrated by the preservation of ATP levels, CrP/Cr ratio, CrP/ATP ratio and energy charge during the pacing and by the maintenance of a subendocardial to subepicardial blood flow ratio above unity. Besides, there were no changes in left ventricular end-diastolic pressure and reactive hyperemia was not observed after restoration of the control heart rate. The protective effect was probably mediated by the activation of ATP-dependent potassium channels because it was abolished by pretreatment with glibenclamide. However, the duration of the protective effect of rapid pacing was brief since much of the protection was lost when 15 min of control heart rate was allowed to separate the 30 min rapid ventricular pacing period from the sustained occlusion. Furthermore, when glibenclamide administration was initiated only after the 30 min of pacing and at the onset of the 15 min of control heart rate before coronary occlusion, residual protection was not further attenuated. This suggested that ATP dependent potassium channels did not contribute to the protection persisting beyond 15 min after cessation of the ventricular pacing.

We observed a reduction of infarct size in dogs preconditioned with five periods of 5 min tachycardia, each followed by a 5 min period of control heart rate [10]. During the tachycardic periods the oxygen consumption of the heart increased by 65%, coronary flow increased by 38% and the coronary arteriovenous oxygen content difference increased by 20%. These changes are similar to those observed in man and in conscious healthy dogs during exercise [11–14]. Induction of myocardial ischemia during the preconditioning tachycardia was excluded by documenting maintenance of coronary vasodilatory reserve across the left ventricular wall including the subendocardium, the absence of changes in left ventricular end-diastolic pressure, segment length dimensions, and lactate extraction by the heart, and the lack of ECG ischemic changes [15]. This protective effect was not related to changes in hemodynamic variables or collateral flow to the ischemic region either. We also observed that preconditioning was abolished by 8-phenyltheophylline [15] administered either before the tachycardic periods or after them (before the prolonged coronary occlusion). These results suggest that the protective effect is mediated by adenosine receptors and that these receptors must be activated during the prolonged ischemia for the protection to express itself. These results agree with the demonstration that adenosine is produced during an increase in myocardial oxygen demand due to rapid pacing and exercise in dogs [16, 17].

Future experiments should investigate whether pre–conditioning by increasing myocardial metabolic rate has characteristics similar to those of ischemic preconditioning such as temporal limitation, second window, tolerance to several repeated preconditioning stimuli, and evaluation of its protective efficiency related to the duration of the sustained ischemia. Furthermore, remote myocardial preconditioning induced by increased metabolic rate of the skeletal muscle is an interesting hypothesis to test because of the demonstration of remote myocardial preconditioning by ischemia of a myocardial region adjacent to the infarct region [18] and by renal ischemia [19, 20]. Is non-ischemic preconditioning as powerful as ischemic preconditioning in protecting the heart? This is another important issue to be investigated since ischemic preconditioning seems to be the most powerful experimental strategy known for protecting the functioning heart from infarction [21]. We have used in dogs a protocol of brief periods of tachycardia analogous to the brief periods of ischemia used by Murry *et al.* [1] in their original experiments. The decrease in infarct size that we obtained was much smaller than that obtained after ischemic preconditioning [10].

Considering that an increase in myocardial oxygen consumption and ventricular distention, in the absence of ischemia, induce preconditioning and assuming that these triggers occur frequently, it could be hypothesized that the heart is being continually preconditioned for ischemia. This hypothesis has not been tested yet, but the effect of exercise on the size of induced myocardial infarction in chronically instrumented animals should be evaluated.

Finally, does non-ischemic preconditioning occur in humans? This is the most important question from a clinical viewpoint, and could have therapeutic implications. Physical activity prior to rupture of a coronary plaque could have a dual effect. On the one hand, it would accelerate the progression of myocardial ischemia because of the higher metabolic demand, but on the other hand, it might attenuate the degree of myocardial cell damage through preconditioning.

In summary, non-ischemic preconditioning is a new field for exploration and could have clinical implications of perhaps greater relevance than ischemic preconditioning. Procedures that increase myocardial oxygen consumption, including exercise, could become important determining factors in the protection of the heart.

References

1. Murry CE, Jennings RB, Reimer KA: Preconditioning with ischemia: A delay of lethal cell injury in ischemic myocardium. Circulation 74: 1124–1136, 1986
2. Przyklenk K, Kloner RA, Whittaker P: Synopsis of ischemic preconditioning: What have we learned since 1986? In: K Przyklenk, RA Kloner, DR Yellon (eds.) Ischemic Preconditioning: the concept of endogenous cardioprotection. Boston/Dordrecht/London, Kluwer Academic Publishers, 1994, pp 153–170
3. Tomai F, Crea F, Gaspardone A: Mechanisms of cardiac pain during coronary angioplasty. J Am Coll Cardiol 22: 1892–1896, 1993
4. Kloner RA, Yellon D: Does ischemic preconditioning occur in patients? JACC 24: 1133–1142, 1994
5. Ovize M, Kloner RA, Przyklenk K: Stretch preconditions canine myocardium. Am J Physiol 266: H137–H146, 1994
6. Szekeres L, Papp JG, Szilvassy Z, Udvary E, Vegh A: Moderate stress by cardiac pacing may induce both short term and long term cardioprotection. Cardiovasc Res 27: 593–596, 1993
7. Vegh A, Szekeres L, Parrat JR: Transient ischemia induced by rapid cardiac pacing results in myocardial preconditioning. Cardiovasc Res 25: 1051–1053, 1991
8. Iwamoto T, Bai XJ, Downey J: Preconditioning with supply-demand imbalance limits infarct size in dog heart. Cardiovasc Res 27: 2071–2076, 1993
9. Koning MMG, Gho BCG, Klaarwater E, Opstal RLJ, Duncker DJ, Verdouw PD: Rapid ventricular pacing produces myocardial protection by nonischemic activation of K_{ATP}^+ channels. Circulation 93: 178–186, 1996
10. Domenech R, Macho P, González A: Preconditioning of the infarct size by tachycardia in dogs (abstr). Rev Chil Cardiol 14: 87, 1995
11. Nelson RR, Gobel FL, Jorgensen CR, Wang K, Wang Y, Taylor HL: Hemodynamic predictors of myocardial oxygen consumption during static and dynamic exercise. Circulation 50: 1179–1189, 1974
12. Khouri EM, Gregg DE, Rayford CR: Effect of exercise on cardiac output, left coronary flow and myocardial metabolism in the unanesthetized dog. Circ Res 17: 427–433, 1965
13. Lochner W, Nasseri M: Uber den venöseen Sauerstoffdruck, die Einstellung der Coronardurchbluting und den Kohlenhydratstoffwechsel des Herzens bei Muskelarbeit. Pflügers Arch 269: 407–416, 1959
14. Restorff W, Holtz J, Bassenge E: Exercise induced augmentation of myocardial oxygen extraction in spite of normal coronary dilatory capacity in dogs. Pflügers Arch 372: 181–185, 1977
15. Domenech R, Macho P, Vélez D, Sánchez G, Liu X, Dhalla N: Tachycardia preconditions the infarct size in dogs. Role of adenosine and protein kinase C. Circulation 97: 786–794, 1998
16. Saito D, Nixon DG, Vomacka RB, Olsson RA: Relationship of cardiac oxygen usage, adenosine content, and coronary resistance in dogs. Circ Res 47: 875–882, 1980
17. Hori M, Kitakaze M: Adenosine, the heart, and coronary circulation. Hypertension 18: 565–574, 1991
18. Przylenk K, Bauer B, Ovize M, Kloner RA, Whittaker P: Regional ischemic preconditioning protects remote virgin myocardium from subsequent sustained coronary occlusion. Circulation 87: 893–899, 1993
19. McClanahan TB, Nao BS, Wolke LJ, Martin BJ, Metz TE, Gallagher KP: Brief renal occlusion and reperfusion reduces myocardial size in rabbits. (Abstr.) FASEB J 7: 682, 1993
20. Gho BCG, Shoemaker RG, Van der Lee C, Sharma HS, Verdouw PD: Myocardial infarct size limitation in rat by transient renal ischemia. (Abstr.) Circulation 90 (suppl I): I-476, 1994
21. Speechly-Dick ME, Baxter GF, Yellon DM: Ischaemic preconditioning protects hypertrophied myocardium. Cardiovasc Res 28: 1025–1029, 1994

Molecular and Cellular Biochemistry **186**: 205–207, 1998.

Index to Volume 186

Developments in Molecular and Cellular Biochemistry

Series Editor: Naranjan S. Dhalla, Ph.D., M.D. (Hon.), FACC

1. V.A. Najjar (ed.): *Biological Effects of Glutamic Acid and Its Derivatives.* 1981 ISBN 90-6193-841-4
2. V.A. Najjar (ed.): *Immunologically Active Peptides.* 1981 ISBN 90-6193-842-2
3. V.A. Najjar (ed.): *Enzyme Induction and Modulation.* 1983 ISBN 0-89838-583-0
4. V.A. Najjar and L. Lorand (eds.): *Transglutaminase.* 1984 ISBN 0-89838-593-8
5. G.J. van der Vusse (ed.): *Lipid Metabolism in Normoxic and Ischemic Heart.* 1989 ISBN 0-7923-0479-9
6. J.F.C. Glatz and G.J. van der Vusse (eds.): *Cellular Fatty Acid-Binding Proteins.* 1990 ISBN 0-7923-0896-4
7. H.E. Morgan (ed.): *Molecular Mechanisms of Cellular Growth.* 1991 ISBN 0-7923-1183-3
8. G.J. van der Vusse and H. Stam (eds.): *Lipid Metabolism in the Healthy and Diseased Heart.* 1992 ISBN 0-7923-1850-1
9. Y. Yazaki and S. Mochizuki (eds.): *Cellular Function and Metabolism.* 1993 ISBN 0-7923-2158-8
10. J.F.C. Glatz and G.J. van der Vusse (eds.): *Cellular Fatty-Acid-Binding Proteins, II.* 1993 ISBN 0-7923-2395-5
11. R.L. Khandelwal and J.H. Wang (eds.): *Reversible Protein Phosphorylation in Cell Regulation.* 1993 ISBN 0-7923-2637-7
12. J. Moss and P. Zahradka (eds.): *ADP-Ribosylation: Metabolic Effects and Regulatory Functions.* 1994 ISBN 0-7923-2951-1
13. V.A. Saks and R. Ventura-Clapier (eds.): *Cellular Bioenergetics: Role of Coupled Creatine Kinases.* 1994 ISBN 0-7923-2952-X
14. J. Slezák and A. Ziegelhöffer (eds.): *Cellular Interactions in Cardiac Pathophysiology.* 1995 ISBN 0-7923-3573-2
15. J.A. Barnes, H.G. Coore, A.H. Mohammed and R.K. Sharma (eds.): *Signal Transduction Mechanisms.* 1995 ISBN 0-7923-3663-1
16. A.K. Srivastava and J.-L. Chiasson (eds.): *Vanadium Compounds: Biochemical and Therapeutic Applications.* 1995 ISBN 0-7923-3763-8
17. J.M.J. Lamers and P.D. Verdouw (eds.): *Biochemistry of Signal Transduction in Myocardium.* 1996 ISBN 0-7923-4067-1
18. E.-G. Krause and R. Vetter (eds.): *Biochemical Mechanisms in Heart Function.* 1996 ISBN 0-7923-4118-X
19. R. Vetter and E.-G. Krause (eds.): *Biochemical Regulation of Myocardium.* 1996 ISBN 0-7923-4259-3
20. G.N. Pierce and W.C. Claycomb (eds.): *Novel Methods in Molecular and Cellular Biochemistry of Muscle.* 1997 ISBN 0-7923-4387-5
21. F.N. Gellerich and S. Zierz (eds.): *Detection of Mitochondrial Diseases.* 1997 ISBN 0-7923-9925-0
22. P.K. Singal, V. Panagia and G.N. Pierce (eds.): *The Cellular Basis of Cardiovascular Function in Health and Disease.* 1997 ISBN 0-7923-9974-9
23. S. Abdel-aleem and J.E. Lowe (eds.): *Cardiac Metabolism in Health and Disease.* 1998 ISBN 0-7923-8104-1
24. A.K. Srivastava and B. Posner (eds.): *Insulin Action.* 1998 ISBN 0-7923-8113-0

Developments in Molecular and Cellular Biochemistry

25. V.A. Saks, R. Ventura-Clapier, X. Leverve, A. Rossi and M. Rigoulet (eds.): *Bioenergetics of the Cell: Quantitative Aspects*. 1998 ISBN 0-7923-8118-1
26. G.N. Pierce, H. Rupp, T. Izumi and A. Grynberg (eds.): *Molecular and Cellular Effects of Nutrition on Disease Processes*. 1998 ISBN 0-7923-8171-8
27. K. Ahmed, E. Chambaz and O.G. Issinger (eds.): *Molecular and Cellular View of Protein Kinase CK2*. 1998 ISBN 0-7923-8208-0
28. M.V. Cohen, J.M. Downey, R.J. Gelpi and J. Slezák (eds.): *Myocardial Ischemia and Reperfusion*. 1998 ISBN 0-7923-8173-4
29. D.A. Bernlohr and L. Banaszak (eds.): *Lipid Binding Proteins within Molecular and Cellular Biochemistry*. 1998 ISBN 0-7923-8223-4
30. R. Alvarez-Gonzalez (ed.): *ADP-Ribosylation Reactions: From Bacterial Pathogenesis to Cancer*. 1998 ISBN 0-7923-8235-8
31. S. Imai and M. Endo (eds.): *Muscle Physiology and Biochemistry*. 1998 ISBN 0-7923-8264-1

KLUWER ACADEMIC PUBLISHERS – DORDRECHT / BOSTON / LONDON

GPSR Compliance

The European Union's (EU) General Product Safety Regulation (GPSR) is a set of rules that requires consumer products to be safe and our obligations to ensure this.

If you have any concerns about our products, you can contact us on ProductSafety@springernature.com

In case Publisher is established outside the EU, the EU authorized representative is:

Springer Nature Customer Service Center GmbH
Europaplatz 3
69115 Heidelberg, Germany

Batch number: 10755910

Printed by Printforce, the Netherlands